FRONT AND BACK ENDPAPER MAPS USED WITH PERMISSION OF MRS. ERWIN RAISZ.

Continental Drift

THE EVOLUTION OF A CONCEPT

Ursula B. Marvin

To learn about the earth—a fundamental preoccupation of mankind—geologists are applying ever more efficient remote sensing and sampling techniques to the ocean basins, to the interior of the planet, and to the moon. Characteristically, the new advances raise new problems. From about 1930 to the mid-1960s continental drift was dismissed as a dead issue by most American geologists; today it is regarded as a key to earth history. Why the sweeping reversal of opinion?

In this comprehensive study Ursula Marvin reviews the early scientific and polemical debates on continental drift, and singles out the rise of solid-earth geophysics as the most important factor contributing to the premature demise of the continental drift theory. But other factors abounded, and in order to place the drift controversy of the 1920s in proper perspective, the book probes far back in time to outline the slow growth of geographical, geological, and geophysical knowledge about the origin and distribution of continents and oceans.

The pivotal period occurred in the early decades of this century when German scientist Alfred Wegener proposed his hypothesis that the continents have not always occupied their present positions on the globe: In the late Paleozoic era they were all part of one huge landmass, Pangaea, which occupied half of the earth's surface; the other half was covered by the primeval Pacific Ocean. This concept ran counter to one of the most deeply held convictions in earth science: the permanence of continents and deep ocean basins, enunciated after much careful consideration by James D. Dana in 1846 and subsequently accepted by American geologists as an established principle.

Wegener's hypothesis was proposed and defended in the same period that saw rapid advances in seismology and in experimental studies of the behavior of rocks under high pressures. Most of the new geophysical evidence suggested that the earth has more than sufficient internal strength and rigidity to preclude the superficial drift of continents.

For almost forty years arguments for continental stability prevailed. Then a dramatic reversal of opinion occurred in the late 1960s, when geologists throughout the world became convinced that crustal plates, both continental and oceanic, have moved over many degrees of latitude and longitude since the Cretaceous period. By 1967 the evidence for crustal motion, which had been accumulating through the efforts of a few, widely scattered investigators finally became overwhelming. A new global model called plate tectonics was proposed, and suddenly the burden of proof shifted decisively from those favoring some form of "drift" to those supporting immovable continents.

Today, although most scientists and laymen appear persuaded of the truth of the new global tectonics, none of the latest pieces of evidence or interpretations has gone unchallenged. An adequate causal mechanism is still being sought and many fundamental problems remain to be solved.

This book is rich in fresh insights into the personalities and ideas that have influenced the development of geology. It does not propose an exhaustive compendium of facts and references, but rather an overall view of the development of knowledge and the ebb and flow of scientific opinion on what has been one of the most controversial issues in earth science.

About the author . . .

A Vermont native, Ursula Marvin is a graduate of Tufts College and received her M.A. and Ph.D. degrees in geology from Harvard University. With her husband, Thomas C. Marvin, an economic geologist, she spent six years in field mapping and studying mineral deposits in many parts of South America and Africa. In 1961 Dr. Marvin joined the Smithsonian Astrophysical Observatory in Cambridge, Massachusetts, to do research on the mineralogy of meteorites and, in recent years, lunar samples. She has published many articles on mineralogy and continental drift, including one for the *Encyclopaedia Britannica.* Dr. Marvin is a vice president of the Meteoritical Society, a Fellow of the Geological Society of America, a member of The American Geophysical Union, the American Association for the Advancement of Science, and the International Association of Geochemistry and Cosmochemistry.

"I didn't grow up with the idea of drifting continents and sea floor spreading. But I tell you, when you look at the way the pieces of the northeastern portion of the African continent seem to fit together separated by a narrow gulf, you could almost make a believer out of anybody."

Astronaut Harrison Schmitt
during the flight of Apollo 17

Continental Drift

THE EVOLUTION OF A CONCEPT

Ursula B. Marvin

SMITHSONIAN INSTITUTION PRESS CITY OF WASHINGTON

Frontispiece: The sphere of the earth as photographed from the Apollo 17 spacecraft in December 1972. The view includes the south polar ice cap of Antarctica, the entire continent of Africa, the island of Madagascar, Iran, and, on the far northeastern horizon, the west coast of India. The East African rift system is clearly delineated from the Indian Ocean northward to the triangular Afar depression at the juncture of the Red Sea and the Gulf of Aden. The course of the Nile River may be traced from the Ethiopian plateau region to the Mediterranean Sea.
(National Aeronautics and Space Administration photo)

Second printing, with corrections, 1974
Library of Congress Cataloging in Publication Data appears on page 240.
Distributed in the United States and Canada by George Braziller, Inc.
Distributed throughout the rest of the world by Feffer and Simons, Inc.
Printed in the United States of America
Designed by Elizabeth Sur

Foreword

This book has its origins in a series of seminars held at the Smithsonian Astrophysical Observatory in the summer of 1966 to discuss the scientific applications of satellite tracking. At that time each of the camera-equipped tracking stations in an equatorial network could be located to an accuracy of plus-or-minus ten meters. The substitution of lasers for cameras was expected to increase the accuracy to the centimeter range and, for the first time, make possible direct measurements of global crustal motions.

As my contribution to the seminars I was invited to assess the possibilities of measuring continental drift. The subject held a special interest for me because, imprinted with a classical education, I had originally viewed continental drift as a geological chimera that had been securely laid to rest. My earliest field experiences, however, were with my husband in South America and Africa where we learned to our astonishment that geologists took continental drift for granted. Neither our laughter nor our serious arguments in the name of geophysics carried any persuasive force with our friends in the southern hemisphere. After returning home we broached the subject to numerous colleagues who, I now remember, laughed heartily but said very little, except to point to the lack of a mechanism.

In preparing for my seminar in July 1966 I reviewed with mounting excitement the recent results of paleomagnetic studies and the matching of dated contacts between Africa and South America. Nevertheless, fair as I tried to be in my presentation, the audience noted a distinct bias against continental drift. But by November, when the manuscript was due, I had switched my allegiance.

Overwhelmed by the evidence—most of which was so new it had never been available to the pro-drifters in the southern hemisphere—I had become a "drifter." That experience led me to wonder very curiously about the early history of the drift controversy and the factors and personalities that had caused acceptance of the hypothesis by some geologists and its total rejection by others from the same body of scientific evidence. In recounting the long, many-faceted history of ideas on continents and oceans I have selected those voices that seemed to me to be the most illustrative of a given point of view. I regret that many an important scientist has been omitted from my story, and I am acutely aware that volumes could be written on the same subject with substantially different dramatis personae.

My researches have altered many preconceptions. They have, for example, inspired a profound respect for Alfred Wegener, whose reputation as a scientist still suffers in North America despite some attempts at rehabilitation. I have also learned to admire the accomplishments of many an earlier scientist who, working without computers or remote sensing equipment, perceived the planet as a whole in terms very similar to our own. Are not those scientists and detectives most brilliant who deduce the nature of things from the fewest good clues?

Ursula B. Marvin, *Smithsonian Astrophysical Observatory, Cambridge, Massachusetts, February 1973*

Acknowledgments

I wish to express thanks to my friends who have helped in the preparation of this book. I especially thank Paul A. Mohr who, red pencil in hand, worked through an early draft of the entire manuscript once and through the section on plate tectonics twice. James W. Skehan also read the manuscript with a critical eye and offered many helpful suggestions, as did James D. Hume and Charles T. Stearns. Owen Gingerich prompted a major revision of the section on the history of astronomy; Clifford Kaye rendered valuable advice on the history of geology; John A. O'Keefe, on geophysics; G. Jeffrey Taylor, John S. Reid, and George Megrue, on geochronology. All of these scientists worked diligently to rid the text of errors and internal inconsistencies. Any that remain are wholly my own.

Eliza Collins, who is not a scientist, read an early draft from a layman's viewpoint and indicated numerous passages that needed clarification. Afterward she paid me the ultimate compliment of reporting that my manuscript had stimulated her to read a textbook of geology. If my book leads any future reader to another book or into a library I will feel the effort was worthwhile.

I appreciate the assistance of Professor Wolf von Engelhardt who arranged for my use of the library at the University of Tübingen where I could examine the early publications of Alfred Wegener.

I am grateful to Ernest Biebighauser and Betty Sur, of the Smithsonian Institution Press, for transforming a manuscript to a book with unfailing good humor while the author performed seemingly endless revisions.

Above all I wish to thank my husband, Thomas C. Marvin, not only for massive aid in library research, in reading drafts and proofs, and in compiling my bibliography and index, but also for maintaining his equanimity through three years when my evenings, weekends, and vacations were devoted mainly to writing. He has even persuaded me to believe that he enjoyed it.

Contents

Introduction

". . . for such a theory as this to run wild"

Alfred Wegener, a young scientist in Germany, had proposed an interpretation of earth history that challenged some of the basic principles of geology and geophysics. "If we are to believe Wegener's hypothesis we must forget everything which has been learned in the last 70 years and start all over again," protested an outraged participant at the 1922 meeting of the Geological Society of America. The prospect was unthinkable.

According to Wegener's hypothesis the continents have not always occupied their present positions on the globe; in the late Paleozoic era they were all part of one hugh landmass, Pangaea, which occupied half of the earth's surface. The other half was covered by the primeval Pacific Ocean. Wegener presented evidence that in the Jurassic period, which was then dated at about 40 million years ago, Pangaea began rifting into fragments, and the yielding of the weaker rock of the ocean floor allowed the continents to move apart like icebergs in water. This concept ran counter to one of the most deeply held convictions in earth science: the permanence of continents and deep ocean basins, enunciated after much careful consideration by James D. Dana in 1846 and subsequently accepted by American geologists as an established principle.

Continental displacement was not a strictly new idea when Wegener first outlined his hypothesis in 1912. A few earlier writers had depicted the splitting asunder of continents by violent cataclysms alleged to have occurred either at the beginning of earth history or at an improbably recent date. None had achieved any serious following.

Wegener made his proposal at a time when many geologists were beginning to find the conventional theories inadequate to explain new observations or to solve some of the old but fundamental problems. He envisioned continental drift as resulting from forces intrinsic to the earth's rotation system, and he presented evidence from such a range of sciences—physical geology, geodesy, geophysics, paleontology, zoology, and paleoclimatology—that his hypothesis could not easily be ignored. It was, indeed, hailed with enthusiasm by some scientists, but at the same time it shocked geological conservatives into an impassioned defense of their positions.

"Can we call geology a science when there exists such difference of opinion on fundamental matters as to make it possible for such a theory as this to run wild?" thundered Professor Rolling T. Chamberlin from his Gothic sanctuary at the University of Chicago.

The theory "ran wild" in the sense that it was hotly debated until about 1930, when the furor subsided with Wegener's hypothesis in general disrepute. So radical a departure from accepted theory would have to be backed by positive proof, and none was forthcoming. Although a few investigators continued to support the idea, they failed to convince the majority that continental drift is necessary in the interpretation of earth history. They also failed to demonstrate the actuality of crustal migration by means of geodetic measurements, or to propose a casual mechanism of anything approaching the required magnitude.

Wegener's hypothesis was proposed and defended in the same period that saw the rise of solid-earth geophysics, with rapid advances in seismology and in experimental studies of the behavior of rocks under high pressures. Most of the new geophysical evidence suggested that the earth has more than sufficient internal strength and rigidity to preclude the superficial drift of continents.

As arguments for continental stability prevailed, the concept of continental drift took on an ever more comic aspect. Ridiculous "matches" between far-flung coastlines and tragic family separations imposed upon the lower vertebrates by the wrenching apart of landmasses provided rich sources of geological humor. In 1944 Professor Bailey Willis, in a short note entitled "Continental Drift, ein Märchen" (a Fairytale), proposed that Wegener's hypothesis be placed in the discard "since further discussion of it merely encumbers the literature and befogs the minds of fellow students." And he added, "Scientists who are not geologists cannot be expected to know that the geology upon which protagonists of the Theory rest assumptions is as antiquated as pre-Curie physics."

During the following two decades new observations seemed to argue more strongly than ever for a permanent distribution of continents. Radiometric age determinations confirmed stratigraphic evidence suggesting that the continents have gradually grown in area as belts of younger rock have been added to their margins. This pattern implied evolution in situ around ancient Precambrian nuclei.

In the 1950s measurements of the rates of heat flow from the earth's interior yielded the totally unexpected information that the average flow from the weakly radioactive oceanic crust is about equal to that from the more highly radioactive continental crust. This result appeared inexplicable unless the suboceanic mantle is richer in radioactive elements than that beneath the continents and unless each segment of crust is permanently coupled with the mantle that lies directly beneath it.

Some of the earliest geodetic satellites launched in the 1960s revealed anomalies in the earth's gravitational field suggesting that the mantle has sufficient strength to support long-term stress differences of 100 to 200 bars, a magnitude far too high to allow for the plastic yielding associated with continental drift.

We have qualified each of these examples with the words "suggested" or "appeared" because in each case other interpretations were offered. There is no doubt, however, that to the majority of earth scientists the bulk of geophysical data gathered before about 1965 militated against the idea of drifting continents.

A dramatic reversal of opinion occurred in the late 1960s, when geologists throughout the world became convinced that crustal plates, both continental and oceanic, have moved over many degrees of latitude and longitude since the Cretaceous period. For American scientists, the change took effect in the winter of 1966. Until then most of them had retained confidence in continental stability; but by the spring of 1967 a reaction was in full swing. The evidence for crustal motion, which had been accumulating through the efforts of a few, widely scattered investigators in paleomagnetism, oceanography, seismology, and geochronology, finally became overwhelming. A new global model called plate tectonics was proposed, and suddenly the burden of proof shifted decisively from those favoring some form of "drift" to those supporting immovable continents.

The term "continental drift" does not accurately describe the motions envisaged in plate tectonics. It has, however, come to be used as a generic title for all versions of continental mobility as opposed to the idea of continental stability. And inasmuch as continental drift was "Wegener's hypothesis" throughout the long decades of scorn and ridicule, it should, in fairness, be remembered today that it was Alfred Wegener who struck the spark, however premature and faltering, of the present revolution in the geosciences.

Today, although most scientists and laymen appear persuaded of the truth of the new global tectonics, none of the latest pieces of evidence or interpretations has gone unchallenged. An adequate causal mechanism is still being sought and many fundamental problems remain to be solved.

In these chapters we will trace the concept of continental drift from its beginnings, and in so doing we will find ourselves reviewing a broad spectrum of ideas on the origin and distribution of continents and oceans. We do not propose an exhaustive compendium of facts and references but rather an overall view of the development of knowledge and the ebb and flow of scientific opinion on one of the most controversial issues in earth science.

This survey of past changes in attitude makes us reticent to predict any that may occur in the future. At present the balance of the evidence favors the idea that the continents have moved vast distances over the surface of the earth. They are probably still moving.

The asymmetrical face of the earth

The continents, by their very existence, pose our first and most fundamental problem. Continents as we know them—ragged in outline and asymmetrical in their distribution on the globe—would not be predicted by physical theory. A large rotating planet with a powerful gravitational field should mold itself into a smooth, featureless spheroid which, given the chemistry of the earth, should be covered by a layer of water about 2.4 kilometers deep. The striking contrast of the earth with such a serene model is one of the puzzles of geology. The surface of the planet, far from being regular and homogeneous, is divided into two morphological provinces—continents and ocean basins—that differ markedly not only in elevation but in chemical composition, petrology, density, and age. Of these two provinces the ocean basins predominate, occupying 60 percent of the 510 million square kilometers of the earth's total surface area. In addition, the ocean waters flood the continental shelves, leaving only 148 million square kilometers, or 29 percent of the surface, unclaimed by the sea.

On casual examination, the configuration of the continents and oceans appears to lack any orderly pattern. No continent is balanced by another on the far side of the globe. On the contrary, 80 percent of the total land area is crowded toward the northern hemisphere in such a way that every continent lies opposite an ocean. In the polar regions the distribution is reversed, with the Arctic Ocean basin opposite Antarctica, the largest landmass in the southern hemisphere and the most isolated of continents. Thus, the earth is divided into a land hemisphere and a water hemisphere as shown in Figure 1.

The concentration of land in the northern hemisphere is enhanced by the fact that four of the continents have broad northerly reaches that taper southward to narrow peninsulas. North America and Eurasia rim the top of the world, almost enclosing the Arctic basin with dry land and shallow-shelf seas; but these landmasses dwindle to a single narrow isthmus without reaching the equator, which passes to the south of Panama City and Singapore. The narrow tips of Africa and South America extend southward to only moderate latitudes. Cape Town is 80 kilometers closer to the equator than is the Strait of Gibraltar; and both Cape Town and Buenos Aires lie near latitude 34° south, a parallel corresponding to the one which, in North America, marks the border between Alabama and Tennessee. Punta Arenas, on the Straits of Magellan, lies near latitude 53° south, the same distance from the equator as Hamburg, Germany.

The broadly triangular shape of the continents began to puzzle men of science in the middle of the 16th century when the first roughly accurate maps of the world were drawn. Francis Bacon remarked upon it as a problem demanding an explanation in his *Novum Organum*, written in 1620. More than 250 years later, in 1885, it was described in the opening sentences of *Das Antlitz der Erde* (The Face of the Earth), the four-volume masterwork of Eduard Suess:

> If we imagine an observer to approach our planet from outer space, and, pushing aside the belts of red-brown clouds which obscure our atmosphere, to gaze for a whole day on the surface of the earth as it rotates beneath him, the feature beyond all others most likely to arrest his attention would be the wedge-like outlines of the continents as they narrow away to the south. This is indeed the most striking character presented by our map of the world, and has been so regarded ever since the chief features of our planet have become known to us.

As the continents taper southward, the oceans, in complementary fashion, taper toward the north from an unbroken ring of water near the antarctic circle. The oceans make the earth unique in the solar system as "the water planet." Traces of water vapor occur in the atmospheres of Mars and Venus, but bodies of standing water are unknown on any other planet or satellite. Perhaps Eduard Suess was mistaken after all when he suggested that the wedge-shaped outline of the continents is the feature most likely to capture

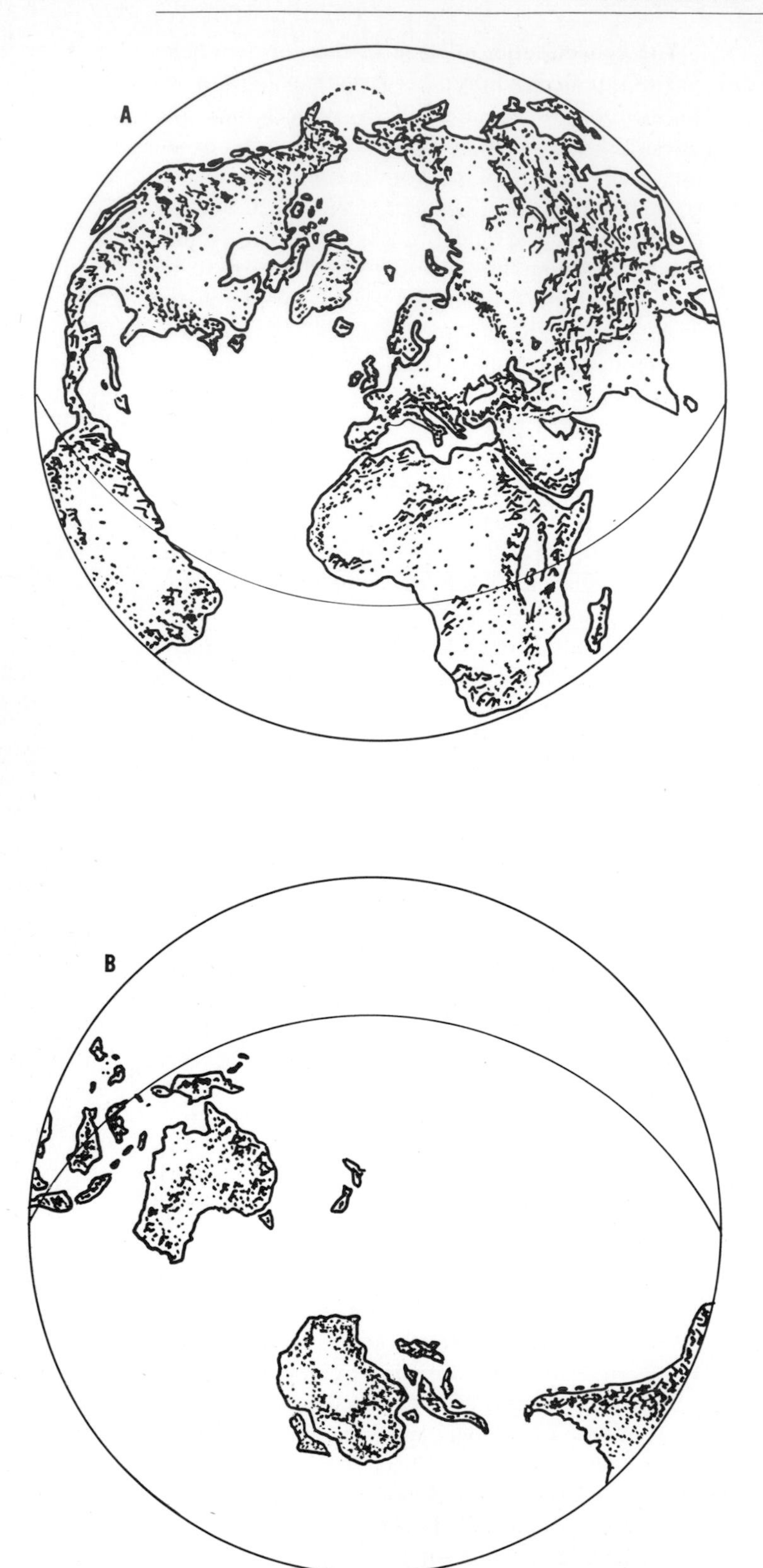

Figure 1. A: The land hemisphere, centered near Nantes, France, includes 82 percent of all the land area of the world; nevertheless, within this hemisphere only 46 percent of the surface is occupied by land. B: The water hemisphere, centered on the Antipodes Islands at longitude 180° east, latitude 50° south, includes 18 percent of the land area and 63 percent of the oceans of the world. Within this hemisphere 88 percent of the surface is occupied by water. Antarctica is shown with its ice cap melted away.

the attention of a visitor from space. Of much more absorbing interest to this hypothetical voyager from a desiccated homeland might be our great undulating oceans of blue-gray water, shining in the sunlight, advancing, retreating, and breaking in long rolling waves against the shorelines of the continents.

The marked opposition of land and sea is due in large measure to the vastness of the Pacific Ocean, which alone occupies one-third of the globe. This unique depression, rimmed with deep trenches backed by chains of folded mountains and interrupted only by groups of volcanic islands and coral reefs, is by far the largest single topographic feature on the earth. It, too, has perplexed scientists for centuries, as it differs from other oceans in its character and, so it was presumed, in its origin.

The distribution of land and sea over the earth is, then, asymmetrical, with the continents occurring as small, high-altitude patches amid a worldwide system of oceans. Why do the continents exist at all? Why are most of them roughly triangular and crowded into one hemisphere? Neither of these questions can be answered with any certainty, but we do know that asymmetry is not "normal" for the surface of a rotating sphere. Calculations show that if the continents had ever been regularly disposed, either in the form of a band around the equator or as one large landmass over each pole, or one in each quadrant, no force intrinsic to the earth's rotation system would throw them out of balance. The present configuration may have resulted from factors that caused asymmetry when the crust first began to form. Otherwise, the asymmetry has been imposed since the beginning by forces of large magnitude and uncertain origin. Thus, on the basis of distribution alone and before we begin to examine the geological framework of the crust, we are confronted with the problem of the permanence of continents and ocean basins as opposed to the possibility of their horizontal displacement.

Throughout the discussions on continental drift, scientists have argued the philosophical question of whether earth history must always be interpreted in terms of the continuous functioning of familiar processes or whether we may admit of occasional, seemingly fortuitous, events. Behind this argument stand the conflicting doctrines of uniformitarianism versus catastrophism. The geological sciences are firmly grounded in the uniformitarian principle—and how firmly we shall see in succeeding chapters. During the slow growth of geology as a science the longest and hardest battles had to be fought, not once but again and again, against catastrophists—particularly those who claimed that, after the original creation, the earth's landscape was modified only once, during the divinely decreed Deluge of Noah. The struggle was so prolonged and bitter that by the end of the 19th century most scientists had come to view as anathema any proposition involving unique events with unknown causes. A major factor that always weighed heavily against the acceptance of Wegener's hypothesis was that to many geologists it appeared tainted with catastrophism. Wegener did not call upon a catastrophic triggering event, but he did postulate one great episode of continental drift, occurring at a late stage in earth history and caused by forces that others perceived were too weak by orders of magnitude.

From the first, therefore, continental drift was a highly charged emotional issue. Some scientists saw it as a legitimate subject for investigation; others rejected it outright as irresponsible trifling with the known facts and established principles of geology. The resulting furor was bitter, but it followed in the tradition of many earlier controversies concerning the earth: its size and shape, its place and motion in the heavens, and the distribution of its continents and oceans. Challenges to the prevailing "common sense" view of the world have always provoked intensely hostile reactions.

In 1922 a critic of Alfred Wegener wrote: "A moving continent is as strange to us as a moving earth was to our ancestors, and we may be as prejudiced as they were." Having paid this lip service to open-mindedness, the critic then launched so vituperative an attack upon Wegener and his hypothesis as to outmatch, prejudice for prejudice, the most uncompromising of his grandfathers.

A brief review of some of the highlights in the stormy development of our geographical and geological knowledge of the earth may help us understand why the concept of a moving continent appeared so strange earlier in this century.

Geographical Speculations

The Distribution of Continents and Oceans

For the first 4,500 years of recorded history, the distribution of continents and oceans remained unknown. Maps were drawn, nevertheless, showing hypothetical continents that influenced geographical thought from ancient times until the early years of the 19th century A.D. When the broad configuration of lands and seas was finally established, after the discovery of Antarctica in 1820, the battle for a rational interpretation of their geological origin and evolution was just beginning. With respect to such fundamental problems as the size and shape of the earth, however, some of the earliest deductions were among the best.

The spherical earth

Several centuries before the time of Christ, Greek scholars concluded that the earth is a sphere. About 500 B.C. the Pythagoreans argued the case philosophically, saying that the earth must possess the most perfect and harmonious of forms. Two hundred years later the case rested less on theory than on observations such as the gradual disappearance of a ship's masts over the horizon and the succession of new constellations that becomes visible as one travels toward the north or south. Aristotle stated, about 330 B.C., that the earth must be spherical because it casts a circular shadow on the moon during lunar eclipses. At the same time, in examining the surface of the earth Aristotle recognized the significance of fossil marine shells in strata occurring high above sea level. He wrote that the same regions do not remain always sea or always land, but all change their condition in the course of time.

About 230 B.C. Eratosthenes of Cyrene, director of the library of Alexandria, calculated the circumference of the earth as 250,000 stades. How long was a stade? No one knows for certain, and the story is retold in numerous slightly different versions. Reviewing the evidence in 1969, Fred Hoyle, the English astronomer, stated that there were three stades in use: an itinerary stade of 157 meters, an Olympic stade of 185 meters, and the Royal Egyptian stade of 210 meters. Pliny, the Roman historian, wrote that Eratosthenes used the itinerary stade. If he did, and if that stade was indeed 157 meters, his result equalled about 39,250 kilometers and was remarkably close to our present value of 40,008 kilometers for the polar circumference. The equatorial circumference is 40,075 kilometers. [Both values were calculated in 1972 from geodetic satellite data.]

Eratosthenes believed that in Syene (Aswan), on the upper Nile, sunlight penetrated to the bottom of deep wells at noon on the day of the summer solstice.

For comparison he measured the length of the shadow of a vertical rod in Alexandria at noon on the same day, and he found that the shadow was equal to one-fiftieth part of a circle (7°12′). He concluded that the earth's circumference is 50 times the distance between the two cities. Eratosthenes thought that the earth is a perfect sphere, that Syene stood on the Tropic of Cancer, and that Syene and Alexandria lay 5,000 stades apart on the same meridian. None of these assumptions is quite true. The earth is an oblate spheroid slightly flattened at the poles, but this fact remained unknown until the time of Isaac Newton in the 1670s. Syene actually lay 3°3′ east of Alexandria and 38′ (about 60 kilometers) north of the Tropic of Cancer (where sunlight would not penetrate deep wells at the solstice). Furthermore, the direct, great circle, distance between the two cities could not be measured by any means then available, but it was certain to be substantially shorter than the overland route which followed the bends of the Nile. Perhaps Eratosthenes appreciated the inaccuracy inherent in this part of his calculations, for he revised his final result upward to 252,000 stades (39,564 kilometers), a figure evenly divisible by the 360 degrees of a great circle. Fortunately, all of his errors cancelled out to yield approximately the right answer.

As one result of Eratosthenes' calculations, Greek geographers learned that they lived on a spherical earth almost four times as large as the area of their known, inhabited world, the Oecumene. Confronted by this problem, some of them could not believe that their familiar landmasses, extending from the Mediterranean northward beyond the islands of Britain, southward to the cataracts of the Nile, and eastward to India, could be the only continental area of the earth. The Stoic philosophers were forced by their love of symmetry to invent three more continents, one in the northern hemisphere and two in the southern, to balance their own. As the spirit of nature loves life, they assumed that these continents were inhabited and they named them and their peoples with reference to the positions they occupied relative to their own: the Perioeci (peoples around the globe from the Oecumene), the Antoeci (peoples below the Oecumene), and the Antipodes* (peoples on the opposite side of the globe). Thus arose the concept of the Antipodes, who were to become the subject of fierce controversy in the Christian centuries.

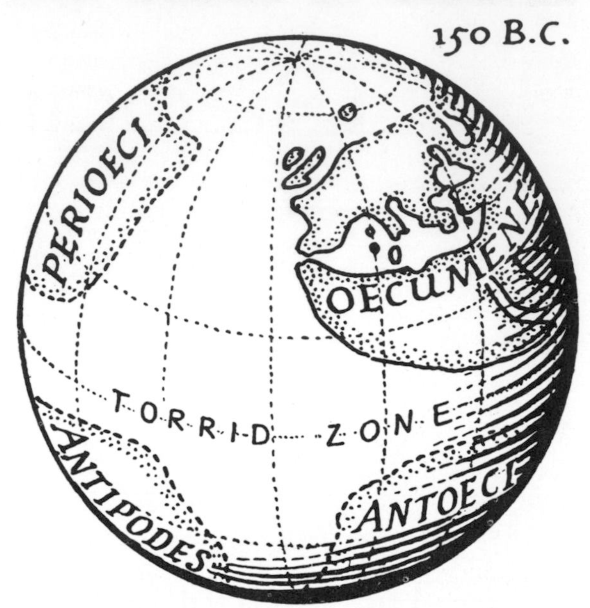

Figure 2. A sketch of the globe of Crates by the cartographer Erwin Raisz. It is not known whether the globe was ever constructed but many versions have been drawn from Crates' description. The Oecumene in this sketch is based on the world map of Eratosthenes. (From *General Cartography* by Erwin Raisz, 1938. Used with permission of McGraw-Hill Book Company.)

In addition to deducing the spherical form of the earth, Greek observers measured the obliquity of the equator to the plane of the sun's orbit and concluded that the globe is divided into climatic zones: frigid near each pole, torrid along the equator, and temperate in between. Nine parallels and eleven meridians, although they were not named as such, were inscribed on Eratosthenes' map of the Oecumene.

The globe of Crates (Figure 2), described about 150 B.C., implies both a craving for symmetry and the assumption that if the area of the unknown is large enough it includes other lands and other men. A similar assumption motivates those of today's astronomers who are searching for orbital clues or radio signals that might suggest the possibility of inhabitable planets circling some of the stars of our galaxy.

Posidonius the Rhodian, who lived a century after

*Anti-podes: literally, people whose feet are placed opposite ours. The term was loosely applied to lands as well as to peoples.

the time of Eratosthenes, is credited with recalculating the earth's circumference by sightings on the star Canopus from Alexandria and from the Island of Rhodes. Such a task was really impossible because no means existed for measuring the distance by sea between the two points. Nevertheless, his result, as reported about 50 B.C. by Cleomedes, equalled about 38,400 kilometers, a figure so close to that of Eratosthenes that some modern astronomers suspect that Posidonius was working the problem backwards—that his actual purpose was to measure the distance from Rhodes to Alexandria by using Eratosthenes' value for the earth's circumference. As Posidonius' own writings were lost very early, we cannot be certain of either his purpose or of his original result. However, his calculations attained the most far-reaching importance when Strabo, about 20 B.C., quoted him as establishing the circumference as 32,918 kilometers. This newer and presumably better figure, which shrank the earth to a sphere only about four-fifths its true size, was accepted not only by Strabo, the historian's historian, but by Claudius Ptolemy, the most influential of ancient geographers, and it remained an article of faith among scholars for the next 1500 years.

The world map of Ptolemy (Figure 3), drawn about A.D. 150, depicted only one landmass, but that one of considerably greater extent than the Oecumene of Eratosthenes. Ptolemy extended Africa southward across the equator to the legendary Mountains of the Moon at the sources of the Nile (a detail that caused endless confusion among explorers until the 18th and 19th centuries when James Bruce discovered the source of the Blue Nile above Lake Tana and John Speke discovered that of the White Nile in Lake Victoria). Ptolemy pictured the African coastline as turning abruptly eastward at about 15° south of the equator and stretching beyond the Golden Chersonese (the Malay Peninsula) where it merges with mainland Asia. Thus, on his map the Indian Ocean is enclosed, much like a huge counterpart of the Mediterranean. The south shore of this ocean is labeled "Terra Incognita."

That egregious error derived from an early Greek conception of land-enclosed oceans, the antithesis of the configuration shown on the globe of Crates. The conception had been championed as early as the second century B.C. by Hipparchus, the astronomer, and in the first century A.D. by the geographer Marinus of Tyre. Ptolemy apparently accepted the opinion of these two authorities despite several lines of evidence that water routes were open between the Indian Ocean and both the Atlantic Ocean and the China Sea. As early as 450 B.C., for example, Herodotus repeated a tale that Phoenician sailors, on orders from King Necho of Egypt, had circumnavigated Africa from the Red Sea on the east, around the southernmost cape, up the west coast to the Pillars of Hercules, and back to Alexandria through the Mediterranean. The trip, which took place about 600 B.C., was said to have required more than two years, with the crews going ashore to sow and harvest crops each autumn, an effort that, in those latitudes, would require 60 to 90 days. The sailors reported that as they rounded southern Africa the sun shone upon them from their right. Today this detail is widely accepted as the strongest evidence that the expedition, while highly exceptional, was an authentic one. That it failed to have any influence on ancient ideas of geography appears doubly strange in view of the fact that King Necho's order seems to imply previous knowledge of the shape of Africa. Herodotus himself displayed his skepticism of man's ability to cross the torrid zone by doubting that part of the story about the sun slanting from the right (north). Later scholars evidently dismissed the story as a fable. Much the same fate befell a passage relating to the waters around southern Africa in a mariners guide titled *The Periplus of the Erythrean Sea*, written about A.D. 60 by an anonymous Greek merchant.

Ptolemy's errors, however gross, dwindle to insignificance beside his accomplishments in placing geography on a firm scientific basis by projecting a sphere on a plane and locating places with reference to a system of coordinates. Ptolemy was the first modern geographer, and he was also the last of the classical geographers with an interest in the broad configuration of lands and seas. By his day the Roman Empire had been flourishing for nearly two centuries, and Roman cartographers were narrowly utilitarian in outlook. They plotted, in highly schematic fashion, the routes of the Roman roads as these spread out from the Mediterranean to the boundaries of the expanding empire. The maps were useful to travelers, both civil and military, but were devoid of the vast richness of new information that Roman technology might have provided about the world as a whole.

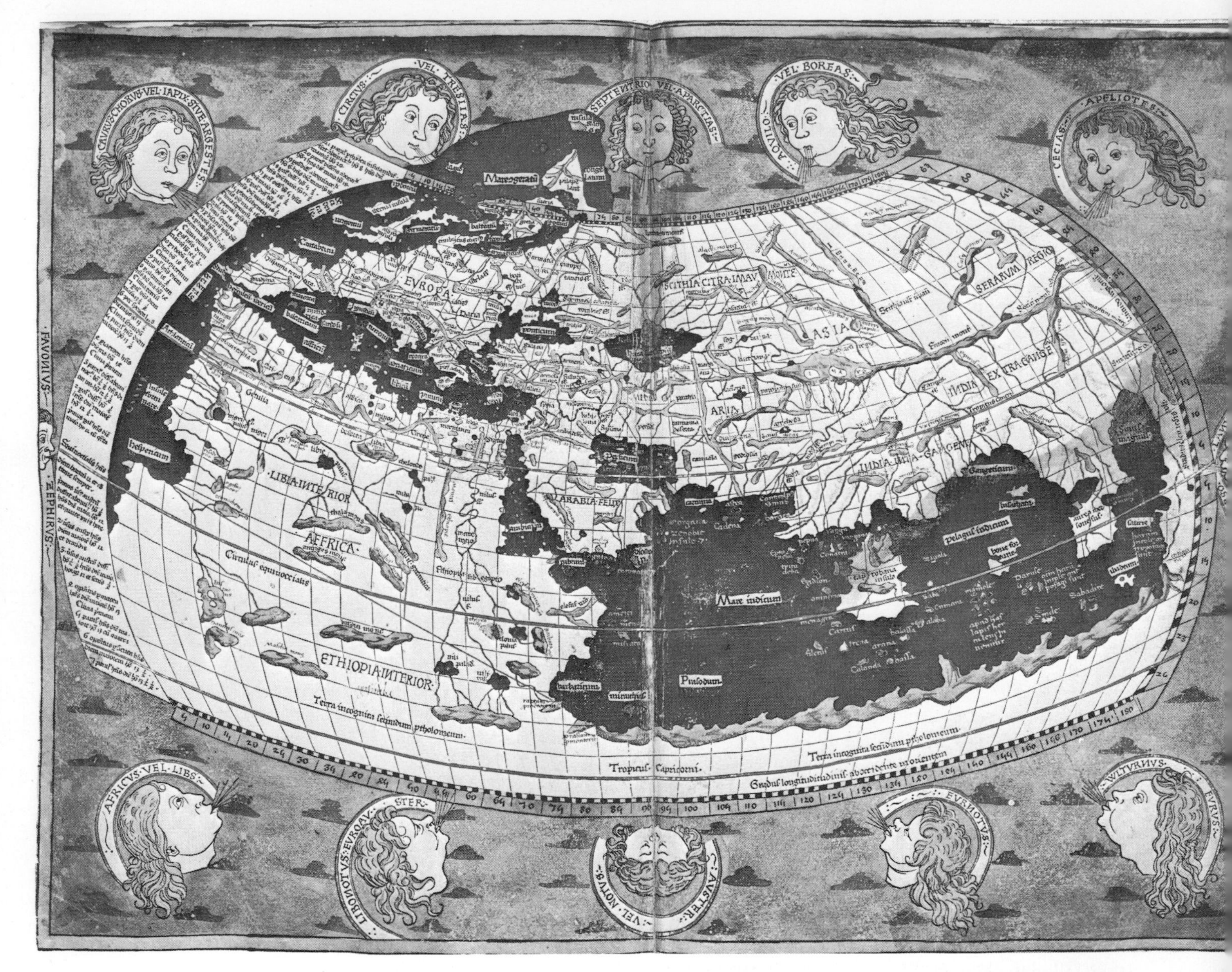

Figure 3. The world map of Ptolemy (about A.D. 150), as reconstructed and updated in the *Cosmographia* published at Ulm in 1482. The Indian Ocean is enclosed on the south by Terra Incognita. India lacks its proper peninsular form, and Ceylon (Taprobane) is shown as a large island straddling the equator. The sources of the Nile are depicted as two lakes fed by streams issuing from a range of mountains at about latitude 20° south. In other editions of the Ptolemy maps, these mountains are named Lunes Montes, the Mountains of the Moon. (Used with permission of the John Carter Brown Library, Brown University.)

The four corners of the earth

The concept of a large, spherical earth with room for more than one habitable landmass was profoundly disturbing to the Christian theologians who began to dominate the thought of the Roman Empire in the 4th century A.D. The scriptures taught that all men are descended from Adam through Noah, and St. Paul had said that God "hath made of one blood all nations of men for to dwell on all the face of the earth." Of those who preached the gospel he wrote: ". . . verily, their sound went into all the earth, and their words unto the ends of the world." How, then, could the Antipodes exist, separated from Christendom by the torrid zone? Some authorities were willing to accept the idea of a spherical earth with extra continents that were habitable but uninhabited. Others argued that if unknown men could not exist neither could undiscovered continents—God would create no landmass in vain. St. Basil, the Cappadocian, declared the whole question to be irrelevant: whether the earth be a sphere, a cylinder, or a disk is of no consequence to Christians intent upon salvation.

Belief in the Antipodes, and hence in a spherical earth, was denounced as absurd, if not heretical, by the Latin father Lucius Lactantius about A.D. 300. He and later theologians came to view scientific investigations in general as unprofitable, unnecessary, and dangerous—as indeed they are to any dogmatic system of thought.

A scriptural interpretation of geography entitled *Topographia Christiana* was written about A.D. 548 by Cosmas of Alexandria. Cosmas "proved" by an appeal to common sense, bolstered by scorn and ridicule, that the men on the underside of a globe would be upside down and the rivers and lakes would be emptied of their waters. Referring to the scriptures, he described the earth as flat, four-cornered, and designed upon the same plan as the ancient Tabernacle of the desert that was constructed by Moses under divine direction. His single inhabitable continent was surrounded by an unnavigable ocean beyond which lay Terra Ultra Oceanus, not a worldly landmass but the abode of man before the Deluge, a land that included Paradise itself where arose the great rivers that water the earth. The apparent rising and setting of the sun was ascribed to the passing of that body behind a great conical mountain in the north (Figure 4). To all who still favored a global earth Cosmas wrote: ". . . your reasonings are capricious, self-contradictory, inconsistent, doomed to be utterly confounded, and to be whirled round and round even more than that unstable and revolving mythical sphere of yours."

Before he settled into a monastic life in Sinai to write his book, Cosmas had traveled widely in the Red Sea and across the Indian Ocean where he must have seen many a ship's mast gradually disappear over

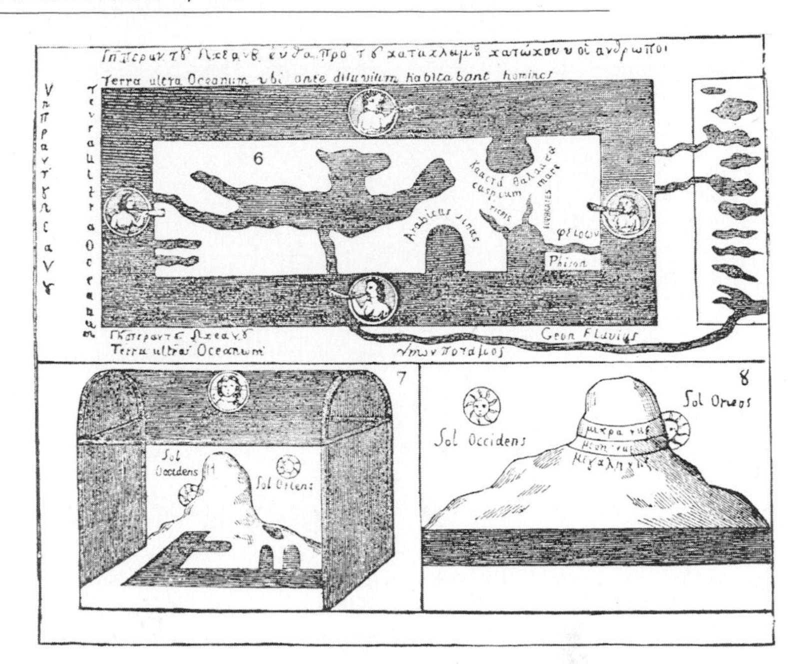

Figure 4. The earliest maps known to be of Christian origin are by Cosmas of Alexandria. They depict the face of the earth (6), the earth enclosed by the Vaults of Heaven (7), and the path of the sun (8). "Cease, O ye wiseacres! prating worthless nonsense, and learn at last, though late, to follow the divine oracles and not your own baseless fancies," wrote Cosmas to the advocates of a spherical earth. On his own rectangular earth the River Gheon flows westward out of Paradise and crosses the ocean to feed the Nile. The Mediterranean is the large body of water below the numeral 6; also shown are the Caspian Sea, the Arabian Gulf (Red Sea), and the Tigris and Euphrates rivers flowing into the Caspian. In 7, the land enclosed by the vaults of heaven is seen to slope southward from the base of a great conical mountain. In 8, the mountain is encircled by three lines which mark the paths of the sun as "he" moves round it at different altitudes, thus making summer nights shorter and winter nights longer. (From *The Christian Topography*, The Hakluyt Society, 1897.)

the horizon. Such visible evidence for a curving earth's surface he rejected in favor of an introspective search for truth with results that, however farcical they seem today, were of deadly effect in their time.

Cosmas' Alexandria was the great center of learning where both Eratosthenes and Ptolemy, at periods separated by nearly 400 years, had been directors of the world-famous Library. After Egypt came under Christian rule the Library ceased to be mentioned in the writings of scholars. Whether it fell into decay

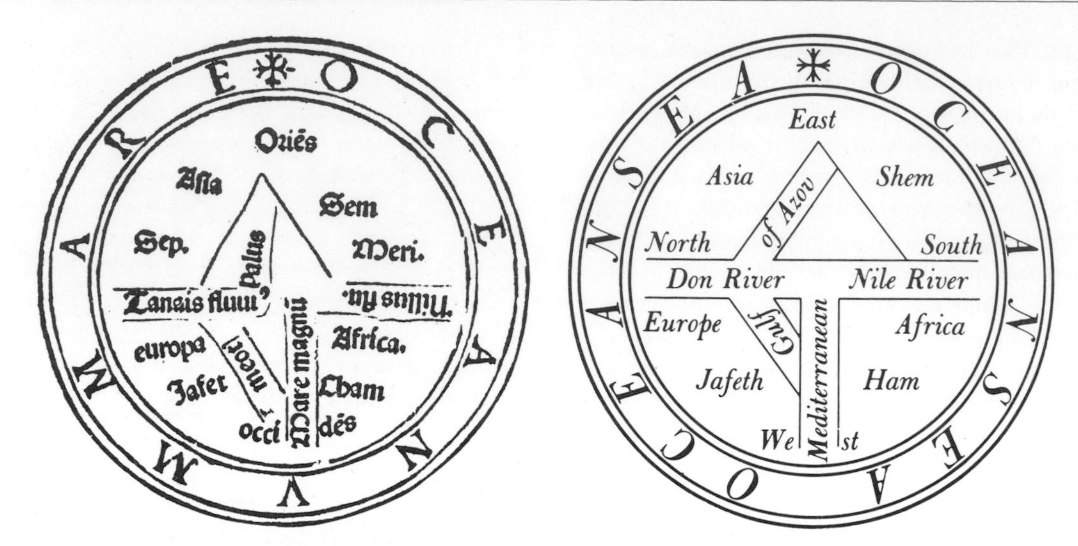

Figure 5. The disk-shaped world map of Isidore, Bishop of Seville (A.D. 570-636), with an English translation. The East (Orient) was at the top in most maps of the Middle Ages. The importance formerly attached to knowing the direction of east is reflected today in our use of the words oriented and disoriented. (Used with permission of The Walters Art Gallery.)

simply from neglect or was subjected to the systematic destruction imposed by some Christian rulers on pagan manuscripts and works of art remains unknown. In any case, the fate of the Library symbolizes the eclipse of learning that settled over Europe for the next thousand years.

. . . the triumph of Christianity [wrote George Sarton in his *Introduction to the History of Science*] was a distinct gain from the point of view of morality, but a loss from the point of view of scientific research. . . . The Greeks had taught the nobility of scientific study and that the pursuit of disinterested knowledge is the greatest purification; the Romans had urged the necessity of applying knowledge to immediate needs; the Christians were now insisting that if we have not charity it profits us nothing. . . . Unfortunately most men are incapable of grasping an idea, unless they exaggerate it to the exclusion of all others . . . most of the people who finally understood that charity was essential did not stop there, but jumped to the conclusion that it was all-sufficient. This led them to consider scientific research not only useless but pernicious. Thus the ruin of science, begun by Roman utilitarianism, was in danger of being completed by Christian piety. It has taken about one millennium and a half to make people generally understand that knowledge without charity and charity without knowledge are equally worthless and dangerous; a great many people do not understand it yet.

Fortunately, as scientific learning languished in Europe it flourished among the Arabs, who actively pursued new researches and also preserved both the *Almagest* (astronomical works) and the *Geographica* of Ptolemy. Farther east, in the Fergana Valley of Central Asia, the Moslem astronomer Alfraganus remeasured the circumference of the earth about A.D. 800, and arrived at the enlarged figure of approximately 43,584 kilometers. Meanwhile, Europeans lost interest in the configuration of landmasses. Their world maps degenerated to schematic disks with Jerusalem at the center and showing Europe, Asia, and Africa transected horizontally by the waters between the Don and the Nile, vertically by the Mediterranean, and ringed by the Ocean Sea (Figure 5). Europeans were proud of their other-worldly world view.

". . . there must be equal dispositions of land and water"

Nevertheless, the tradition of a spherical earth was never wholly lost in Europe. "It does not lie in the destinies of mankind that all should equally experience mental obscuration," commented Alexander von

Humboldt when he reviewed the ideas of the Middle Ages from the perspective of the 19th century.

In England, about A.D. 700, the Venerable Bede described the earth as occupying the center of the universe like the "yelk" of an egg. He believed in the possibility of unknown continents but not in the Antipodes. Churchmen who held similar views could argue that the inhabitants of any lands beyond the torrid zone would not be men but subhuman creatures called Antipods.

In 1268, Roger Bacon, an English Franciscan friar, sent to Pope Clement IV his *Opus Majus* in which he described the earth as a sphere, outlined what was known of the geography of Europe, Africa, and Asia, and lamented the great deficiency of information on the world as a whole. Referring to the *Almagest*, Bacon said that Ptolemy, among all who have lived since the Incarnation, gave the most accurate information, and he agreed with Ptolemy's conclusion (subsequently rejected by most scholars) that habitable land lies in more than one quadrant of the earth. Arguing once again from the requirements of symmetry, Bacon stated that if we

> consider natural processes in accordance with natural philosophy, we shall find that [the other] quarters are not covered with water as the majority of mathematicians estimate . . . there must be equal dispositions of land and water in our quarter and in the quarter beyond the equinoctial circle toward the other pole; and likewise in the [other two] quarters.

Bacon quoted Aristotle as saying that the Atlantic is a narrow ocean separating Spain on the east from India on the west. Aristotle had so concluded because of the abundance of elephants in the region of the Atlas Mountains of North Africa (which he regarded as part of Spain) and in India, but not in the intervening lands of the Near East. Similar species suggested similar habitats and implied that the Atlantic is a narrow water gap which elephants could cross by swimming more easily than they could walk across the broad landmass of Asia. Aristotle may have been the first philosopher to indulge in the pastime—which became very popular in our own century—of arranging continents and oceans on the basis of animal distribution. To illustrate his own concept of the Atlantic Ocean, Roger Bacon drew a diagram (Figure 6) showing water occupying both poles and a narrow strip between Europe and Asia.

The world map which Bacon included with his *Opus Majus* has been lost, but his description of the globe with its narrow Atlantic Ocean was transcribed almost verbatim in *Imago Mundi*, a geographical treatise compiled in 1410 by Cardinal Pierre d'Ailly of Cambray. This work, in turn, was studied in detail by Christopher Columbus who carried it on his voyages and quoted from it in letters he wrote to King Ferdinand and Queen Isabella in 1498 from the island of Hispaniola.

Roger Bacon was one of the two or three greatest men of science in pre-Renaissance Europe. He prac-

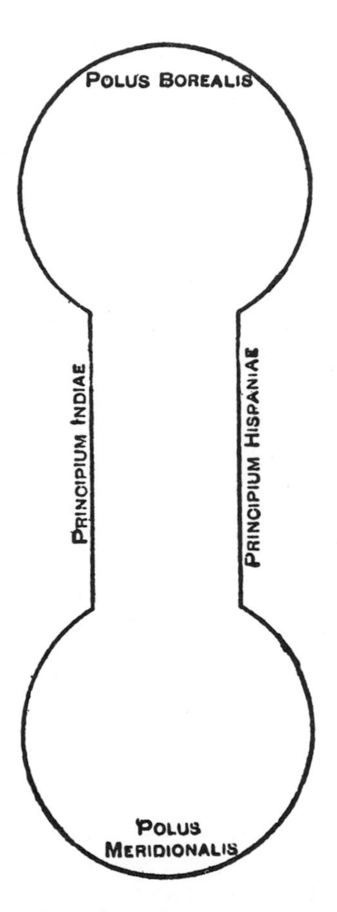

Figure 6. Roger Bacon's mathematical sketch (in 1268) of the Atlantic as a narrow ocean connecting two circular basins at the north and south poles (Polus Borealis and Polus Meridionalis) of a spherical earth. Spain (Principium Hispaniae), on the east coast, lies opposite India (Principium Indiae), on the west coast. (From *Opus Majus*, by Roger Bacon, translation by Robert Belle Burke, 1928; used with permission of University of Pennsylvania Press.)

ticed both experimental and observational techniques but believed that science ultimately should be subordinated to ethical purposes of the highest order, which he was confident resided in the teachings of the Church. His justification for urging upon the pope a complete and accurate survey of world geography was not, he said, a craving for knowledge of the human race and its habitat but for the defense of the Church, which vitally needed information on the whereabouts and habits of the evil races Gog and Magog (mentioned in Ezekiel 38-39 and Revelations 20) who were destined to come forth in the days of Antichrist and lay waste to all Christendom. These evil races (among whom Bacon counted the ten lost tribes of the Jews) had, according to Christian tradition, been imprisoned behind the Caspian Gates by Alexander the Great, who, although he lived 300 years before the time of Christ, evidently was regarded as a champion of the Christendom to come. In a fascinating compound of geology, mineralogy, and myth Bacon commented that unable to conquer them Alexander prayed to God:

> Although he was not worthy to be heard, yet God of his goodness for the preservation of the human race commanded that there should be a great earthquake, and the mountains distant a stadium approached each other so that the width of only one gate separated them. Alexander then built at the bottom of the pass bronze columns of wondrous size and erected gates and covered them with bitumen which could be dissolved neither by fire, water, nor iron. He sought this bitumen from the islands of the sea. These gates could be destroyed in no manner except by an earthquake; and now they have been destroyed.

Like modern men, medieval Europeans lived with a sense of imminent destruction. An early end of the world was always a tenet of Christianity, justifying a wholehearted focusing of concern on the life to come. In a more immediate sense, the legends of Gog and Magog came true with a vengeance for all peoples living east of Vienna in the days of Genghis Khan and his son Ogadai. Roger Bacon was 27 years old when the Mongols crossed the Danube and the Vistula Rivers to invade western Europe. They turned back that time not because of the fortified Caspian gates or Christian armies with improved geographical knowledge but because, on the death of Ogadai Khan in 1241, they had to go home to Karakorum to choose a successor. Bacon was wise in warning the pope of destructive hordes from the east, although he clearly lacked knowledge of their history, identity, and organization. His scientific views, however, which had won for him the protection of one pope, aroused the enmity of two other popes who regarded him as a heretic. He spent the last decades of his life in exile. Nothing came of his recommended survey of world geography until 200 years later when the ideas of Aristotle and Ptolemy (transcribed by Bacon and retranscribed by Cardinal d'Ailly) cast a spell on Christopher Columbus.

News from the Antipodes of the west

A broad interest in geography began to revive in Europe in the 15th century when the *Geographica* of Ptolemy was translated into Latin. As the Turks encroached on Byzantium, Greek scholars and manuscripts began filtering westward. Greek and Arabic versions of the works of Ptolemy had been copied and recopied for centuries with surprisingly few additions and revisions. They showed no scriptural reorganizations of landmasses and none of the fanciful medieval monsters that had become so common a decoration on European maps.

In 1406 a Latin translation of a Byzantine version of the *Geographica* was completed in Florence and handmade copies were widely disseminated, first without maps but later with one world map and 26 regional maps. These maps were soon updated by the addition of new information on the configuration of northern Europe, western Africa, and the vast territories of China, Japan, and southeastern Asia as described by Marco Polo, who had completed his travels in 1295. Although Ptolemy's map was clearly an improvement over the maps of the Middle Ages, it was, as we remember, based on a very serious error. Following Strabo, Ptolemy had limited the circumference of the earth to about 32,918 kilometers, only four-fifths of the actual circumference.

The distance from Cape St. Vincent at the southwestern tip of Portugal to the far side of Asia, according to Ptolemy, was 180° of longitude, or 16,460 kilometers reckoned by Ptolemy's short degrees of 91.4 kilometers. This was an overestimate of about 50 percent. The addition of China and Japan extended Asia some 4,780 kilometers farther east, or two-thirds of the way around Ptolemy's small globe. The implications of this new geography proved irresistible to Christopher Columbus, who found ways to shrink the globe and extend Eurasia even farther eastward. Columbus singled out the calculations of Alfraganus, the 9th-century Moslem astronomer, as the best measurement of the size of the earth. In fact, Alfraganus' globe was too large, but Columbus concluded that the astronomer must have used the short Roman mile of 1,480 meters rather than the Arabic mile of 2,137 meters. The substitution of Roman miles in Alfraganus' calculations reduced the earth's circumference to only 30,180 kilometers, about 10 percent below the figure of Ptolemy and 25 percent below the true value.

On this diminutive globe, Columbus stretched Eurasia for some 283° of longitude by adding China and Japan not to Ptolemy's estimate of 180° but to an earlier estimate of 225° made in the first century A.D. by Marinus of Tyre. This brought the coast of Japan all the way east to the true longitude of Havana, Cuba. Columbus convinced himself that along latitude 28° north, where degrees are shorter than they are along the equator, he had only 4,320 kilometers to sail between the Canary Islands and Japan.

Armed with this vision, Columbus spent eight years trying to persuade the monarchs of Portugal and Spain to sponsor him in a voyage to the Indies, which he was certain he could reach by sailing westward. For all his single-minded determination, Columbus would never have contemplated crossing the vast span of over 19,000 kilometers that actually separates western Europe from eastern Asia. And he would have been fully aware of the true distance if the measurement of the earth's circumference made by Eratosthenes in the third century B.C. had been accepted as correct, or if any European scholar, ancient or medieval, had shown enough scientific curiosity and skill to try a new measurement and get it right.*

Columbus' general view of the world is illustrated on a globe that was completed in 1492 by Martin Behaim, a merchant of Nuremburg and amateur chartmaker who had lived for several years in the Azores and had himself commanded an exploratory voyage in 1489. Behaim's globe (Figure 7), 51 centimeters in diameter, was commissioned by the burghers of Nuremberg for the decoration of their town hall, a circumstance that testifies to a widespread and then probably very modern belief in the sphericity of the earth. Behaim departed from Ptolemy by showing Africa as a peninsula and China as thrusting far to the eastward. He also placed numerous large islands between the Azores and Japan. Columbus and Behaim clearly drew many of their ideas from the same sources.

Columbus was not, of course, the first navigator to sail westward into the Atlantic. By 1427 Prince Henry the Navigator, of Portugal, began sending ships westward as well as southward along the coast of Africa, and over the next four years his captains discovered seven islands of the Azores. Earlier still, Norsemen had sailed to North America, not in their sleek, dragon-headed fighting ships but in their tubby knarrs which carried colonists and trading goods between Norway, Iceland, and Greenland.

According to Samuel Eliot Morison in his book *The European Discovery of America*, the first European to cross the Atlantic and return to tell the tale was probably a Norse trader named Biarni Heriulfson. In the autumn of A.D. 986, Biarni left Iceland for Greenland but strayed southward off-course until he sighted a level land covered with forests. Knowing this could not be Greenland, he turned northward, sighted another forested land, and then sailed northeast until he reached the Norse settlements on Greenland's west coast. Biarni never put ashore on the unknown western lands, but his description of heavy forests prompted Leif Ericsson, in the summer of 1001, to retrace Biarni's course in search of timber for the Greenland colonies. Leif landed at sites that almost certainly were on Baffin Island, in Labrador, and in Newfoundland where, at L'Anse aux Meadows (Vinland), he spent the winter and where subsequent Norse parties attempted colonization.

The Vinland settlements were abandoned within a few years, and those in Greenland died out at some unknown time in the 1400s long after all contact with Europe had been lost. Although the Greenland and Vinland colonies were both mentioned in the *Hamburg History of the Church*, written by Adam of

*The first recorded measurement in Europe of a degree of a meridian was made in 1528 by Jean Fernel, a French physician who counted the revolutions of his carriage wheels on a trip from Paris to Amiens and calculated a value very close to the correct one of 111.13 kilometers.

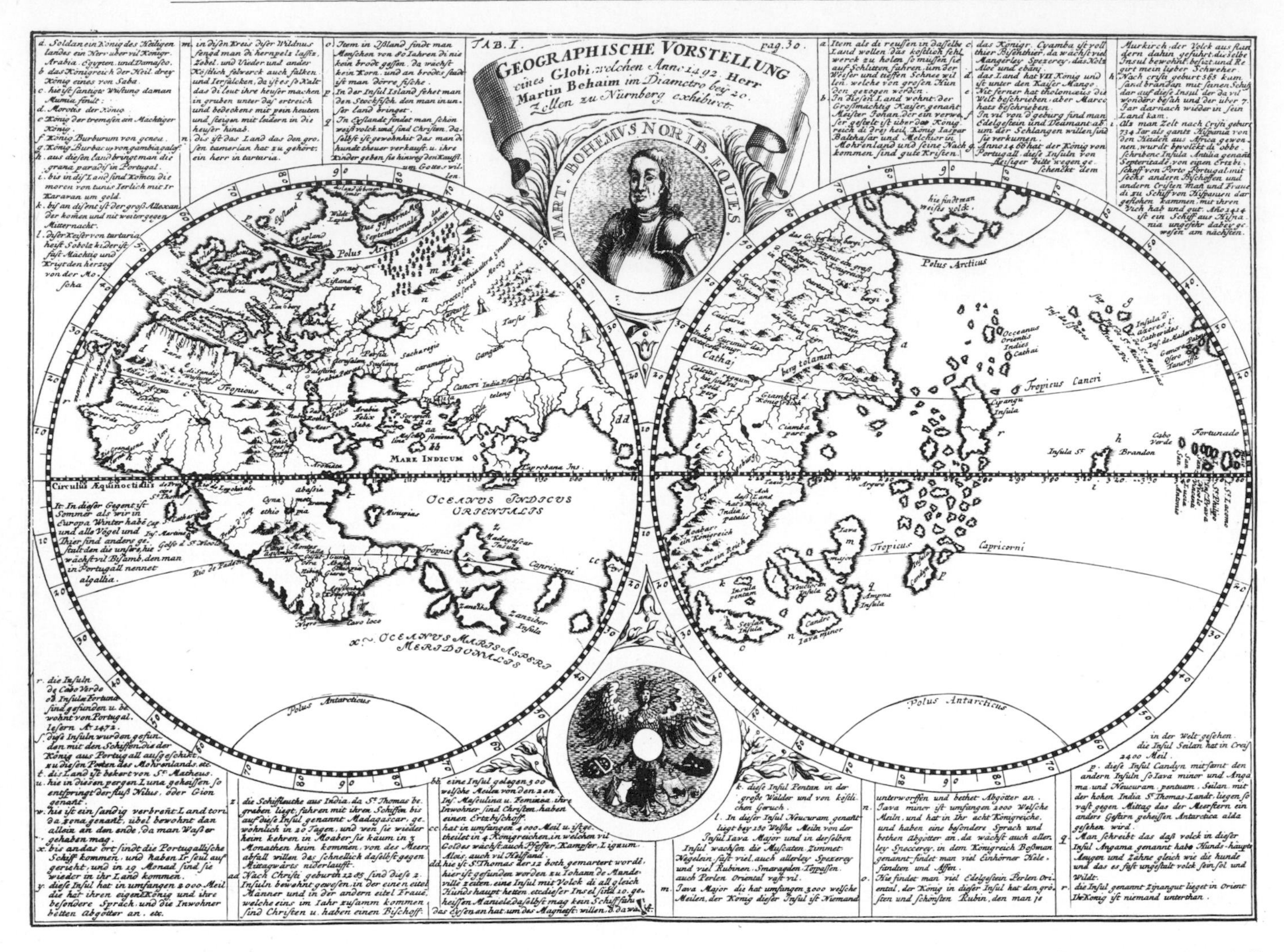

Figure 7. An 18th-century engraving by Johann Doppelmayer of the map used by Martin Behaim on his Erdapfel ("earth-apple," or globe) of 1492. Although Africa bends eastward in a fashion reminiscent of Ptolemy, the Indian Ocean is open and the south polar region is free of land. Behaim's Erdapfel, the oldest terrestrial globe in existence, is preserved in the Germanisches Nationalmuseum at Nuremberg. (Used with permission of Universiteits-Bibliotheek, Amsterdam.)

Bremen in 1074, the popes showed no effective concern for the fate of their far northern communicants. In time, Greenland itself was so completely forgotten that it was mapped as a new discovery when a Portuguese ship landed at its southernmost cape in 1500. Thus, remarkable as the Norse achievements had been, they seem to have added nothing to the geographical working-knowledge of the rest of maritime Europe.* Columbus himself is reported by his son Ferdinand to have visited Iceland in 1467. Ferdinand does not mention Vinland, however, although he tended to emphasize the extent of his father's knowledge and experience. If Columbus went to Iceland, despite the doubts of serious historians—or if he ever heard of Vinland,

*This is also true of the hosts of other voyagers who are alleged to have crossed the Atlantic before Columbus. If they came—those Phoenicians, Egyptians, Romans, Hebrews, Welsh, Irish, and others—they vanished utterly from contact with the Old World and so had no influence on the growth of geographical knowledge. Also, they left no positively identified trace in the New World.

which is equally doubtful—he clearly viewed it as a place to avoid on his route to the warm, gold- and spice-laden Indies.

Columbus' proposed expedition to Asia benefited not only from his underestimate of distances but also from a new, inquisitive intellectual climate in Europe. A course into the open ocean had been out of the question as long as people believed that the earth was flat and the ocean teemed with monsters or was bounded by the mystical reaches of Paradise, and as long as they used galleys requiring banks of oarsmen and frequent stops for provisions. The revival of geographic learning, the development of small, decked vessels capable of tacking into the wind, and the development of the mariner's compass, however crude, all took place in the 12th to 15th centuries and helped toward the solution of these problems.

Columbus' conviction about his mission never faltered. Throughout his four voyages he believed that the lands he visited were outlying parts of Asia. On his first landfall at San Salvador in the Bahamas he had sailed close to his calculated longitude for Japan. On his second voyage he bound his officers in an oath that they had landed on the Malay Peninsula. He was forced to turn back from his fourth and final voyage when he believed himself a short distance from the mouth of the Ganges. Columbus died in 1506 without conceding that he had discovered a "new world," which he had, and in the conviction that he had opened a westward sea-route to the Indies, which he had not. As one result of his obsession, based on too small a globe and too large a Eurasian landmass, the descendants of a northern Asiatic people who had crossed the Bering Straits some 25,000 years earlier and inhabited the two western continents acquired the name Indians—to their own confusion and to that of a wholly unrelated people, the Hindus, who refer to their homeland not as India but as Bharat.

If Columbus was too intent upon his inward vision to perceive the true nature of his discoveries, he was, unlike Cosmas of Alexandria, a man of action whose accomplishments were to prove valuable to others. The excitement of the time is manifest in the letters of Peter Martyr, an Italian scholar at the court of Spain:

May 14, 1493. Within these last few days has arrived a certain Christopher Columbus from the Antipodes of the West; a man of Liguria (Genoa). . . . Although sent forth by our sovereigns he could but with difficulty obtain three ships, since what he said was regarded as fabulous.

September 13, 1493. The wonders of this terrestrial globe round which the sun completes a circuit in the space of four and twenty hours have, until our time, been known only in regard to one hemisphere, from the Golden Chersonese to our Spanish Cadiz. The rest has been given up as unknown by cosmographers, and any mention of it has been slight and dubious. But now, O blessed enterprise . . . .

November 1, 1493. Columbus, he who discovered the New World, the Sea-Chief of the Indian Ocean. . . .

October 31, 1494. Hear what things have lately been discovered at the Antipodes in the Western Hemisphere. Marvellous things are related . . . these I would describe but the impatience of the messenger will not suffer it.

Peter Martyr coined the phrases "new world" and "western hemisphere." Even so, for some time he believed that Columbus had sailed to Asia, and he titled him the Sea-Chief of the Indian Ocean. He also depicted the sun as daily encircling the earth.

Terra Australis

The outlines of islands and larger landmasses discovered by Columbus and by the explorers that immediately followed him soon began appearing on maps. For several years these lands were plotted as though they were eastward-thrusting promontories of Asia, and some mapmakers continued that practice until 1533. More realistic maps were drawn, however, by officers who sailed with Columbus, Vespucci, and others.

In 1507, Martin Waldseemüller, in Alsace, drew the first world map showing the newly discovered western lands clearly separated from Asia by a large ocean. He outlined a southern continent and labeled it "America" in honor of Amerigo Vespucci, who had charted the coasts of Venezuela and Brazil and thus, according to Waldseemüller, had added a fourth continent to those known to Ptolemy. Although this credit clearly belongs to Columbus, who had also charted the coast of Venezuela (and concluded that the Orinoco arose in Paradise), it is an honor Columbus would have emphatically rejected in his lifetime.

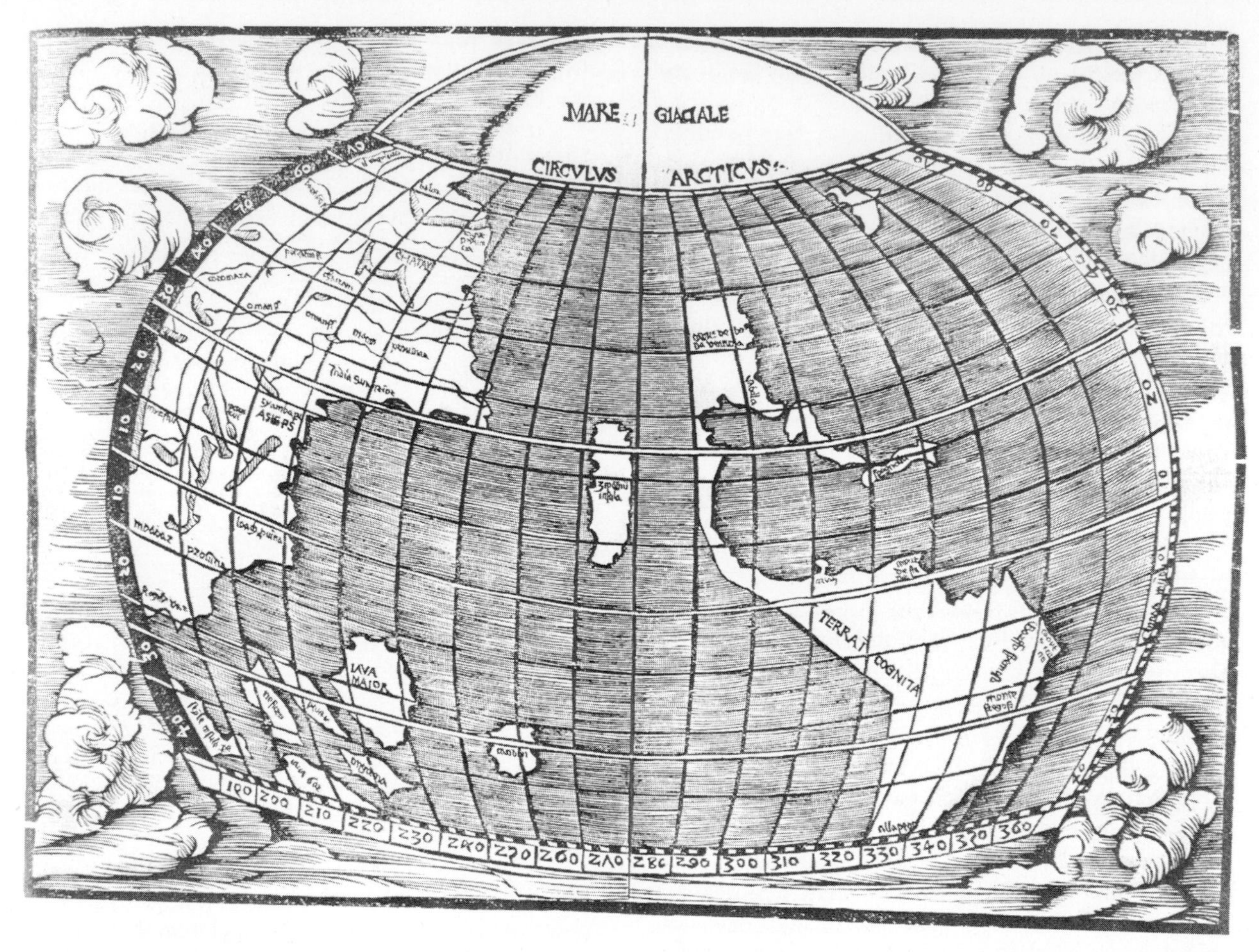

Figure 8. Martin Waldseemüller's depiction of the western continents separated from Asia by an ocean appeared in a marginal inset to his large world map of 1507. The name "America" appears on the large map but not on the inset; the narrow isthmus appears only on the inset. Japan (Zipangu) lies closer to the isthmus than to Asia. This print is from a copy of Waldseemüller's inset published in Cracow in 1512 by Johannes de Stobnicza. (Used with permission of the John Carter Brown Library, Brown University.)

One curious feature of Waldseemüller's map is a small, marginal inset (Figure 8) of the western hemisphere in which the isthmus of Panama and the *west* coast of South America are represented schematically but with haunting accuracy. Although Columbus had been told by the inhabitants of the Panama region that a large sea lay a short distance to the south, he had assumed it was an Asiatic gulf. Not until six years after the drawing of Waldseemüller's map was the isthmus crossed and the ocean discovered, one September morning, in the event immortalized in John Keats' stanza:

Or like stout Cortez, when with eagle eyes
He star'd at the Pacific—and all his men
Look'd at each other with a wild surmise—
Silent, upon a peak in Darien.

Poetic license notwithstanding, the explorer who made this discovery was not stout Cortez but brave Balboa, and "all his men" numbered 190 Spaniards and 1,000 Indians. The ocean was not named the Pacific until seven years later by Admiral Magellan.

After Ferdinand Magellan (or rather the 18 men who were left of his original crew of 275) completed the circumnavigation of the world—from Seville in

1519 and back to Seville in 1522—the Ptolemy map became totally obsolete and world maps began to show roughly the right configuration of landmasses, at least in the region of the equator. But the polar regions remained mysterious. The discovery of the Americas had confirmed the ancient concept of the western Antipodes—what, then, of the equally venerable tradition of a great continent in the south?

The Greek concept of a spherical earth with four symmetrically disposed landmasses was, as we have seen, outlawed by early Christian authorities. In the Middle Ages, however, two cycles of maps that were in circulation showed a southern landmass separated from the northern world of Europe, Asia, and Africa by the impassable torrid zone. One map cycle derived ultimately from the writings of Cicero who, in the first century B.C., defended the idea of a spherical earth and the Antipodes in his *Dream of Scipio*. Cicero's work was expanded upon in a commentary written about A.D. 410 by Aurelius Theodosius Macrobius. The map in Figure 9 was drawn in the sixth century to illustrate a manuscript copy of Macrobius' commentary. The other map cycle, called the Beatus cycle, derived from the following sentence in the *Etymologiarum* of St. Isidore of Seville, written about A.D. 600:

Figure 9. A map based on the ideas of Macrobius. The known world, centered on Jerusalem (Hierusa), is balanced by a large southern continent, the unknown land of the Antipodes. (From the first printed edition of Macrobius, *In Somnium Scipionis exposito*, Brescia, 1483. Used with permission of the Annmary Brown Memorial, Brown University.)

> Moreover beyond [these] three parts of the world, on the other side of the ocean is a fourth inland part in the south, which is unknown to us because of the heat of the sun, within the bounds of which the Antipodes are fabulously said to dwell.

Macrobius was a pagan, and so of questionable authority, and Isidore was not consistent in his descriptions of the earth. (Elsewhere in the *Etymologiarum* Isidore favored the three-continent disk map shown in Figure 5.) Nevertheless, these two scant sources kept alive the tradition of at least one undiscovered continent until 1488 when Bartolomew Dias proved conclusively that open water lay south of Africa. As a result of Dias' voyage the globe of Martin Behaim (Figure 7), made in 1492, showed unbroken ocean occupying the region of the south pole.

A 2,000-year old idea does not die easily, however. In 1515 Johannes Schöner of Nuremberg completed a globe on which there appeared a large southern landmass labeled "Brasilie Regio." Schöner's globe displayed two other interesting details: a strait through the Isthmus of Panama, where none exists, and a strait between the southern landmass and South America, where none had been discovered. Six years after Schöner's globe was completed, Magellan sailed through a narrow passage separating South America from a mountainous land, dotted with fires, which he named Tierra del Fuego. Magellan did not claim that Tierra del Fuego was a continent but its existence fanned imaginations in Europe and revived in full force the belief in the Antipodes of the South. The first map to depict Magellan's strait was drawn about 1529 by Franciscus Monarchus of Antwerp. That map shows a large continent centered on the south pole and a smaller one at the north pole. Thus was the earth, in its position at the center of the universe, assured of the proper balance.

The great southern continent was named Terra Australis (Land of the South) on a map drawn in 1531 by Oronce Finé, a professor of mathematics at Paris. Terra Australis, sometimes described as Incognita (Unknown) or Magellanica (Magellan's land), remained on maps where it inspired voyages of exploration for the next 250 years. Its northern

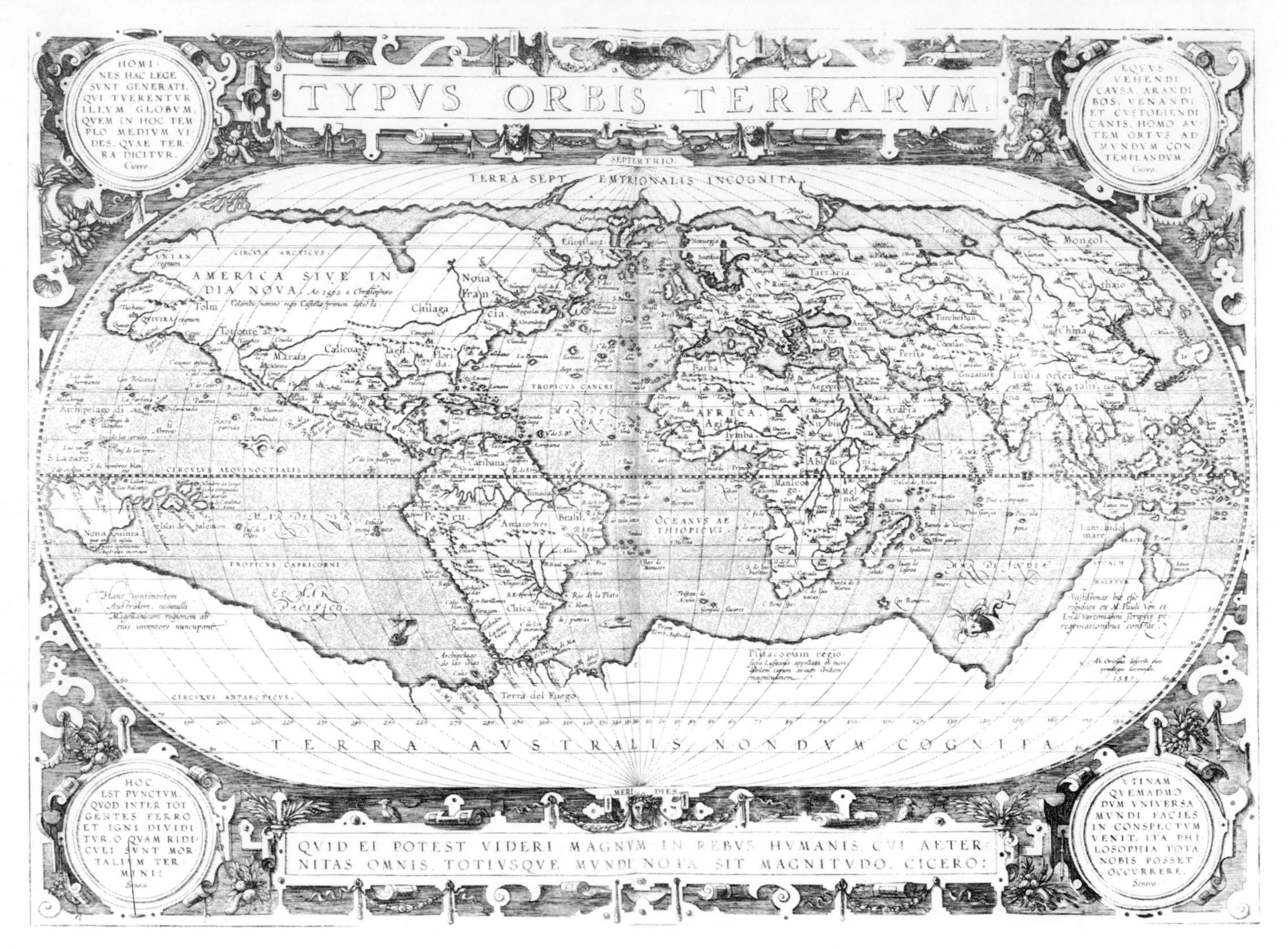

Figure 10. The world map of Abraham Ortelius dated 1587. On this map the two polar continents, Terra Australis and Terra Septemtrionalis, reached their greatest extent. (Used with permission of the John Carter Brown Library, Brown University.)

counterpart, lying beyond the tundras and glacial seas of Eurasia, was less alluring as an object in itself. But in 1538 when Gerhard Mercator, one of the master cartographers of all time, published a world map showing North America separated from the arctic landmass by a narrow strait he sparked the search for a northwest passage to Asia. Mercator's map is famous for being the first to name both of the western continents "America." It should also receive credit for luring many a vessel into the dangerous waters of the arctic archipelago. Mercator's intuition was not wholly wrong: a precarious passage is possible through the islands of northern Canada. The first commercial vessel to use this passage, however, was not en route to the spice islands of Asia. It was an oil tanker, the *S.S. Manhattan*, which made the voyage from New York City to Point Barrow, Alaska, in 1969, 431 years after publication of the Mercator map.

The maps of the 16th century, drawn by Mercator and others, showed both polar continents growing in area until a limit was reached in 1587 with Abraham Ortelius' *Typus Orbis Terrarum* (Figure 10). On this map the unknown land of Terra Septemtrionalis

Figure 11. A sketch of the southern hemisphere taken from Buffon's "Carte magnétique des deux hemisphères" of 1778. Buffon wrote that the record of the second voyage of the celebrated Captain Cook was not published in France until after completion of the volume containing this map, but "I have seen to my great satisfaction my conjecture confirmed by the facts . . . that the region of the arctic pole is occupied by ice and that of the antarctic pole by even more ice." (From Buffon, *Supplément a l'Histoire naturelle*, Volume 5.)

occupies the north, and Terra Australis extends almost to the tropics in all three southern oceans. Maps such as this one prompted Francis Bacon, in 1620, to write:

Even in the very figure of the World in its greater parts *Conformable Instances* are not to be neglected. As Africa and Peru with the continent which stretches to the Straits of Magellan: for each region has similar Isthmuses and similar Capes, which is no mere accidental occurrence. So too the New and the Old World are Conformable in this, that both Worlds are broad and extended towards the North, but narrow and pointed towards the South.

No mere accidental occurrence—for this observation Francis Bacon is often cited as an early advocate of continental drift.

In 1738, Pierre Bouvet, a French officer, persuaded the French East India Company to send him into the South Atlantic to prove or disprove the existence of Terra Australis. He sighted a promontory rising above a fogbound coast 2,250 kilometers south of Cape Town at latitude 54° south, and, turning eastward, he skirted fields of pack ice crowded with seals and penguins for about 320 kilometers. Bouvet believed that he was close to land throughout his trip along the margin of the ice. His mate disagreed, and later was proved to be right. The land they sighted is now called Bouvet Island.

Thirty-two years later, in 1770, Captain Cook charted the eastern coast of a landmass lying southeast of Java on the Tropic of Capricorn. He named this land New South Wales and claimed it in the name of King George III. Dutch navigators had charted the western coasts 150 years earlier and had named the land Nova Hollandia, but the British claims to the entire area prevailed. In 1810 a London cartographer inscribed the new continent with the name Australia, which came into general usage and displaced all earlier names, including "Ulimaroa," the melodious name used by the Maoris of New Zealand. To Englishmen, whose homeland lies close to the opposite end of a diameter of the earth, Australia and New Zealand have, since their discovery, been regarded as the Antipodes.

Australia, however, was too small a continent and too far north to be equated with Terra Australis. Therefore, in 1772, Captain Cook sailed southward again on what one of his officers, Charles Clerke, called a continent-hunting expedition.

On January 17, 1773, Cook made the first recorded crossing of the antarctic circle. He circumnavigated the south pole at latitudes ranging as high as 71°10′ south, but he never sighted land. Cook succeeded handsomely in disproving the existence of any huge continent like Terra Australis, and for the next 45 years many of the maps drawn in Europe resembled the 1492 globe of Martin Behaim by showing the south polar region as a vast expanse of empty ocean (Figure 11). Yet, a continent does lie over the south pole, and it is larger than either Europe or Australia. Antarctica was first sighted in 1820 when an English vessel under Captain Bransfield charted the waters around Trinity Island. The first landing on the continent was made a year later by men sailing under Captain John Davis, a sealer from New England.

Perhaps the most exciting moment in Antarctic exploration came in 1841 when the men of the English vessels *Erebus* and *Terror*, commanded by James Clark Ross, penetrated the ice packs to the edge of the continent and suddenly caught sight of the belching fumes of an active volcano at latitude 77°30′ south, longitude 167° east. This spectacular mountain of fire in a land of ice they named Mount Erebus, after their ship, which had borrowed the Greek name for the region of darkness at the entrance to Hades.

Observing Ross and his men from the cliffs along the shore were packed ranks of black and white birds standing erect and greeting the arrivals from Europe with hoarse, excited cries. Here were the authentic Antipods, nonhuman creatures inhabiting a land with a climate too harsh for men.

The broad configuration of the earth's continents and oceans was known at last.

Terrestrial Motion and Magnetism

Geography burst out of the Ptolemaic and Christian traditions about the turn of the 16th century as ships from the ports of Europe embarked on global explorations. The first stirrings of new ideas about the earth's place in the universe and on the processes altering its surface features occurred about the same time, but the establishment of modern astronomy was still more than a century away, and that of geology more than three centuries away.

". . . like any other planet"

In 1514, five years before Magellan sailed from Seville to circumnavigate the earth, Nicolaus Copernicus completed his *Commentariolus* in which he stated: "We revolve about the sun like any other planet."

Copernicus was quietly challenging the authority of Aristotle, of Ptolemy, and of Christian tradition. Aristotle's views carried great weight after his works were reintroduced to Europe by St. Thomas Aquinas in the 13th century. Aristotle had said that the earth and the universe had the same center. About 40 years after the death of Aristotle, this view had been challenged by Aristarchus of Samos who placed the sun at the center of the universe and described the daily rotation of the earth. Aristarchus also perceived the planets and stars as extending outward for distances far vaster than those envisioned by any other early astronomer. Aristarchus' heliocentric system could not be proved correct by any means available in his time, nor did it provide a more reasonable explanation of observed planetary motions than the "common sense" idea of an earth-centered system. It was rejected, therefore, long before Ptolemy worked out his elaborate system in which, with epicycles superimposed on cycles, the sun and planets revolve about the earth. Ptolemy would have been surprised indeed, according to Fred Hoyle, had he been told that, after his *Almagest*, no significant advance in astronomy would be made for the next fourteen hundred years.

By the time of Copernicus certain difficulties had become apparent, including troubles with the calendar. To most investigators, however, none of the difficulties seemed so serious as to warrant the overthrow of the Ptolemaic system. And church tradition, long since divested of belief in a flat earth, still embraced the concept of the earth as a special creation, immovable in the center of the universe and not to be regarded as "like any other planet."

Copernicus, a Catholic, was not at heart a rebel. Indeed, at least one historian of astronomy, Owen Gingerich of Harvard University, has suggested that Copernicus designed his heliocentric system mainly to achieve greater symmetry and esthetic appeal rather than to reform the science. In any case, for thirty years Copernicus pursued his calculations at Frauenberg, Poland, where he was a canon of the cathedral. He might never have attempted to publish his final results had it not been for the urging of a young disciple, the Lutheran astronomer Joachim Rheticus. Copernicus was old and ill when he finally completed his manuscript, and, through Rheticus, he entrusted it to Protestant publishers. Due to a mischance, Rheticus could not be present to see it through the press and so, in 1543, when the first printed copy of *De Revolutionibus Orbium Coelestium* was placed into the hands of Copernicus on the day he died it included a preface, inserted by the publisher without authorization, stating that the new system was not necessarily true or even probable. Thus, Copernicus, who is now generally regarded as the founder of modern astronomy, failed to bring about any change in his science during his own lifetime.

The opposition to Copernicus was not solely theological. Many astronomers remained unconvinced. Copernicus' calculations embodied no actual proof that the earth moves, nor did they predict the motions of heavenly bodies more rigorously than had those of Ptolemy. The arguments that had been used against Aristarchus of Samos could still be raised in the time of Copernicus. Realizing this, Copernicus removed all reference to that ancient astronomer from his final manuscript. The insights of Copernicus were rejected outright by Tycho Brahe, the greatest astronomical observer of the 16th century, because Tycho could not see the stellar parallax that was predicted by Copernicus' model. Not until new calculations made by Johannes Kepler, some 80 years later, indicated the superiority of Copernicus' fundamental idea did astronomers begin to reorient their thinking in favor of a sun-centered system.

"The immovability of the earth is thrice sacred"

The change did not take place in time to affect Kepler's contemporary, Galileo, who was charged with heresy when he reaffirmed the theory that the earth moves around the sun. Galileo, a devout Catholic and long-time friend of the man who had become pope, was both aggressive and indiscreet in the publication of his ideas. In 1633 he was forced to recant: "I, Galileo . . . aged 70 years, arraigned personally before this tribunal and kneeling before you . . . having before my eyes and touching with my hands the Holy Gospels . . . do abjure, curse, and detest the errors and heresies . . . that the Sun is the center of the world and immovable, and that the Earth is not the center and moves."

Galileo spent the last eight years of his life under house arrest but during that time he carried out experiments with the pendulum and developed a fundamental theorum of dynamics which probably constituted his greatest and most original contribution to science.

If the earth were a planet, and only one among several planets, no such great things can have been done for it as those taught by the Christian doctrine. Since God creates nothing in vain, other worlds, if they exist, must be inhabited. But how could their peoples be descendants of Adam and of Noah? How can they be counted among those redeemed by the sacrifice of Our Lord? These questions were asked by the Church fathers of Galileo's time from the deep shelter of a culture still maintaining the rightness of such exquisite logic.

In 1633, in a ringing denunciation of the Copernican doctrine, Father Melchior Inchofer wrote:

> The opinion of the earth's motion is of all heresies the most abominable, the most pernicious, the most scandalous. The immovability of the earth is thrice sacred; argument against the immortality of the soul, the existence of God, and the Incarnation, should be tolerated sooner than an argument to prove that the earth moves.

E pur si muove. "Yet it does move." These words, so often attributed to Galileo, murmuring as he rose from his knees, obviously could not have been pronounced audibly—nor could they fail to have been in his thoughts, secure as he felt in the ultimate rightness of the knowledge he had gained from experiment and observation.

Before the end of the 17th century new knowledge of the shape and motion of the earth was deduced by Isaac Newton in England. From a study of its rotation and precession, Newton calculated that the earth should be an oblate spheroid, slightly flattened at the poles. This idea (which was steadfastly opposed by the Paris Observatory's first three directors, Domenico Cassini and his son and grandson) made sense of the observation that pendulums had to be shortened in order to keep accurate time in the region of the equator. Newton also worked out the law of universal gravitation which governs the motions of the planets and satellites of the solar system, of falling bodies on the earth, and of the waters of the oceanic tides. Gravity not only impels the fall of an apple to the ground but keeps the moon in orbit, constantly "falling" toward the earth, otherwise the moon's simple forward motion would have carried it straight off into the reaches of space after a close brush with our planet.

Certainly one of the geniuses of Western civilization, Newton was an introspective man who fought bitterly to establish the priority of some of his ideas but who was indifferent to the publication of others. The *Principia* (1687) was written at the insistence of the astronomer Edmond Halley, who appreciated the great import of Newton's results the moment he heard of them on a casual visit to Newton. Although far from a wealthy man, Halley bore in full the costs of publication and carried out the prodigious labor of seeing the work through the press. Thus Halley played the indispensable role of communicator, for without him Newton's greatest work might have been lost.

Outside a small circle, such loss would have occasioned no immediate regret. To many scientists, Newton's picture of gravitational influences acting across the vast emptiness of space smacked uncomfortably of the occult. Catholics and Protestants charged him with atheism, although he was passionately religious and he was to spend his last years speculating about the nature of God. In Newton's time there was no voyage of exploration that could refute both common sense and sacred history and "prove" the reality of such abstract notions as a moving earth or a universal law of gravitation.

To improve the knowledge of the longitude and variations of the compass

The magnetic properties of certain black rocks have intrigued men since ancient times, and the north-seeking behavior of magnetized needles mounted on a swivel, or stuck into straw and floated on water, has been used as a navigation aid since at least the 12th century. The first references to the subject are found in a Chinese manuscript written by Chu Yi about 1113 and in one by Alexander Neckam, a monk of St. Albans, in England, about 1187. About 1200, Guyot de Provins, a French poet, expressed the wish that the pope were as fixed and true a guide in spiritual matters as is the pole star to mariners, who can find that star even in storm and cloud by the use of a needle rubbed on a loadstone. No connection between the Chinese and European sources is known, but all the sources indicate that the magnetized needle was a well-known and widely used instrument.

The insight that the earth itself behaves as a huge magnet with a dipolar field is ascribed to William Gilbert, who was distinguished in his time as physician to Queen Elizabeth I but is more famous today as the author of *De Magnete*, published in Latin in 1600. Gilbert, a stout defender of the Copernican system, believed that the compass direction is "forever unchanging" unless catastrophes occur—such as the "great break-up of a continent and annihilation of countries, as of the region of Atlantis, whereof Plato and ancient writers tell." In this observation Gilbert erred. The earth's magnetic field is subject to constant minor variations that were noticed by mariners at least as early as the 13th century. In the belief that the field should be stable, the variations were commonly ascribed to faulty compasses until the observations of Columbus (who carried two compasses of different design) and of those who followed him across the Atlantic persuaded navigators that the fluctuations were real. By 1635 sufficient measurements were available for H. Gellibrand to announce that the compass direction at London had changed 7° in 54 years. From that time, records kept in London and Paris have showed continual variations, differing in each city, of both the declination (deviation from true north) and inclination (departure from the horizontal) of compass needles. These changes, called secular variations, are very rapid on a geologic time scale. Their predominating trend appeared to be a westerly drift of the line of zero declination, which passed through London about 1660 (the year of the coronation of King Charles II) and temporarily made work easy for English surveyors. But a westward drift could not be the whole story because Columbus reported crossing a line of zero declination in the vicinity of the Azores in 1492. A better understanding of the earth's magnetic field clearly required measurements taken from many sites on the globe over long periods of time.

In order to obtain much-needed observations from the southern hemisphere, Edmond Halley, who had already spent two years (from 1676 to 1678) on the Island of St. Helena compiling a comprehensive catalog of the southern stars, petitioned his government for a ship to sail the South Atlantic. As a result, H.M.S. *Paramour Pink*,* the first seagoing vessel to be built and commissioned specifically for scientific investigations, sailed from Deptford, England, on October 20, 1698, under Halley's command. In effect, Halley had written his own instructions by which the Admiralty ordered him to proceed so as to improve the knowledge of the longitude and the variations of the compass along Atlantic coasts south of the equator, and "if the Season of the Yeare permit . . . stand soe farr into the South, till you discover the Coast of the Terra Incognita, supposed to lye between Magelan's Streights and the Cape of Good Hope. . . ." The Royal Navy was not fully professional at that time, and it was a common enough practice for captain's commissions to be granted to landsmen or their heirs for reasons of politics or prestige, generally with no expectation that they would actually sail with their ships. Halley, however, strode on board and took command.

Halley made three scientific voyages in the *Paramour Pink* between 1698 and 1701. In the course

*A pink was a small, sharp-sterned ship, originally of Dutch design, that had a shallow draft and bulged at the water line so that it could sail into shallow waters and have room to store provisions for long voyages. The *Paramour Pink* was 54 feet long and 18 feet wide, with a depth of 7 feet 7 inches and a displacement of 89 tons.

of these voyages he sailed as far as latitude 53° south, where, instead of Terra Incognita, he approached three enormous icebergs and found his ship in danger. He was arrested as a pirate by the English consul at Pernambuco, Brazil, and was fired upon as a pirate in Canadian waters. Throughout his first voyage Halley met with such insolence from his first lieutenant that he finally confined the man to quarters and he, personally, navigated the ship home from Newfoundland to England.* (What better navigator could a ship obtain than a future Astronomer Royal?) Despite all his adventures and difficulties, Halley produced what were for many years the best charts showing longitudes of shore stations and lines of equal magnetic declination (isogonic lines, often called Halley's lines) in the North Atlantic and South Atlantic oceans. Eventually he expanded the data to a magnetic chart of the world but with isogonic lines lacking over most of the Pacific Ocean. His voyages successfully completed, Halley turned the *Paramour Pink* back to the Admiralty, which, in 1706, sold the world's first scientific vessel for the equivalent of $245.

Halley formulated the theory that the earth is concentrically layered with a core that rotates at a rate different from that of the innermost shell. The core and inner shell each has its own dipolar magnetic field, thus giving the planet four magnetic poles, the relative motions of which produce the observed field variations. Halley's ideas sound very advanced indeed for his time; therefore it seems highly appropriate that it was in large part from measurements of magnetism (although of a totally different type from Halley's) in the ocean floors that led, nearly 300 years later, to the current tectonic model of the earth.

*Halley's first lieutenant felt he had an old score to settle. He had once written a small book proposing a method for making determinations of longitude from a ship at sea and it had been unfavorably criticized by Halley. Such determinations were impossible at that time and remained so until the development of an accurate chronometer by John Harrison in the 18th century.

Geological Speculations

The Origin of Continents and Oceans

In his introduction to the first issue of the *Geological Magazine*, published in London in July 1864, T. Rupert Jones, the editor, wrote that geology began because men were faced with the necessity of accounting for the presence of sea shells on dry land. The first problem was to establish the organic origin of the shells. Given this premise, it followed that the lands and seas are subject to changes in level; and when some of the shells were shown to represent extinct species, it followed that changes in organic forms have taken place over periods of time. Eventually the principle was established that since the earth's formation its surface has been undergoing continuous change as a result of the natural processes that are seen in operation today. The latest advance invokes the results of experiments carried out in the laboratory to duplicate processes that cannot be directly observed.

In Jones' résumé the development of geology as a science appears simple and obvious; but in fact each phrase represents an idea that was embattled for centuries. To establish the organic origin of the shells: authorities throughout Europe held that the "shells" either were the debris of the Noachian Deluge, inorganic curiosities—"sports of nature"—or decoys of the Devil implanted to confound the intellect of mankind. Repeated exchanges between land and sea and the concept of extinction of species seemed to challenge the skill and judgment of the Creator. Church authorities, recognizing that change is dependent upon time, saw intuitively that Biblical history was not long enough to embrace the gradual alteration of the earth's surface by the imperceptible processes operating today. They maintained, therefore, that the earth as we know it, with its continents and oceans and all its living things, is a recent product of special creation, as described in *Genesis*, and has been modified only by sudden overwhelming catastrophes.

In the hope of establishing a definite limit to the age of the earth, numerous scholars attempted to count the years chronicled in the Bible. Finally, in 1650 and 1654, a two-part treatise, *Annales Veteris et Novi Testamenti*, was published by James Ussher, Archbishop of Armagh and Anglican Primate of all Ireland. Ussher reckoned that heaven and earth were created by the Trinity upon the entrance of the night preceding Sunday, October 23, 4004 B.C. On Tuesday the waters under the heaven were gathered together unto one placed and dry land appeared. On Friday, man was created. The Deluge occurred 1,656 years later when "all the fountains of the great deep [were] broken up and the windows of heaven were opened.... Fifteen cubits upward did the waters prevail; and the mountains were covered." According to Archbishop Ussher, Noah entered the ark on

Sunday, December 7, 2349 B.C., and emerged on Wednesday, May 6, 2348 B.C.

The research of Archbishop Ussher fascinated not only men of his own church but men throughout Christendom. In London, in 1701, the margins of a Bible were inscribed with his dates which consequently acquired much of the force and authority of the King James version itself. Ussher's figure for the age of the earth is one of the shortest ever proposed, but it is of the same order as others of his era. "The poor world is almost six thousand years old," said Rosalind in *As You Like It*. But Rosalind did not learn this from Ussher; she sprang full-grown from the brow of Shakespeare fifty years before Ussher published his time scale. The main result of Ussher's chronology was to invest catastrophic doctrine with more authority and influence than ever at a time when observational science was showing signs of rejuvenation throughout Europe.

Water Enough; and Time

One of the first investigators to penetrate a millenium of intellectual darkness and to recognize the signs of change mentioned by the classical scholars was Leonardo da Vinci, who saw for himself the significance of uplifted fossiliferous strata. In the late 1400s, when Columbus was charting a westward course for the Indies, Leonardo wrote that fossil shells are surely organic in origin not only because they duplicate living forms but also because they occur in groups of species that inhabit the same waters and represent individuals in all stages of development. He worked out an entire cycle from stream erosion, sediment transport, and the burial of shells in shallow marine waters to uplift and exposure by renewed erosion. Leonardo's insights on these and many other aspects of science were fundamentally correct. Most of them, however, he committed to mirror writing in his notebooks. Although some critics have argued that publication of his ideas would have advanced science and technology by centuries, it seems more probable that, in view of the intellectual climate of his time, publication would have led to the destruction of the man himself and of all his works.

About 1570, a century after the time of Leonardo, Bernard Palissy, a French ceramicist and naturalist, was denounced for heresy and presumption when he gave public lectures on the organic origin of the beautifully preserved fossils of the Paris Basin. Palissy also observed that the earth's surface rocks are wasting away and the lands soon would be exhausted were it not for the ability of new rocks to grow and replace the old. In later years the wearing down of the continents became a serious moral issue: How, if God is good, can the abode of Man be destroyed? This problem became so troublesome in the 18th century that some geologists denied outright the visible evidence of soil erosion. Others solved the problem by concluding that a balance is maintained with new continents arising from the floor of the oceans—an idea which, shorn of any moral component, harks back to Aristotle.

Waters from the depths of the earth

In the 17th century scientists were ready to speculate on the origin of the earth in terms of natural processes. The first to do so was the French scholar René Descartes. While Galileo was living under house arrest in Florence, Descartes was working on *Le Monde*, a volume which described the infinitude of the universe and the revolution of the earth around the sun. He was planning publication in 1634, but in seeking to check his own work against Galileo's he learned that Galileo had been condemned and most copies of his book burned. As a loyal Catholic, Descartes then put aside *Le Monde*, which was not published until after his death. Many of his other works were published in his lifetime, but in each one, however penetrating his observations, he always found ways to reconcile his conclusions with church doctrine.

Descartes believed that the earth originated as a glowing mass like the sun and eventually cooled and contracted into a layered body with a self-luminous nucleus enclosed by a shell of opaque rock. That shell was covered by a thick watery layer beneath a smooth outer crust. The rough topography we observe on the earth today he attributed to collapse of crustal segments into the watery layer. Here, in the *Principia Philosophae* of Descartes, published in 1644, are found the germ of the contraction hypothesis of a cooling earth—a hypothesis which dominated geological thought in the 19th and the first half of the 20th centuries—and of numerous interpretations of geologic history involving collapsed ocean basins or waters from the depths of the earth.

Cyclic exchange between land and sea

Descartes' English contemporary, Robert Hooke, rejected the idea that the configuration of continents and oceans was wrought in one or more great catastrophes. Viewing the earth's surface with keen insight, Hooke perceived evidence of cyclic exchange between land and sea reoccurring over a long span of time. The presence in English strata of fossils, which he never doubted were the remains of marine organisms, persuaded him that Great Britain was once covered by the sea and "had Fishes swimming over it." The present land surface is constantly being eroded and its substance carried back to the seabed where, Hooke conjectured, new strata are consolidated by any of four processes: fusion by subterranean heat, cementation by saline substances, hardening of bituminous matter, or hardening by long-continuing cold and compression. Hooke attributed the uplift of marine strata and the sinking of continental areas to seismic convulsions which always lead to a balance between continents and oceans. Thus, he commented, Atlantis may have sunk as Great Britain rose.

The observation that some of the fossil shells of England were tropical varieties led Hooke to suggest that it "may perhaps to some seem not impossible that the center of gravity or Method of Attraction of the Globe of the Earth may change and shift places and if so then certainly all the fluid part of the Earth will conform thereto and twill follow that one part will be covered and overflowed by the Sea that was before dry, and another part be discovered and laid dry that was before overwhelmed." One of the processes Hooke named, a shift in the center of gravity of the earth, has been proposed repeatedly since that time as a possible cause of polar wandering—defined as the reorientation of the globe with respect to its rotation axis. Hooke believed that by this mechanism every point on the earth has passed through the full range of climatic zones. Many modern scientists are in substantial agreement with this, but it no longer is supposed that a redistribution of the earth's mass necessitates a wholesale exchange of land and sea.

Hooke's geological ideas were presented in his lecture course at Oxford and most of them appear in his *Discourse on Earthquakes* that was completed by 1668 but not published until 1705 in a posthumous volume of his writings. In certain of his conceptions, including those concerning the transformation of sediments to solid rock, Hooke was more than a century in advance of his time. Yet, Hooke exerted no discernible influence on the development of earth science. All of his arguments appear to have been forgotten by the time he died. His eclipse resulted partly from the wide divergence between his views and those of his contemporaries and partly because of a long, vituperative battle over priorities in gravitational theory which he waged with a powerful adversary, Isaac Newton.

Sedimentation and the Flood

In striking contrast to the fate of Robert Hooke was that of Niels Steensen, a Dane who changed his name to Nicolaus Steno when he took up residence in Italy in 1665. Steno studied medicine at the University of Copenhagen and authored numerous tracts on anatomy and physiology. After pursuing anatomical research in Leyden and Paris he moved to Florence where he was appointed physician-in-ordinary to the Grand Duke Ferdinand II. The landscapes of Tuscany aroused his interest in geology and in 1669 (the year after Robert Hooke had written *Discourse on Earthquakes*) Steno published a dissertation we usually call

Prodomus although the word simply means "Introduction" and the full title of his work was *De Solido intra Solidum naturaliter contento Dissertationis Prodomus* (The Introduction to a Dissertation Concerning a Solid Naturally Contained with a Solid).

The solids within solids were fossils, but in *Prodomus* Steno also described crystals and laid the foundations of the sciences which, in later centuries, would be called crystallography and paleontology. In studying quartz crystals Steno observed that the angles between equivalent pairs of faces were always the same regardless of whether the crystals were large or small, regular or deformed. This observation led, in the next century, to the fundamental *Law of the Constancy of Interfacial Angles*, often called Steno's Law. With respect to fossils, Steno (after examining the same region studied by Leonardo da Vinci nearly 200 years earlier) concluded that they are the remains of living organisms, and he described how to distinguish marine from freshwater types. In *Prodomus* Steno also set forth several of the principles by which we still interpret the history of sedimentary strata. One of these was the principle that beds are deposited in horizontal layers by the settling of particles in water. Tilted strata, therefore, record subsequent deformation. Another of Steno's principles was that younger beds overlie older ones, and so an orderly succession of strata represents a time sequence.

Although Steno's principles are accepted today as broad generalities, he concluded *Prodomus* by applying them specifically to the events described in the Scriptures. He wrote that the earth during the Creation was covered by a primeval ocean which precipitated a series of fine-grained, unfossiliferous sediments. The first continents to emerge from the waters were smooth platforms undermined by huge subterranean caverns. In time the collapse of the cavern roofs opened valleys which were inundated by the ocean waters of the Biblical Flood. As the Flood postdated the creation of life, the sediments it deposited bore fossils. When new continents emerged they too were undermined by caverns and, according to Steno, a second episode of crustal collapse, not mentioned in the Scriptures and unaccompanied by a great flood, wrought the earth's present topography.

The longer text for which *Prodomus* was meant as the introduction was never written. Steno was converted to the Catholic faith in 1667 and soon afterward he began to lose interest in science. In 1675 he took holy orders and the following year was appointed Apostolic Vicar of northern Germany and Scandinavia. Steno assumed his duties with the utmost dedication. He sold or gave away all of his possessions for the benefit of the poor, wore ecclesiastical garb of the roughest serge, and obeyed a self-imposed regime of abject poverty and much fasting. His behavior appeared so eccentric that in Hamburg some of his fellow Catholics threatened to cut off his nose and ears, and even to kill him. Steno hoped to return to Italy but in 1685 he accepted a missionary post at Schwerin, in Mecklenburg, where he died a year later from the effects of chronic malnutrition.

Steno was buried in the church of San Lorenzo in Florence. His tomb was the object of a pilgrimage made in 1881 by the participants of the Second International Geological Congress, held that year in Bologna. That event was commemorated by the erection of a medallion portrait, encircled by a marble wreath, and a Latin inscription which refers to Steno as "a man of surpassing distinction among Geologists and Anatomists." In 1953 Steno's remains were disinterred, carried ceremoniously through the streets of Florence, and committed to the small, newly renamed Steno Chapel in San Lorenzo. Critical as we tend to be today of the churches in stifling scientific development, the fact remains that most of the brilliant men of science from the Dark Ages to the 19th century were devout Catholics or Protestants. Clearly the churches possessed their own appeal for men of towering intellect.

The backbones of continents

The felicity with which *Prodomus* combined science with the Mosaic record and time scale made it widely read and admired. In 1671 *Prodomus* was translated into English by Henry Oldenburg, who had attended some of the lectures of Robert Hooke. The suggestion has been made that Steno may have been familiar with the ideas of Hooke through correspondence with some third party such as Oldenburg, but for this there is no positive evidence. In any case, despite the work of both Hooke and Steno, England, in 1681, was fertile intellectual ground for a book called *The Sacred History of the Earth* by Thomas Burnet.

Burnet's book is of interest because it carries echoes of Descartes and also tells a story much like that which would be recounted two centuries later by Antonio Snider, one of the earliest advocates of continental displacement.

Contemplating the Biblical Deluge, Burnet wrestled with the problem of deriving enough water, without resorting to a miraculous source, to fulfill Moses' description of inundating the whole world and all the highest hills to a depth of 15 cubits. The Flood was prodigious when Moses supposed that the earth consisted of the Fertile Crescent and Egypt—whether he measured by the Egyptian great cubit of 55.6 centimeters or any lesser cubit. But for the Flood to cover the 500 million square kilometers and all of the mountain ranges of the globe as it was known in the 17th century was another problem altogether. Moses, speaking in tones of thunder to his own people in the 13th century B.C., unknowingly created enormous difficulties for nations of gentiles living 3,000 years later.

Burnet concluded that, aside from gentle hills, the earth had no mountains before the Flood, and at the time of Eden it had a smooth, fair skin enclosing a thick layer of primeval water. The baking heat of the sun opened deep fissures in the crust and began to heat the waters below. Suddenly, at the moment when God in his wrath chose to punish sinful mankind, the waters burst forth in successive tidal waves and great blocks of the unsupported crust crashed into the depths, shortening the radius of the earth. The force of this cataclysm was so great as to tilt the axis of the planet and condemn forever the sons of Noah to the yearly succession of the seasons. In due time, some of the waters remained in the great collapsed basins of the Atlantic, Pacific, and Mediterranean, while the rest drained back into underground voids beneath piles of upended crustal slabs that make up our continents and mountain ranges (else no dry land would be exposed). No significant change has occurred since then: our landscapes are a great "ruine" appropriate to man's condition after the Fall. Burnet's most lasting contribution to geology was a phrase in which he described mountain ranges as the backbones of continents, as though they were the framework on which the rest is built. This idea persisted for nearly two centuries before geologists discovered that mountains are among the youngest rather than the oldest features of the continents.

The saltiness of the ocean

Not all Englishmen accepted Burnet's florid account of earth history. By the beginning of the 18th century, Edmond Halley, whom we have already followed as he charted longitudes and magnetic variations in the Atlantic, proposed that the oceans might hold the clue to the age of the earth. Along with the widespread belief that the earth is only a few thousand years old there was a minority opinion that the earth is eternal. Neither view made sense to the acute scientific imagination of Halley, who wrote, in 1715, *A Short Account of the Cause of the Saltiness of the Ocean, and of Several Lakes that emit no Rivers; with a proposal, by help thereof, to discover the Age of the World.* Halley supposed that the ocean and numerous inland lakes and seas grow progressively saltier as they receive very weak saline solutions from rivers and lose fresh water by evaporation. How helpful it would have been, he commented, if the Greeks of 2,000 years ago had left accurate records of the salt contents of the Atlantic, the Mediterranean, the Caspian Sea, and other bodies of water. By remeasuring the same waters we could determine the rate at which they have grown saltier, and, by reckoning back to the time when the ocean was fresh, we would be very close to the date of the beginning of the world. Halley proposed that for the benefit of scientists in future centuries the Royal Society should forthwith institute a program to record the salt contents of the oceans, seas, and salt lakes. (A forward-looking proposal indeed, but one that was never carried out.) Halley conceded that the ocean may have been somewhat salty in the beginning, but that would only shorten the estimated age of the earth below the calculated maximum. His main argument, he said, "is chiefly intended to refute the ancient Notion, some have of late entertained, of the Eternity of all Things; though perhaps by it the World may be found much older than many have hitherto imagined."

Two centuries later, in 1898, the Irish scientist John Joly attempted to calculate the age of the ocean from its sodium content. Joly, who had none of the figures that Halley had proposed to place on record for him, estimated the average sodium content of crustal rocks, the worldwide rates of erosion, stream run-off, and numerous other factors. He concluded that the ocean is 80 to 89 million years old.

"The spoils of the ocean are found in every place"

In the century after Descartes, one of the most eminent scientists of France was Georges LeClerc, the Comte de Buffon, who believed that the earth began as a white-hot mass torn from the sun by the impact of a comet. Buffon pictured the rocky crust as honeycombed with large caverns, but–unlike Descartes and Burnet–he thought it was initially covered, not underlain, by water. Buffon believed that the triangular shape of the continents and the linearity of their mountain ranges were caused by submarine currents before cracks in the floor of the universal ocean allowed enough of the water to flow into the subterranean caverns to expose continents and islands. He wrote: "It was no sooner suspected that our continent might formerly have been the bottom of the sea, than the fact became incontestible. . . . the spoils of the ocean are found in every place."

Buffon was convinced by the low-lying topography of eastern America and its multitude of lakes and sparsity of human population, among other lines of evidence, that the Americas had been drained of the ocean waters later than had Europe, Africa, and Asia and were, therefore, a "new world" in geologic age as well as in time of discovery.

Buffon's earliest description of the origin of the earth was denounced by the theological faculty of the Sorbonne. As a result, he agreed to abandon everything in his book which ran counter to the record of Moses. But 30 years later, in 1778, he published *Epochs of Nature*, in which he traced earth history through seven geological epochs unmarked by theological influences. The first epoch saw the earth ripped from the sun and the moon from the earth. The succeeding epochs witnessed the condensation and partial withdrawal into caverns of the universal ocean, the early development of marine life, the formation of mountains, volcanoes, and mineral deposits, the coming of animals, and the emergence of the present continental areas. Finally, the seventh epoch celebrated the advent of man. The stratigraphic record of marine life struck Buffon as representing a vast span of time, perhaps as long as 3 million years. Such a figure seemed so indefensible, however, that he removed all reference to it from his final manuscript.

In order to judge accurately the time required to reduce the earth from a state of incandescence to its present temperature, Buffon measured the cooling rates of red-hot globes. From his experiments he concluded that the earth was molten 74,832 years ago, and that living nature orginated 38,849 years ago and will come to an end with the planet sheathed in universal ice 93,291 years hence–with all dates reckoned from 1774. Buffon's subdivision of events into dated epochs, based in part on experimental evidence, was the first serious attempt to formulate a historical geology divorced from the Scriptures.

Buffon puzzled at length over the mystery of elephant remains in the subarctic. He satisfied himself that the ivory found there was identical with that of modern, tropical elephants, but apparently he never heard evidence that mammoths were hairy creatures adapted to cold temperatures. He concluded that the northern latitudes have undergone a drastic change in climate, but he expressly rejected the suggestion of polar wandering–a tilt in the rotation axis altering the obliquity of the ecliptic. Variations of obliquity occur, he said, but they are governed by the planets and are small and oscillating. Therefore, he suggested that the entire surface of the earth formerly was warm because of the high rate of heat flow from the molten interior and that as the crust cooled and thickened those animals adapted to warm climates migrated toward the equator where tropical conditions are maintained by solar heat. We shall hear much more of polar wandering, shifting climatic zones, and terrestrial heat flow, but is is doubtful that we will find a more ingenious interpretation of the evidence than this one of Buffon's.

Like many another philosopher before and after his time, Buffon discussed the lost continent of Atlantis.* He wrote that the lands swallowed up by

*Francis Bacon also discussed Atlantis. He suggested that the Americas are remnants of that vanished land, and numerous later writers have done likewise, but the subject is not pursued here because Atlantis is not a geological problem. According to Plato, the inhabitants of Atlantis lived about 9600 B.C. and waged aggressive war on a race of proto-Athenians who were noble, healthy, handsome, and living according to the tenents of Plato's ideal Republic–as formulated much later. Disaster overtook both races simultaneously. Mythographers have clear priority. Geologists, remembering that 9600 B.C. was about when the last ice cap began to retreat, would be no more likely to look in the real world for vestiges of high civilizations of that vintage than they would for the Kingdom of Lilliput or for the Slough of Despond.

the waters were perhaps those which united Ireland to the Azores and the Azores to America, "for, in Ireland, there are the same fossils, the same shells . . . as appear in America, and some of them are found in no other part of Europe." The sinking of this part of the North Atlantic took place, according to Buffon's chronology, in the sixth epoch of the earth's history, about 10,000 years ago.

The migrating oceans

In 1802 there appeared in France a small volume by J. B. Lamarck entitled *Hydrogéologie*. Lamarck was an independent thinker, famous in his time and today for his researches in paleontology and biology. Looking at the fossil record, Lamarck thought he saw evidence of the slow transformation of living organisms over immense periods of time. He worked out a family tree for the vertebrates in which the principle of evolution, although not named as such, was implicit. Lamarck believed that organs and appendages are strengthened or atrophied depending on their use, and that changes, which come to be inherited, may be impressed upon organic forms by their external environment. This view of Lamarck's was subject to extensive but unrewarding research in the 20th century by T. D. Lysenko, a biologist of the Soviet Union. During his lifetime, Lamarck's hardest battles in biological science were fought against the doctrines of the immutability of species and of universal extinctions championed by Georges Cuvier, the French paleontologist and catastrophist.

Turning to geology, Lamarck, who seems to have been influenced only by Buffon, saw the earth itself as subject to gradual change over an almost "inconceivable" span of time. The heading of Chapter 3 of *Hydrogéologie* poses several questions: "Was the ocean basin always where it is now? And if we find proof of the sea's former presence in places where it no longer exists, why was it there before and why is it not there still? " Lamarck's answer was that the oceanic waters, under the tidal influence of the moon, are continuously migrating around the globe, scouring away at the eastern margins of the continents and withdrawing from their western margins. As a result, every point on the earth has been occupied by the ocean and has also been exposed as a continental area "not only once, but several times." At present, the waters impinging upon the Americas and China are diverted southward, causing a buildup of water and of marine sediments in the southern hemisphere and a concurrent exposure of more and more land in the northern. The resulting change of mass over different earth radii will, Lamarck predicted, lead to a shift in the earth's center of gravity and a reorientation of the crust with respect to the north and south poles. By this means, Lamarck postulated (with no priority conceded to Robert Hooke) that all parts of the earth's surface have already migrated through the full range of climatic zones.

Lamarck estimated that the ocean basin creeps westward at a rate of about 4 meters per century, a rate which brings it around the earth in 9 million centuries. He added that any given area will be exposed as a continent during about a third of each complete revolution of the ocean basin, or 3 million centuries. Lamarck stated that the ocean has already completed several revolutions. If by "several" he meant four or five, this process would have continued for 3,600 to 4,500 million years, which is the same order as our present age measurements of the continents and the earth. He was wrong in this line of reasoning and wrong in his mechanisms for evolution, but he was right to an impressive degree in his insistence upon change by natural processes and in his conception of geologic time.

Lamarck ended his book with a broad pronouncement against all those who, fearing to witness the fall of an idol, would not open their eyes to reality: "Let those who prefer to be deceived be thoroughly deceived." To Lamarck's dismay, his contemporaries apparently preferred to be deceived. He could find neither a publisher nor an audience for *Hydrogéologie*, so he printed 1,025 copies at his own expense. The book fell into almost complete oblivion. Despite his rank as a paleontologist, no one listened when Lamarck wrote of the earth itself and time.

"That which we call the Atlantic Ocean is only a valley"

A description of the Atlantic Ocean as a huge erosion valley was published in 1801 by the German scientist

Alexander von Humboldt. Von Humboldt remarked upon the parallelism of the Atlantic shorelines, saying that the ocean between latitudes 23° south and 70° north has the form of a longitudinal valley in which the advancing and retreating angles are opposite each other. He also emphasized the geological similarities of the Old and New Worlds. Some of the mountain chains of America, he said, can be traced eastward into the old continent; the primitive mountains of Brazil correspond to those of the Congo; the immense plain of the Amazon faces the plains of lower Guinea in Africa; the Mississippi lowland, under water at the time of the opening of the Gulf of Mexico, lies opposite the Sahara Desert. He wrote: "That which we call the Atlantic Ocean is only a valley excavated by the force of the waters; the form of the seacoast, the salient and re-entrant angles of America, of Africa, and of Europe proclaim this catastrophe."

From the pyramidal form of the continents, with apexes pointing south, and the greater flattening of the globe at the south pole, von Humboldt concluded that the waters rushed up from the south, were diverted northwestward by the mountains of upper Guinea, and northeastward again (completing the zigzag of the Atlantic) by the American cordilleras. "It would be very interesting for the geological science," he remarked, "if in a navigation made at government expense, one tried to find the direction, inclination, and attitude of the beds which make up the coastlines of the Atlantic."

Alexander von Humboldt, like Francis Bacon, often is cited as an early advocate of continental drift, but it is difficult indeed to extract this meaning from the writing of either man. Von Humboldt was clearly envisioning a sudden, catastrophic scouring-out of the Atlantic basin by the force of the water. He listed several smaller seaways formed more recently in the same manner: the Straits of Calais (English Channel), the Pillars of Hercules, the Hellespont, and the Sunda Strait. He did not picture the floors of such seaways as foundering or their coastlines as drifting apart.

In later years von Humboldt drew attention to aspects of symmetry shared by continental coastlines of the northern and southern hemispheres. For example, the pronounced easterly bend which begins just north of the equator on the west coast of Africa has a counterpart just south of the equator on the Pacific coast of South America; and the fiord-fringed coasts of Norway and Scotland are duplicated in the archipelagoes of southern Chile. Von Humboldt believed that such repetitions of form were important guides to the factors controlling global deformation—among which factors he never listed drifting continents.

Many men who were neglected in their lifetime are honored posthumously. With von Humboldt the situation is reversed. He was the first scientist to explore South America with an open notebook, describing the people, languages, and customs as well as the archaeology, geology, flora, fauna, and climate. He also traveled widely in North America, Asia, and Europe where he is said to have been the most famous man of his time, except for Napoleon Bonaparte. His greatest accomplishment was the writing of *Cosmos*, a book in which he tried to encompass all the known facts and intelligent speculations on the universe. Yet, von Humboldt has been largely forgotten by modern scientists, and when his works are read they are often misunderstood. Ironically, his proposal that the Atlantic Ocean was scoured out by marine erosion has sparked a renewal of interest in von Humboldt as one who favored the idea of continental drift early in the 19th century.

Time unlimited

The first effective change in the conceptions of geologic time and processes was initiated toward the close of the 18th century by James Hutton of Edinburgh, Scotland. A marker placed on Hutton's grave in 1947 designates him as "The Founder of Modern Geology"; yet, his ideas might well have been lost to science except for the happy circumstance that his disciple, John Playfair, rescued his writings from oblivion and transcribed them into a book of luminous prose.

Hutton was a medical doctor, farmer, and natural scientist troubled by the moral problem posed by the wasting away of the continents. In considering all ramifications of this issue he concluded that, by preordained purpose, erosion and decay are counterbalanced by deposition and uplift so that new continents raised from sea floors always replace the old. To this scarcely original idea Hutton added two concepts that proved of fundamental importance to geology. He extended the time scale indefinitely, and he envisioned a dynamic earth wherein structural change is caused mainly by internal heat instead of a static earth altered chiefly by superficial waters. Hutton stressed the importance of heat in accomplishing three phenomena: the transformation of sediments (sand to sandstone, limey deposits to limestone and marble, muds to shales and schists); igneous processes, including the extrusion of basalt and the intrusion of granite as molten rocks; and the faulting, folding, and uplift of marine strata to form new continents.

Hutton was right, in principle, about thermal metamorphism of sediments but wrong when he postulated that this takes place in seabeds where volatile gases are trapped and the process is enhanced by the weight of the overlying water. His services to geology were so considerable, however, that he should not be faulted for failure to envision geosynclines and orogenesis half a century before any other scientist did so.

Hutton's estimate of geologic time was orders of magnitude longer than any previously contemplated (except by Lamarck, who could not make himself heard, and by Hutton's fellow Deists, who saw the earth as eternal). In fact, Hutton left his time scale open. He wrote that, to nature, time is "endless and as nothing," and that "the progress of things upon this globe . . . cannot be limited by time, which must proceed in a continual succession." Focusing his attention on the evidence of slow, cyclic change, he made it his task to clear the geological Augean Stables of the encrusted catastrophist doctrines of over 1,000 years. Given unlimited time, massive alterations of landscapes can be accomplished by small increments without the aid of catastrophes. Hutton ascribed geologic phenomena to what he called "actual causes," meaning those which can be witnessed in operation. In accounting for change he decreed that "no powers [are] to be employed that are not natural to the globe, no action to be admitted of except those of which we know the principle, and no extraordinary events to be alleged in order to explain a common appearance."

Hutton had the rare and exhilarating experience of predicting at least two geological relationships which he later found exposed in the field. If he was right that worn-down continents become sea floors and these are later uplifted to form new continents, then somewhere in Scotland horizontal strata should be found lying upon an eroded surface of ancient continental rocks. And indeed they are. At Siccar Point, on the Berwickshire coast, he found a dramatic exposure where tilted and contorted Silurian schists are beveled flat and blanketed by the horizontal strata of the Devonian Old Red Sandstone, a type of contact that ever since has been called a geological unconformity. His other triumph involved evidence of the intrusive nature of granite. If this rock were once molten, it should be found intruding older strata in irregular tongues and veins. Such a relationship was found in Glen Tilt, Scotland, where a dark gray schist is laced by cross-cutting dikes of a pink granite.

Hutton supposed that he was proving granites to be igneous "plutonic" rocks that rise as molten magmas from the depths of the earth. At that time he was challenging the "Neptunist" school of geology led by Abraham Gottlob Werner at Freiburg, Saxony. Werner, another staunch advocate of natural causes of geologic phenomena, taught that the primeval earth was covered by a deep ocean oversaturated with chemicals that deposited the rocks of the crust as a series of precipitates. In his view, granites—the earliest primary precipitates—form the world-encircling bedrock which is overlain by stratified transition rocks such as limestones, shales, coal, and basalt, and these, in turn, are covered by unconsolidated sands and clays. Werner doubted that internal heat was important in crustal evolution and, like numerous philosophers before him, he saw volcanoes as accidental occurrences resulting from the ignition of coal or other combustibles at shallow depths. He was persuaded that basalt is a chemical precipitate baked, in some localities, by the heat of burning coal seams.

As a scientist Werner was one of those very rare individuals whose fame depended mainly upon his inspiring qualities as a teacher. His publications were few but his influence was enormous, and he built an unrivaled center of learning at Freiburg where young men, including Alexander von Humboldt, became ardent naturalists and disciples of his Neptunist school of geology. Today Werner is credited by specialists for his system of classifying minerals and for his emphasis on an orderly succession in geological strata. Most students of geology, however, think of him as the man on the losing side of two fundamental issues (although both arguments were actually "lost" after his lifetime). These issues were the aqueous origin of basalt, which was proved by Nicolas Desmarest, in France, to be volcanic, and the precipitation of granite as the universal primary rock, which was proved by James Hutton to be intrusive and therefore younger than the surrounding wall rock. Hutton's view of granites as plutonic rocks was challenged in the 19th and 20th centuries by petrologists favoring the "granitization" of preexisting sedimentary rocks. That some granites represent sediments transformed and mobilized into igneous-looking rocks is now generally conceded, but no scientist today would dispute Hutton's conclusion that the granites of Glen Tilt were emplaced as intrusives.

". . . on the same footing with the system of Copernicus"

Hutton's views were colored by the fact that he—like George Washington, Thomas Jefferson, and numerous other able contemporaries—was a Deist, one who believes in God as revealed through nature and reason, without divine revelation. He was accused of atheism, and worse, by religious conservatives, and his scientific ideas were bitterly opposed by dissenting geologists. He never renounced his long time scale for the earth, but he did concede that he could agree with Moses that man himself is a newcomer to the world. Hutton's geological ideas were presented in 1785 to the Royal Society of Edinburgh and published as an essay in the Transactions of that society in 1788. In response to some fierce attacks he prepared a long work called *Theory of the Earth with Proofs and Illustration*, two volumes of which were published in 1795. The third and final volume remained unpublished when he died in 1797.

Although he was a persuasive talker, Hutton's writing was notoriously dull and his book did little to gain new adherents, much less to counter his critics. The case was stated graphically in 1969 by Gordon L. Davies in his history of British geomorphology, *The Earth in Decay*:

> All told, Hutton's presentation of his own theory could hardly have been worse. Mistitled, lacking in form, drowned in words, deficient in field evidence, and shrouded in an overall obscurity, the theory's chances of finding general acceptance were seriously prejudiced.

In 1802, however, John Playfair righted the situation by publishing *Illustrations of the Huttonian Theory of the Earth*, an eminently readable review of Hutton's ideas. Playfair, a professor of mathematics at Edinburgh University, carried out numerous researches of his own but he is remembered solely as Hutton's disciple who helped to establish the principle of uniformitarianism. Referring to the sacred books, Playfair wrote:

> . . . nor does it appear that their language is to be understood literally concerning the *age* of that body [the earth] any more than concerning its *figure* or *motion*. The theory of Dr. Hutton stands here precisely on the same footing with the system of Copernicus; for there is no reason to suppose that it was the purpose of revelation to furnish a standard of geological any more than of astronomical science . . .It is but reasonable, therefore, that we should extend to the geologist the same liberty of speculation which the astronomer and mathematician are already in possession of; and this may be done by supposing that the chronology of Moses relates only to the human race. This liberty is not more necessary to Dr. Hutton than to other theorists. No ingenuity has been able to reconcile the natural history of the globe with the opinion of its recent origin.

Unique fossil assemblages

While Hutton was battling to revolutionize geological theory, William Smith, a surveyor and self-educated naturalist, was spending the last decades of the 18th century working out the succession of unique fossil assemblages by which the sedimentary formations of England could be identified and placed in a relative time sequence. In 1799 Smith dictated the first table of British strata, and by 1815 he had compiled the world's first geological map covering a large area. The map, at a scale of five miles to the inch, included England, Wales, and part of Scotland. Smith's methods and stratigraphic succession were so definitive that they later were transferred with little difficulty to the rest of the globe. Through the efforts of his successors, the ancient names of places and peoples in Wales and southern England—Cambria,

Devon, the Ordovicii and Silures—are found on every continent and island where early Paleozoic strata occur. Smith is called "the Father of English Geology," although the worldwide importance of his contributions would seem to merit something more. Those geologists who today are hesitating to interpret the broad geology of the moon on the basis of a few rock and soil samples might recall the accomplishment of William Smith who, working alone in the 18th century on one area of one small island group, established a firm basis for interpreting a significant portion of the geologic history of a vastly more complicated planet.

Uniformitarianism

Hutton's principle of seeking "actual causes" of geological events was eventually adopted on the European continent under the French name "actualisme." In England the principle was embraced and transformed in a subtle way by Charles Lyell who named it uniformitarianism. Lyell, whose *Principles of Geology* went through eleven editions between 1830 and 1872, waged total war against catastrophists, seeking, as he put it, to drive them out of the Mosaic record. In 1829 two meetings of the Geological Society of London were devoted to his debates with advocates of the Deluge. The Geological Society of London had been founded in 1807 by a few men of science whose stated wish was to dispense with nonsense and pursue the study of the mineral structure of the earth. Nonsense, however, is never easily dispensed with, and Lyell, by pushing the uniformitarian principle too far, eventually generated some nonsense of his own. Lyell interpreted earth history in terms of present processes always operating at their present rates to produce an endless succession of cycles—changing regions from land to sea and back again—ever maintaining a harmonious balance. To Lyell the earth was a steady-state, perpetual-motion machine in which reversible "thermo-electric" changes at depth preserved a constant flux of internal heat. His view of earth history was, for a time, almost as effective as that of religious fundamentalists in opposing the concept of the progressive evolution of either living organisms or of the planet itself.

Nevertheless, in their *Source Book of Geology*, published in 1939, Kirtley F. Mather and Shirley L. Mason stated that "Sir Charles Lyell probably accomplished more in the advancement of geological knowledge than any other one man." This opinion reflects the fact that, from Lyell's time to this day, geologists throughout the English-speaking world have called themselves uniformitarians in the belief that this is the only right and proper name for all who reject old-fashioned catastrophism. Today, however, most American and English geologists accept without question the proposition that rates of change have varied from one geological period to another and that earth history has been linear rather than cyclic in character. Such a viewpoint identifies a geologist as an actualist, however strongly he may profess to be a uniformitarian.

In the 20th century uniformitarianism settled into a broadly statistical approach to earth science, making it difficult for many scientists to accept any object or event as either new to scientific knowledge or in any way unique. This attitude seriously impeded research on at least two major issues. One such issue is the idea that the earth may bear the scars of meteorites impacting from space. Controversy on this subject began about the turn of the century. Seven decades later the principle of uniformitarianism was extended beyond the earth to the solar system. The hypothesis of continental drift is the other issue in which the uniformitarian outlook discouraged serious investigation. As the two controversies have certain characteristics in common, a brief review of the fate of the impact hypothesis will help set the stage for the longer and more vituperative battle over continental drift.

Uniformitarianism in the solar system

In 1896 a crater 1.3 kilometers wide and 200 meters deep was discovered in a series of sandstones and limestones a few miles southwest of Winslow, Arizona. The crater rim and surrounding plains were strewn with thousands of iron meteorite fragments. This situtation fired the imagination of Grove Karl Gilbert, of the United States Geological Survey, who had just completed a paper entitled "The Moon's Face" in which he argued that the lunar craters were formed by meteorite impact. Gilbert was wondering where the equivalent craters on the earth were to be found. After a two-week field examination, however, he supported the conclusion of Douglas Johnson—the first geologist he sent to the site—that, despite the absence of volcanic rocks in the immediate vicinity, the Arizona crater must have been formed by some unusual type of volcanic steam explosion. With great reluctance Gilbert decided that the meteorites must be coincidental.

Volcanism is a well-known crater-forming process that can be witnessed in operation. Every meteorite seen to fall in historic times has been too small a body to produce more than an insignificant pock-mark in the soil. The idea that an object from space could blast a gaping hollow in the surface of the earth was too radical, and the event too catastrophic, to be countenanced: "No powers to be employed that are not natural to the globe, no action admitted of except that of which we know the principle, no extraordinary events alleged to explain a common appearance"—the voice of James Hutton reverberating down the 40,000 days and nights since the end of the 18th century!

Once Gilbert had made up his mind, geologists dismissed the idea of meteorite impact and had nothing but ridicule for those "amateurs" who kept alive the concept during the first thirty years of the 20th century. Opinion began to change in 1929 when a Russian expedition, returning from a site on the Tunguska River in Siberia, reported that a forest had been flattened, with the trees pointing radially outward from the epicenter of a fireball explosion that had occurred in 1908. The revelation that an extraterrestrial body *could* mark the surface of the earth, even if only in the vegetation, encouraged geologists to take a second look at the crater in Arizona. Today, after worldwide field investigations together with research on artificial impacts, so many meteorite craters and scars have been identified as to give rise to a whole new branch of geology. In a book called *Shock Metamorphism of Natural Materials* that was published in 1968, Bevan French, one of the editors, comments that while explosive meteorite impact is a catastrophe by anyone's standards, the present data on the number and orbits of bodies in space near the earth demonstrates that our ideas of uniformitarianism actually *require* a significant number of large impact structures on the earth during geologic time: "The increasing recognition of such structures on the earth, far from contradicting the Principle of Uniformitarianism, indicates that the Principle may be as valid for processes taking place in the solar system as it is for processes taking place on and within the earth."

Solar system uniformitarians, however, find our moon an acute embarrassment. Not only is the moon the only satellite of the earth but its size, density, and orbital characteristics render the earth-moon pair unique among the bodies of the solar system. No theories of the moon's origin have been proposed that do not call for some nonuniformitarian event that applied to this pair of bodies and none other.

The hypothesis of continental drift appeared to be so antiuniformitarian in spirit that many geologists reacted as they might to a personal insult. Hutton and Lyell, among other scientists, had postulated the uplift of new continents from sea floors. The concept of moving the same continents from one part of the earth to another was novel enough when Alfred Wegener forced the geological community to consider it; much more shocking was the implication that such movement has occurred only once, late in earth history. Continents have always moved or they have never moved is the only permissable uniformitarian view. Given the present "rock-ribbed" appearance of stability, the concept of continental drift becomes catastrophic by definition. It seemed particularly unacceptable to geologists who believed in the doctrine of the permanence of continents and ocean basins.

The Permanence of Continents and Ocean Basins

A powerful challenge to the idea of cyclic exchange between continents and oceans was delivered in the mid-19th century by James D. Dana, professor of geology at Yale. Dana declared that the continents and oceans were delineated in their main outlines at the beginning of geologic time and are permanent features of the earth's crust. Although one of the original problems in geology involved marine fossils in strata comprising dry land, and various investigators had postulated the uplift of new continents from the sea floors, Dana perceived that these uplifted marine formations were all shallow-water sediments of the types found today on the continental shelves. He concluded that local, intermittent flooding of continental areas by seas has occurred often throughout earth history but that no alternation has taken place between the continents and the truly deep ocean basins.

Following Descartes and many later scientists, Dana believed that the earth is cooling from a molten state, and he was convinced that any melted globe, whether it be of iron, lead, or rocky material like that of the earth, would cool and contract unevenly, forming permanent depressions in those areas which cooled last. The continents, he observed, are highlands that have been essentially free of volcanic fires since before the Silurian; the oceans are rimmed with volcanic mountain chains or dotted with volcanic islands that are still active. The continents, therefore, cooled first, and consequently the contraction was most effective in the remaining parts of the crust. He wrote:

> The great depressions occupied by the oceans thus began, and for a long period afterward continued deepening by slow though it may have been unequal progress. This may be deemed a mere hypothesis; if so, it is not as groundless as the common assumption that the oceans may have once been dry land, a view often the basis of geological reasoning.

Dana expressed these views in a long article entitled "On the Volcanoes of the Moon" that appeared in the November 1846 issue of *The American Journal of Science*. Having described the moon's volcanism at length, he used the lunar analogy for the light it might shed upon the earth.

Dana pointed out that before the depression of the ocean floors had made much progress the basins would have been too shallow to contain the seas and the entire surface of the earth would be under water, a situation which he believed obtained in the early Silurian. After that, with each succeeding epoch, greater subsidence of the ocean floors resulted in a retiring of the seas from the land—interrupted by temporary rises in water level whenever local reheating occurred at depth and caused uplift of crustal segments. He concluded that all geological features, including faults, folds, and contortions of strata as well as subsidence and elevations (both real and apparent), result from inequalities in the contraction of the cooling globe.

Dana's strong view that the present distribution of continents and oceans, with all its asymmetry, dates back to earliest geologic times soon became known as the doctrine of permanence of continents and ocean basins. Reduced to the simple slogan "Once a continent, always a continent; once a basin, always a basin," the doctrine was widely accepted in North America and by some European geologists. In 1910 it was reiterated in ringing terms by Professor Bailey Willis of Stanford University: "The great ocean basins are permanent features of the Earth's surface and they have existed where they now are with moderate changes of outline since the waters first gathered."

Meanwhile, some investigators, including Clarence Dutton in America and the Reverend Osmond Fisher in England, had developed serious doubts that contraction from cooling would be sufficient to cause the large amount of crustal shortening indicated by such features as the great thrusts and folds of the Alps. Dana himself was less than wholly satisfied with this aspect of his hypothesis. In 1864 he said that he had adopted the concept that contraction was due to cooling because no other force at all adequate had been suggested.

An alternative cause of contraction was put forward in 1904 by T. C. Chamberlin and F. R. Moulton in their planetesimal hypothesis. These authors proposed that the earth accreted from a cloud of small bodies that were ripped from the sun by a passing star; therefore, the earth was not necessarily homogeneous in composition and was never completely melted. Eventually, however, the compaction under pressure of the planetesimals caused high temperature at the center of the new planet. Molecular collapse and other large-scale rearrangements of the materials of the interior were set in motion and recurred in cycles throughout the earth's history. One result was a greater decrease in total volume than could be achieved by simple cooling.

The concept of global contraction from one or more causes—cooling, compaction, or the extrusion of magmas and water from the interior—dominated geological thought through the 19th and much of the

20th centuries. The contraction theory emphasizes the shortening of the earth's radii as the predominant mechanism in tectonics. The immediate crustal response is subsidence, the dropping of crustal segments under the pull of gravity. Folding, thrusting, and other signs of horizontal compression also occur as all parts of the crust are forced to adjust to a shrinking interior. At least two major problems confront this concept: (1) Why are the effects of compression limited to a few long, linear belts in widely separated parts of the earth rather than being arranged more uniformly like the wrinkles on a drying apple or prune? (2) Why does mountain-building occur in short, violent epochs ("geological revolutions") of worldwide effect that are separated by long periods of relative quiescence? One of the greatest mountain-building episodes began in the Tertiary, only 60 million years ago, and resulted in our spectacular Andes, Rockies, Alps, and Himalayas. Why should so great a contraction, from any cause, occur so late in earth history? Troublesome as they appear, these problems are not fatal to the contraction hypothesis, which explained so many other observations realistically.

"What we are witnessing is the collapse of the world," wrote Eduard Suess about 1900. Given such a model, the massive horizontal displacement of any given block relative to another would clearly be out of the question.

Catastrophic Displacement of Continents

Despite the implications of the contraction theory, or possibly in ignorance of them, two 19th-century Americans published treatises advocating continental displacement. The first was Richard Owen, a medical doctor and professor of geology and chemistry at the University of Nashville, in Tennessee. In his book *Key to the Geology of the Globe*, published in 1857, Owen described the inorganic world as subject to processes similar to reproduction, growth, and decay as observed in living matter. The book is subtitled "An essay designed to show that the present geographical, hydrographical, and geological structures observed on the earth's crust were the result of forces acting according to fixed, demonstrable laws, analogous to those governing the development of organic bodies."

The tetrahedral earth

Owen may have been the first writer to suggest that the earth is not a sphere but a tetrahedron, a figure bounded by four equilateral triangles which, for a given surface area, encloses the least possible volume. He thought that the first land to break the surface of a primeval sea was a thick crust over the northern portion of a tetrahedral nucleus which was rotating about a vertical axis. Erosion of the land surface combined with sedimentation and luxuriant organic growth in the "tropics" caused a change in surface load and this, in turn, shifted the center of gravity of the whole mass. Before this was accomplished the crustal masses that were to become our present continents were in part superimposed (as shown in Figure 12), with South America resting upon northern Africa and Australia overlying Arabia. According to Owen, the change in center of gravity was accomplished by violent convulsions that tilted the spin axis which "would at the same time change the relative position of sea and land, especially if, as supposed, the internal forces were sufficient to attenuate the crust, increasing its total equatorial diameter, separating the more solid mountain masses and continents by extension of the intermediate plastic materials, producing thereby wide channels such as the Atlantic into which the waters flowed, most likely at the final convulsive shocks. . . ."

Owen suggested that the planet underwent one final expansion so great that the moon *may* have been thrown into space "as a terrestrial ovule" from the region of the Mediterranean. This event he compared

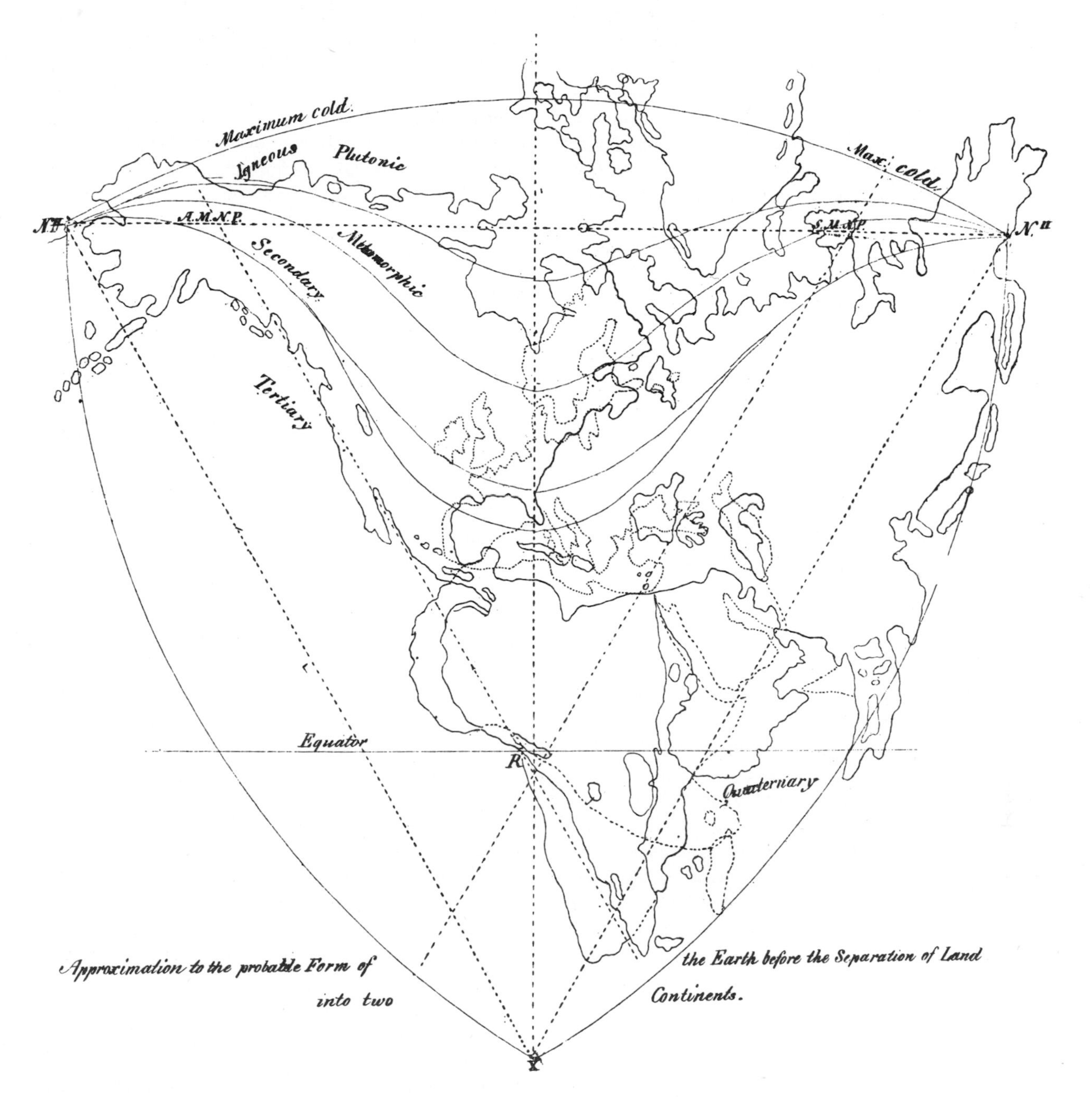

Figure 12. The tetrahedral earth of Richard Owen. Owen's caption reads in part: "The land is supposed to be replaced somewhat as it was before the separation into the present continents: the *dotted* lines indicating those portions supposed farthest submerged, and covered by the parts drawn in *full* lines. Thus, the layers composing the South American Continent are supposed once to have rested on the layers of submerged Africa, particularly in the region of the Sahara. So also Australia is imagined, in some of the earth's early phases, to have been superposed on Arabia, while the points of Cape Comorin and of Cape Horn dove-tailed into its sinuses. Perhaps New Guinea occupied the region of the depressed Caspian Sea. . . . This approximation, although involving great doubt, is given because it may facilitate the working of the problem which all desire to solve, regarding the earth's development." (From Owen, *Key to the Geology of the Globe*, Diagram 1, 1857.)

with the rupture of a Graafian follicle and expulsion of a seed.

Various writers have since associated continental rifting and the opening of the Atlantic with catastrophes involving the moon or with global expansion. Owen was one of the first to propose the splitting of crustal blocks in unambiguous terms, and he ascribed it to both agencies. Taken out of context, as is done here, his hypothesis sounds better than it does in his original text where, to a modern reader, it is hopelessly admixed with medical terms. In his time, however, Owen was a highly respected scientist, and he was one of the earliest fellows of the Geological Society of America. Owen was writing before any sharp distinction was defined between organic and inorganic chemistry. Behind him lay a hoary tradition of belief in the ancient Greek concept of the microcosm and the macrocosm: man, a dimunitive replica of the universe, which is a vast, vital being. Throughout Europe before and after the Renaissance and well into modern times men saw the sun and moon as the cosmic equivalent of man's eyes; the oceanic tides, of his pulse; the rocks, of his bones; the vegetation, of his hair, and so on. This concept colored the thinking of many philosophers including, in his youth, James Hutton.

Owen's concept of the tetrahedral earth was widely adopted by geologists, although most of them believed it originated with William Lowthian Green and so gave no credit to Owen. William Lowthian Green, an English merchant and minister to the king of the Sandwich Islands, made extensive observations of the volcanism in Hawaii and this inspired his book *Vestiges of a Molten Globe*, published in 1875. He attributed the antipodal relationship of continents and oceans to the earth's tetrahedral form resulting from gravitational collapse of the cooling globe. Ocean waters have covered each of the four faces of the tetrahedron, leaving the edges and the four corners dry. (Once again we hear of the four corners of the earth.) With Antarctica occupying the southern tip of the tetrahedron and Eurasia and North America the three northerly tips, Green reasoned that the preponderance of land in the northern hemisphere would cause this sector to lag behind the southern as the earth rotates eastward. As a result, the southern hemisphere has twisted 30° to the east, as on the twin plane of a gigantic crystal, offsetting the southern continents with respect to the northern. The torsion plane is marked by the Mediterranean and Caribbean seas and the east-west mountain belt of Eurasia. "Some things are very interesting even if they are not true," commented B.K. Emerson of this twin-plane analogy, in 1900.

Although it had numerous approximations and imperfections, the tetrahedral hypothesis was without doubt the best strictly geometrical model ever devised to explain the triangular shapes and antipodal arrangement of continents and oceans. It became so widely favored that, in 1938, Walter Bucher, professor of geology at the University of Cincinnati, objected to the hypothesis of continental drift because—among many other failings—it violated the "time-honored" concept of the tetrahedral earth. Yet the concept always suffered from problems, such as how the corners could stay high under the pull of gravity, and it finally had to be abandoned when geodetic measurements showed the earth to be an imperfect ellipsoid rather than an imperfect tetrahedron.

La Création et ces mystères dévoilés

In 1858, one year after the publication of Owen's book, Antonio Snider, an American resident of Paris, published a complete cosmogony entitled *La Création et ces mystères dévoilés*. Snider was a man of wide interests who also wrote brochures on such topics as the temporal power of the pope, the origin of the Sahara Desert and the means of rendering it useful, Austrian justice with respect to Trieste, and the raison d'être of comets. His account of the earth's history, as deduced from the Mosaic record, champions global contraction as a catastrophic event. His book is illustrated by ten superb engravings, two of which (reproduced here as Figure 13) have become classic illustrations in the literature of continental drift.

Snider outlined the course of events between the Creation and the Deluge in terms of days (epochs). During the first day the freezing of a crust over a hot liquid interior caused extreme pressures to build up until the violent explosion of millions of volcanoes took place all at once, blowing the moon out of the earth. Four more epochs followed, each ending with its unique cataclysm until, at the close of the fifth

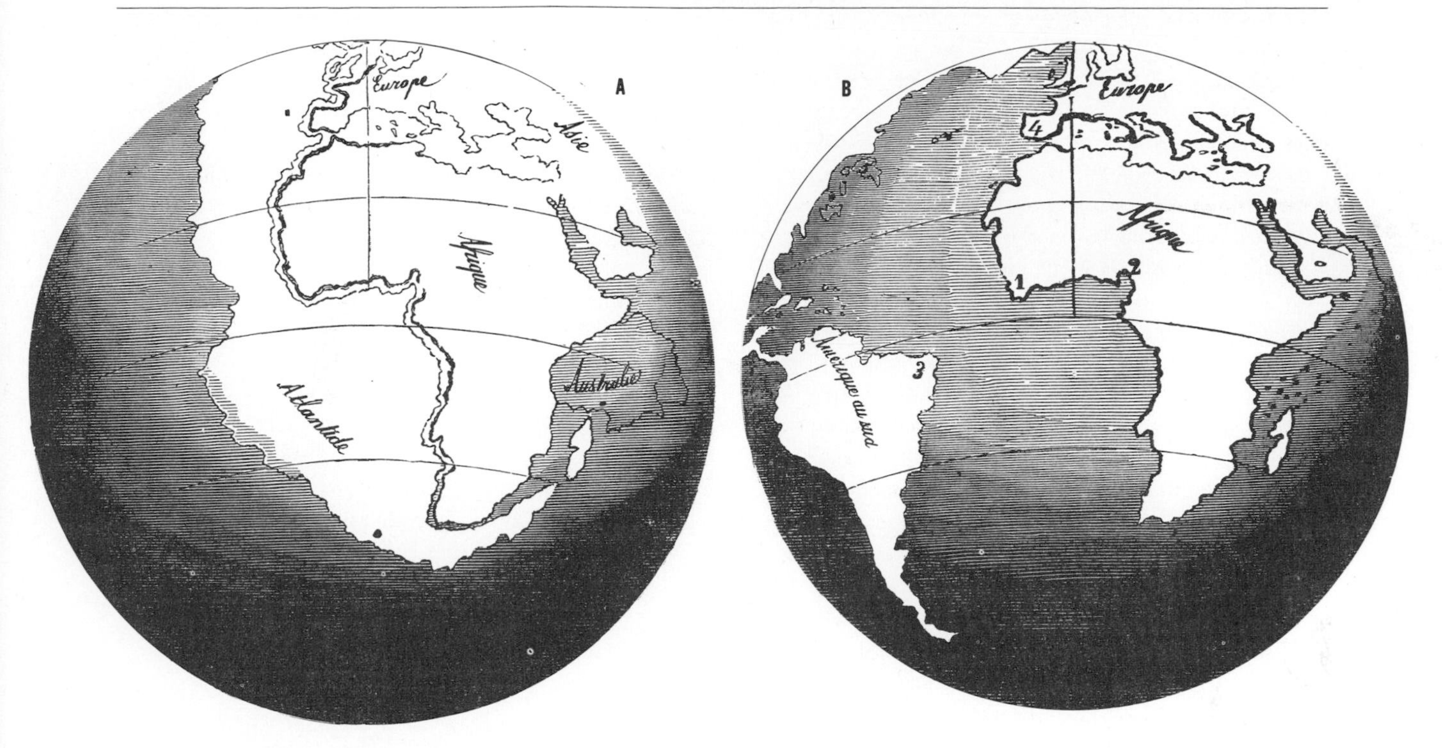

Figure 13. Antonio Snider's engravings (1858) of the terrestrial globe: **A**, as it appeared "Before the Separation," in the interval from Adam to Noah; **B**, "After the Separation," which took place during the Deluge.

day the globe would have appeared as in Figure 13A, with the lands in one large, unstable mass laced by a deep north-south fissure. The sixth day was the time of the Deluge. In a last great buildup of internal pressure, volcanic vapors vented through the fissure, blasting the continents apart and causing a sudden contraction of the earth's circumference from 12,000 to 9,000 leagues. Waters rolled over all the landmasses except the Americas, whose inhabitants (the Atlanteans) escaped destruction and thus are descendants of Adam but not of Noah. After the Deluge the earth would have appeared as in Figure 13B. It has changed only superficially since that time.

Geophysical Speculations

The Nature of the Earth's Interior

The evolution of the earth's surface features cannot be understood without adequate knowledge of the planet's interior. Clues to the nature of the interior accumulated slowly during the 19th century from observations of the earth's orbital characteristics, gravity, heat flow, and crustal structures.

The early concept that the earth has a thin crust over a molten interior followed logically from theories of a high-temperature origin. It appeared to be confirmed by eruptions of flowing lavas, the boiling of hot springs, and the increase in temperature of about 3° centigrade per 100 meters of depth in mines and bores. In 1839, however, William Hopkins of the University of Cambridge set out to test this concept. First he considered the possible effects of conduction, convection, and pressure on the cooling of a molten globe. He concluded that the state of the earth's interior could not be deduced solely on cooling theory and must be tested by measurements of the earth's precession and nutation. His results showed that the crust must be solid to depths of at least one-fourth or one-fifth of the earth's radius (800-1,000 miles). In addition, he argued that as pressure must have been important in the deep interior, the earth probably has a large solid nucleus that crystallized from the center outward as well as a thick, solid crust that cooled from the surface inward. There may still be a thin, imperfectly fluid layer between the two solids. Hopkins concluded that in any case volcanoes do not tap a residual nucleus of primeval fluid; their sources are near-surface reservoirs of limited extent—"subterranean lakes, not a subterranean ocean." Hopkins supposed that the source rocks of volcanic lava must, for some reason, be more easily fusible than the surrounding materials.

Later in the century, Lord Kelvin deduced from the earth's response to lunar-solar tidal forces that it is an elastic medium more rigid than glass. Meanwhile, geodetic measurements on several continents were revealing some puzzling anomalies.

"It is supposed that the crust is floating in a state of equilibrium"

In 1737 Pierre Bouguer, a member of a ten-year geodetic expedition to the Andes for the purpose of measuring an arc of a meridian at the equator, discovered that in the region south of Quito, Ecuador, a surveyor's plumb line was deflected surprisingly little toward the nearby volcanic peak of Chimborazo. From this and a host of similar observations he concluded that the mountains are strangely deficient in gravitational attraction.

Commenting upon Bouguer's observations, a Serbian geophysicist, the Reverend R. J. Boscovitch of the Society of Jesus, wrote, in 1755, that the mountains probably are lifted up due to thermal expansion of material at depth so that the lowered density "compensates" for the overlying mass.

Geodetic measurements in India made during the Great Trigonometrical Survey of that subcontinent from 1838-1843, under the direction of George Everest, produced results of the same kind as those reported by Bouguer in the Andes. This time the discrepancies between the theoretical and the measured deflections of the plumb line due to the gravitational pull of the Himalayas were put on a quantitative basis by the Venerable John Henry Pratt, Archdeacon of Calcutta. Pratt discovered that the deflections measured by Everest were less than one-third as great as calculations predicted they should be. He communicated his results to the Royal Society in London in 47 pages of calculations and discussion that were read at a meeting on December 7, 1854. Pratt's only suggestion as to the cause of the anomaly was that the meridian over India is more curved than the mean ellipticity of the earth. The increase in curvature he ascribed to elevations and depressions "arising no doubt from chemical and mineralogical changes in the mass" during cooling.

The following month, January 1855, George Airy, Astronomer Royal, presented an alternative explanation of Pratt's "singular" results to the Royal Society. At first these results had taken him by surprise, he said, but on reconsideration he saw that they "ought to have been anticipated." Airy proposed (in four short pages) that the crust is not strong enough to support the entire weight of large topographic surpluses such as the Himalayas; that the highlands must be supported by "the downward projection of a portion of the earth's light crust into the dense" substratum. Airy used the word "lava" for the substratum, but stated explicitly that he did not mean it literally. He suggested that the substratum is solid but yielding, in the manner of the rock walls of deep mines. "It is supposed that the crust is floating in a state of equilibrium," he said. He likened the crust to a raft of timber floating in water: if the surface of one log floats higher than the others, we may be sure its undersurface lies deeper.

Airy closed with the observation that the equilibrium obviously is not perfect; that it does not apply to small features such as single mountains and does not account for the changes in elevation which bring submarine strata to the tops of mountains. Airy's proposed explanation of topographic balance has come to be called the "roots of mountains" hypothesis.

Four years later Pratt replied, demolishing Airy's hypothesis on three counts: (1) Airy presumed the solid crust is only a few tens or hundreds of miles thick, whereas Mr. Hopkins of Cambridge says it must be at least 800-1,000 miles thick; (2) Airy assumes that the crust is floating on a denser substratum, yet, as it is cold and solid, it must be heavier than a liquid of the same composition; (3) if tablelands have thickened roots, then hollows (such as oceans) must be underlain by thinner crust, "thus leading to a law of varying thickness which no process of cooling could have produced."

Pratt had discarded his ellipticity hypothesis and was now testing the possibility that the negative effect of seawater south of India was canceling the positive pull of the mountains to the north. He found, as did surveyors in France at that time, that a plumb line is deflected away from the ocean basins by a smaller angle than would be predicted. Gravitationally, mountains appeared to be hollow, and oceans appeared to conceal mountains.

Pratt could not resolve the measured anomalies by any combination of topographic effects so he ascribed them to "hidden causes" in the crust. In 1861 he concluded that the hidden causes are density variations of small magnitude resulting from expansions and contractions of the upper part of the crust during cooling. He pictured the crust as chemically homogeneous and having everywhere the same mass per unit area.

Pratt's three arguments against Airy constitute classic examples of how logical deductions from seemingly reasonable premises can lead to spurious results. His careful consideration of a cooling globe did not allow for the density layering or irregular crustal thicknesses that Airy assumed must be present—and that have since been confirmed by geophysical measurements. Pratt postulated very slight horizontal density variations due to thermal effects. He never supposed that crustal segments have different chemical compositions. And he did not picture the 1,000-mile thickness of the crust as "floating" on the interior. Pratt drew no diagrams of

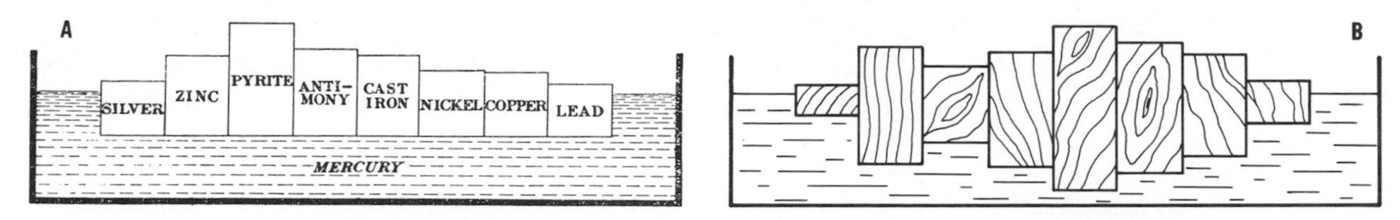

Figure 14. Isostatic compensation. A: Sketch illustrating the concept of equilibrium commonly attributed to John Henry Pratt: Metal blocks of equal mass but different densities are depicted as floating to a uniform level in mercury. (From Bowie, 1921, page 236.) B: George Airy's concept of equilibrium: Wooden blocks of equal density but unequal mass are depicted as floating to different depths in water. (After Holmes, 1965, page 126; used with permission of The Ronald Press, Inc.)

his concept of compensation. From the tone of high seriousness in his papers, one can imagine the distaste with which he would view the diagrams (Figure 14A) now drawn in his name: columns of lead, antimony, and other noxious metals all floating to the same depth in pans of mercury.

Geosynclinals

The question of how the crust responds to the changing distribution of load occasioned by the erosion of rock from highlands and its deposition in seas was addressed by Sir John Herschel, the English astronomer, in a short letter to Charles Lyell dated February 20, 1836. Herschel suggested that an adjustment takes place: the sea floor is depressed and the highlands are pushed upward as materials at depth move out from beneath the loaded sector and come to rest beneath the lighter one to restore an "equilibrium of pressure." He emphasized that he was envisioning movement in imperfectly fluid materials.*

Herschel was theorizing. Geological evidence that his idea was substantially correct came two decades later, in 1859, as a result of a monumental study of the northern Appalachians by James Hall. Hall observed that the mountains are built of sedimentary strata of much greater thickness than flat-lying deposits of the same age and character farther west. Ripple marks and plant fossils occur throughout the stratigraphic section, suggesting that every layer of a series 12 kilometers thick was deposited in shallow water. The sediments that were to become the Appalachian Mountains were therefore laid down in a basin that slowly subsided as it filled. According to Hall, the sinking of the floor caused folding and buckling of the strata along a great synclinal axis which marked the course of the main sediment–transporting current. The deposits occupying the deepest portion of the basin were heated, metamorphosed, and invaded by igneous rocks. Hall pictured the present continental area as a shallow sea floor and the sourceland of the sediments as lying to the east, where we now see only the Atlantic Ocean and the coastal plain. Finally, a large-scale reversal took place: the deeply eroded sourceland subsided below sea level while the continent was uplifted as a unit, and, as the waters drained away, the thick belt of crumpled sediments was exposed as a mountain range. (Figure 15A) Hall rejected the notion that the uplift was confined to the mountains. He wrote that mountains are not elevated as mountains, but as part of the continental movement.

Old continents are engulfed by the sea and new ones are raised from the sea floor—the dictum of Hooke, Hutton, Lyell, and many others since Aristotle. But the birth of a specific mountain range from a sediment-filled trough was a new insight into the problem. Hall believed that mountain ranges throughout the world were formed in the same manner as the Appalachians. He assumed that the increasing load of the sediments caused the sinking of the basins. The continental uplift he ascribed to the cause cited by "Babbage and Herschel," a compen-

*Herschel's letter was published in 1838 in a book entitled *The Ninth Bridgewater Treatise*, edited by Charles Babbage. From that date forward, the concept of adjustment to changes in load has been ascribed to Babbage and Herschel as though they were collaborators—despite the protests of Herschel and others to the contrary.

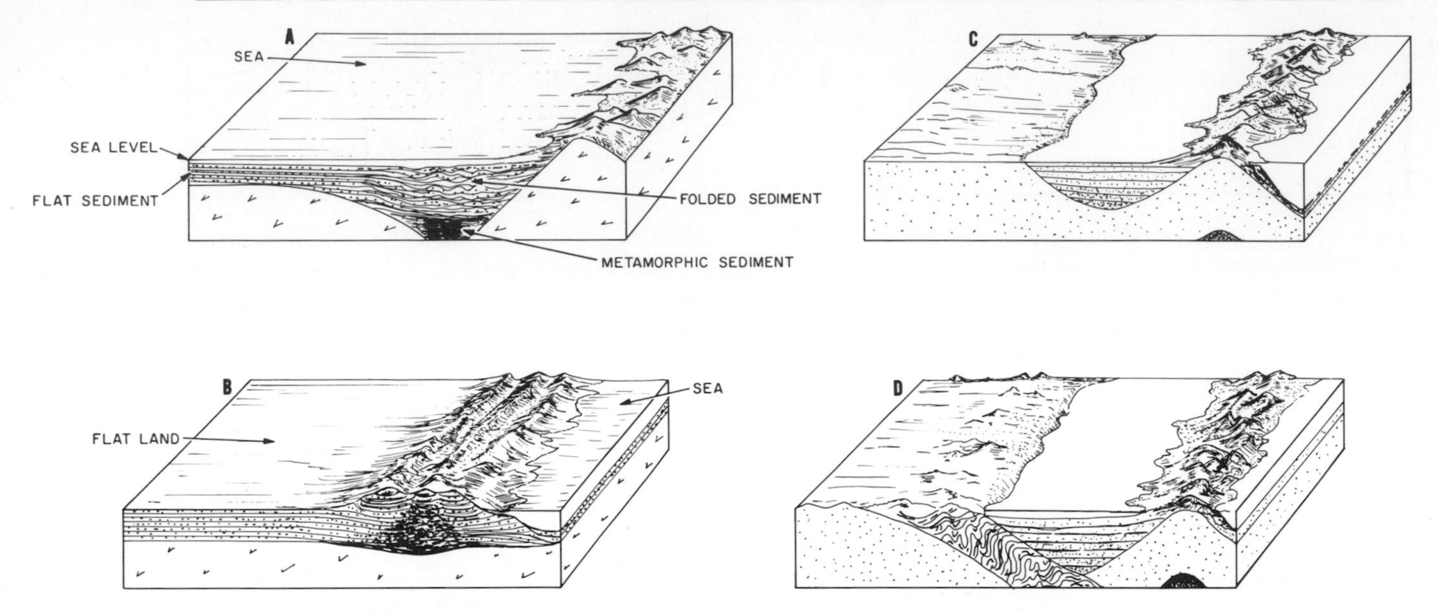

Figure 15. A, B: Schematic representation of James Hall's concept that sediment-filled basins are transformed to mountain ranges by the uplift of continental platforms. C, D: Schematic representation of James D. Dana's conception of continental accretion. A geosynclinal depression, filled with sediments eroded from an offshore geanticline, is compressed against the continental margin. The cycle is repeated by the initiation of a new depression and geanticline.

sating flow of subcrustal material due to loading of the adjacent ocean floor.

With reference to this explanation, J. D. Dana wrote: "Mr. Hall's hypothesis has its cause for subsidence, but none for the lifting of the thickened sunken crust into mountains. It is a theory for the origin of mountains, with the origin of mountains left out."*

Dana himself discussed these sediment-filled basins (which he called geosynclinals) in 1873 in an article entitled "On Some Results of Earth Contraction from Cooling." As sediments are porous rocks, soaked with water, they are less dense than the underlying crystalline basement and are not likely to cause profound depression. Dana interpreted the geosynclinals as tectonic basins forced downward by compression and filled with shallow-water sediments, not as a cause but as a result of subsidence. When Dana put forward the doctrine of permanence he described the ocean floors as more contracted and hence denser than continental bedrock. Global contraction would, he believed, compress the ocean floors against the continents, forcing the crust into great folds: a downfold (geosynclinal) adjacent to the land, bordered on the seaward side by an upfold (geanticlinal). Erosion of the geanticlinal, and not that of the continental area, would, according to Dana, furnish all the sediments that filled the geosynclinal.

Dana pictured the crustal folds as pressing landward in great waves. The geosynclinal belt and its sedimentary load eventually is forced upward into a broad arch and welded to the continent as a coastal mountain range while the eroded geanticlinal subsides beneath the ocean and a new basin and welt form offshore (Figure 15B). Dana's idea that continents grow in area by the addition of a succession of geosynclinal belts gained wide acceptance over the next century. So, too, did the notion of a vanished sourceland, which was to assume great importance in the 20th-century controversy over continental drift.

Geosynclines—sediment-filled basins that turn into

*To which Hall later retorted that he had set out to describe a remarkable occurrence of fundamental geological importance and not to formulate a theory of elevation. Furthermore, he did not have much use for theorizing, his own or anyone else's.

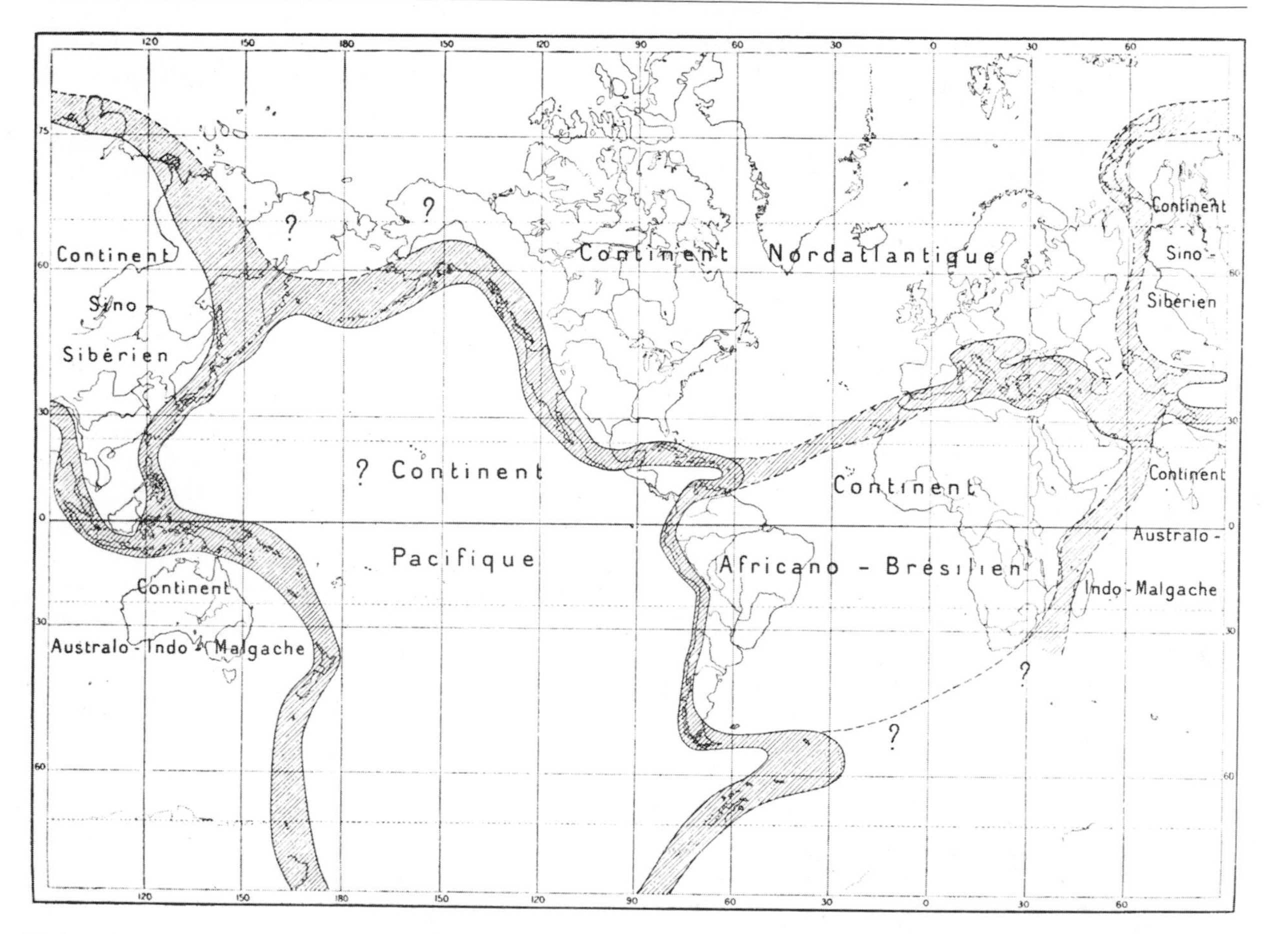

Figure 16. The Mesozoic world of Emile Haug consisted of five vast continents separated by geosynclines. Haug inserted question marks in doubtful areas. (From Haug, *Traité de Géologie*, 1907.)

mountain ranges—were a revolutionary concept which soon became fundamental to geological thinking. In time, complications arose as various investigators defined geosynclines differently. From the beginning, Americans, inhabiting a landmass with mountain ranges paralleling both coastlines, tended to view geosynclines as slowly subsiding basins bordering continents. Europeans were faced with contrary evidence, as many of their mountain ranges cross continental interiors. In 1900 Emile Haug, in France, redefined geosynclines and departed from Hall and Dana in at least two important particulars. He pictured geosynclines as linear troughs, standing open or partially filled not necessarily with shallow-water (neritic) sediments but with muds from the deeper bathyal zone. Haug also removed geosynclines from the border area between continents and oceans and described them as occurring between two continental platforms. By his definition geanticlines were viewed no longer as uparched highlands separating geosynclines from the deep oceans but rather as linear ranges formed within geosynclines by the compression and folding of the sediments. He regarded the Atlantic Ocean as a geosyncline and the Mid-Atlantic Ridge as a newly initiated geanticline. Haug's idea that mountain ranges result from the compression of geosynclines between continental platforms led him to postulate that a vast continent formerly existed in the area of the Pacific Ocean and contributed to the formation of the Tertiary mountains bordering that ocean. Haug's world map (Figure 16) depicts the only Pacific continent that was invented primarily as a structural rather than a biological necessity. Moderately sized Pacific landmasses had previously been visualized by Charles Darwin and Thomas H. Huxley to explain the foundations of coral atolls and the distribution of certain plants and animals. Haug's continent, designed to crumple the coastal ranges, reduced the Pacific Ocean to narrow canals. One can only ponder the rapid evolution of ocean waters since the recent foundering of Haug's landmass.

Erosion proceeds and is irreversible

Despite speculations on compression and deep-seated thermal effects, the cause of the uplift of mountains and continents remained problematical. Uplift violates all preconceptions of what should occur on a cooling, contracting earth. A number of scientists, including Eduard Suess, doubted that uplift—in the sense of a net lengthening of an earth radius—occurs at all. Others, while accepting the actuality of uplift, observed the broken and contorted strata of mountain ranges and ascribed it to catastrophic events.

John Wesley Powell concluded otherwise when he made his pioneer boat trip through the spectacular canyons of the Colorado River in 1869. In the sheer walls rising above him Powell saw evidence that a whole region had risen more than a mile above sea level with no crumpling or tilting of the flat-lying strata. The uplift had been gradual, imperceptible; and as the land rose, the river, loaded with abrasive sediment, had worn through each successive stratum exposed in its bed. As a result of the continuous interaction of uplift and erosion, canyons thousands of feet deep had been cut in the rising plateau by a river that was never more than a few hundred feet above sea level.

Powell viewed erosion as a universal process of base leveling that tends to reduce all landscapes to featureless plains lying near sea level. Where elevations are highest, erosion is most rapid; where they are low, erosion proceeds more slowly. Erosion proceeds, nonetheless, and is irreversible. Why, then, were not all of the continents worn down to sea level far back in the distant past? There are two answers to this: first, the continents very nearly are low plains worn down toward sea level—and we shall hear more on this subject from Alfred Wegener. Second, uplift tends to renew eroded areas, and the rise of rugged highlands has been relatively recent. All mountains, said Powell, are young.

The attribution of the earth's present landscapes to weathering and stream erosion is such a fundamental part of geologic thinking today that it is hard to realize how revolutionary was this conception of Powell's—particularly to British readers, most of whom found it hard to believe that the familiar streams and rivulets of their countryside could possibly have created the valleys they occupy. Even Lyell, the great uniformitarian, believed that our present continental topography was wrought by marine erosion.

Among geophysicists, Powell's description of the uplift of the Colorado plateau stimulated a strong preference for Pratt's explanation of isostatic compensation rather than for Airy's. How could a region of undeformed strata, bearing no signs of compression, suddenly acquire a root of low density rock? Airy's mechanism seemed to require a prodigious transfer of material at depth, with cubic miles of granitic material sweeping in from surrounding areas and accumulating beneath the plateau. Heating and expansion (due to whatever cause) of the underlying materials surely constituted a more reasonable explanation for the vertical rise of the surface.

Fresh insight on the problem of uplift was provided in 1865 and again in 1882 by T. F. Jamieson, who described how the weight of the Pleistocene ice caps had depressed the rocky crust of northern Europe and how release from the load is allowing a rebound. Jamieson argued, as had Herschel, that a weak layer at depth was squeezed away from beneath the loaded area, causing a basin rimmed by a bulge. Uplift is now reestablishing the preglacial level with, geologically speaking, a very short time lag.

Jamieson's evidence for the rebound of Scandinavia was later found applicable to the Hudson's Bay region of Canada and to the basin of the great glacial Lake Bonneville of western North America. Geophysicists have used the rate of postglacial rebound as a meter stick to measure the strength of the interior. A few, however, have doubted the efficacy of ice, which has a density of only 0.92 g/cm^3 and is the very lightest of all solid "rocks" to bow down the earth's crust. Scandinavia and Canada are indeed rising, they argue, but so is every other shield area of the world.

If shields did not rise they would not have remained continental throughout geologic time.

This minority voice has never prevailed with respect to ice caps. Most geologists are convinced that ice 2-4 kilometers thick piled as an extra load on the surface will cause a depression and that the rate of rebound of 1 centimeter per year, (measured in Fennoscandia by precise leveling) is significantly faster than the rise of shields. On the other hand, few, if any, of these geologists believe that the weight of a layer of sediment 12 kilometers thick (density, 2.0-2.5 g/cm^3) is sufficient to cause the sinking of geosynclines. The problem in both cases is one of pushing lighter materials down into denser ones, but the addition of new material to ice caps results in the increased height of the uppersurface as well as the deeper depression of the undersurface. The uppermost sediments of geosynclines remain at or below sea level and the entire mass sinks far deeper than it should if the densities of rocks were the prime consideration. Dynamic factors, the nature of which is still in dispute, must be involved.

Isostasy

In 1889, Clarence Dutton, a protégé of Powell's, addressed the problem of adjustment and found a partial answer in the force of gravity as it controls the figure of the earth. Even if the planet were as strong as steel, he reasoned, its powerful gravitational field and rapid rate of rotation would force it into the form of a spheroid. Why, then, does gravity stop short of perfection and leave the surface so irregular? Gravity stops short of nothing, Dutton concluded. The earth is grossly inhomogeneous and its equilibrium figure is dictated by the densities of its materials. Continents are higher but lighter; ocean basins are lower but denser. Mass per unit area is everywhere equal, or moving toward equality. Dutton proposed that at moderate depth beneath the earth's surface there exists a certain level, all parts of which are under equal pressure from the overlying rocky masses. For this condition of global equilibrium he coined the term "isostasy," from Greek roots meaning "in equipoise."

Working to unbalance the equilibrium are at least two important factors: erosion and sedimentation. According to Dutton the crust is sufficiently strong to support some departures from equilibrium, but beyond a certain limit a wholesale adjustment that involves viscous flow in the substratum is set in motion.

Dutton combined the insights of numerous earlier workers into a broad theory which, he pointed out, could be tested by worldwide gravity measurements. In a famous epigram he rejected outright the hypothesis that thermal contraction has governed the configuration of the earth's surface: "As I have no time to discuss the hypothesis further I dismiss it with the remark that it is quantitatively insufficient and qualitatively inapplicable. It is an explanation which explains nothing which we want to explain."

Dutton's term "isostasy" entered the literature where it took on a variety of different meanings, some of which were far more narrow and specialized than his original definition. Disputes over whether isostasy is a condition or a process and whether it is applicable to various types of features continued into the 1930s—and, to some extent, continue to this day. Eduard Suess opposed the concept, which, despite its broadly international development, he stigmatized as an American idea.

Over the years since Dutton defined isostasy, monumental tables of gravity measurements have been compiled. These tables show that the earth is in a state very close to Dutton's picture of equilibrium. Large features—continents and ocean basins—are fully compensated. Geodetic satellites, while analyzing the gravity field to low harmonic orders, do not "see" continents and ocean basins but pass over their margins unperturbed. The gravitational figure of the earth (the geoid) is not a smooth spheroid, however. It has humps and hollows that appear, on first glance, to have no association with the surface topography. These positive and negative gravity anomalies therefore have been ascribed by some geophysicists to density contrasts deep within the earth's mantle or to irregularities in the core-mantle boundary.

William M. Kaula, of the University of California at Los Angeles, reexamined the data in 1969. After subtracting the low-order harmonics he found residual anomalies at higher harmonics, and he believes that these can be correlated with geological features. The strongest anomalies appear to be associated with zones of relatively recent volcanism or tectonic deformation. Kaula's analysis suggests that the earth is nearly in equilibrium except in those areas where tectonic activity has occurred within the past 60 million years or so. This suggests that 60 million years is the approximate amount of time required for disturbed zones of the earth's surface to achieve isostatic adjustment.

Earth fission and continental drift

How would the earth have responded if the moon had been ripped away early in geologic history? In 1879 the English physicist George H. Darwin postulated such an event from calculations of resonance effects between the oscillations of a slightly distorted planet and the period of the solar tides. A year later he showed that the heat of internal friction due to the planet's slowing rate of rotation would indicate an age for the earth of 3,560 million years. Rather than defend so long a time scale, however, he found ways to reduce that estimated age by two orders of magnitude and to date the separation of the moon as taking place 57 million years ago.

Speculating on this event, Robert Ball, Royal Astronomer of Ireland, wrote in 1881 that "the wound of the earth would soon be healed . . . and at last it would not retain even a scar to testify to the mighty catastrophe." The Reverend Osmond Fisher, rector of Harleton, England, and author of *Physics of the Earth's Crust* (published in 1881 and sometimes cited today as the earliest textbook of geophysics), doubted that so vast a crater would heal so completely. He believed that before the great catastrophe occurred the earth would have become largely solid, with only a thin layer of molten material remaining between the crust and the nucleus. He therefore proposed, in a short article in *Nature* in 1881, that the unique Pacific basin is the scar left by the ripping away of the moon. This cavity, he thought, would begin to fill by the rise of high-density lava from depth. What was left of the light, granitic crust would break into fragments—represented by the continents and islands—and these would slowly float toward the depression. "This," he wrote, "would make the Atlantic a great rent, and explain the rude parallelism which exists between the contours of America and the Old World."

Fisher added that the fissioning-off of a large portion of the earth's crust could also explain the anomalously low density of the moon, which is only about 3.3 grams per cubic centimeter whereas that of the earth is about 5.5 grams per cubic centimeter. And the separated crustal mass would be rich in aqueous solutions which would account for the explosive volcanism that Fisher believed has cratered the lunar surface. "But," added Fisher with exemplary caution, "the difficulties surrounding terrestrial vulcanism are so great that one is hardly tempted to add the lunar to them."

Here is a clear, consistent exposition which seeks to explain the unique form of the Pacific basin, the high density of its floor, the rifting and separation of blocks of light, continental crust, and the parallelism of the Atlantic coastlines as well as the low density and volcanism of the moon. In 1889, in the second edition of his book, Fisher added that the great islands of Sumatra, Borneo, Australia, New Guinea, and New Zealand may have broken away from the region of the Indian Ocean and floated slowly toward the Pacific cavity, leaving behind a trail of fragments. Fisher did not picture the buoyant drift of landmasses as continuing today, but he described mountain-making and other crustal deformations as due in part to currents in the molten substratum that rise along the median lines of the ocean basins and sink beneath the continental margins.

Curiously enough, certain reviewers who remember Richard Owen and Antonio Snider (in spite of their eccentricities) as early advocates of continental drift have overlooked Fisher, although he published in *Nature*, a reputable and still extant scientific journal. H. B. Baker, one critic who did remember Fisher, commented in 1933 that Fisher sounds as though he had Owen's work in the back of his mind but could not, as often happens, quite place the source. To most readers this veiled charge of plagiarism would seem absurd. It is hard indeed to believe that had Fisher ever read Richard Owen's geomedical text he could have forgotten it, much less have borrowed any ideas from it.

"Who it was that first suggested that the Moon originated in the Pacific is unknown. The idea seems to be a very old one." So wrote W. H. Pickering, an American astronomer, in the *Journal of Geology* in 1907—when the idea was 26 years old. Pickering supported this "old idea," commenting:

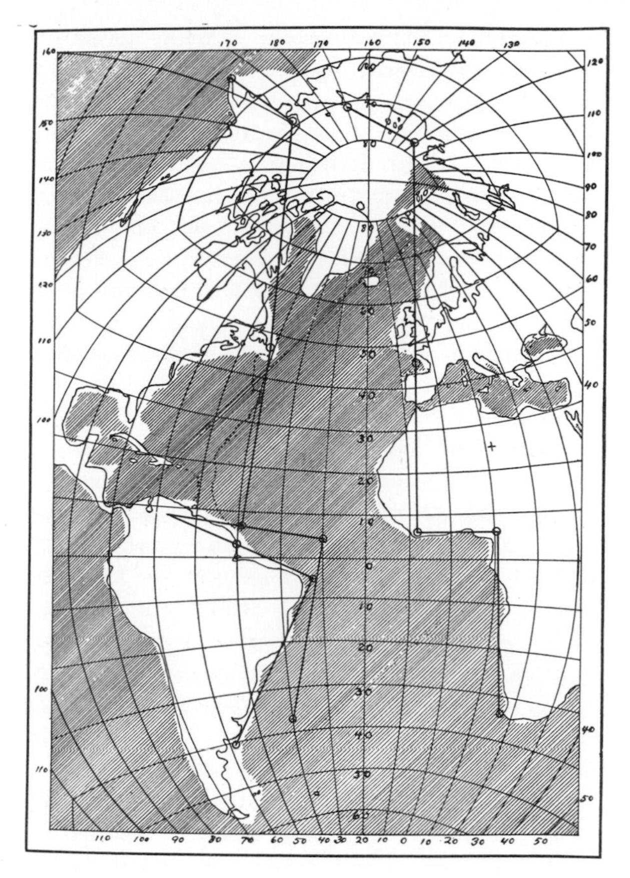

Figure 17. W. H. Pickering's diagram illustrating the parallelism of the Atlantic coastlines. The straight lines bounding part of South America suggest that that continent has rotated clockwise a few degrees while drifting away from Africa. (From Pickering, in *Journal of Geology*, 1907; used with permission of University of Chicago Press.)

> A curious feature of the Atlantic Ocean is that the two sides have in places a strong similarity. . . . When the Moon separated from us, three-quarters of this crust was carried away and it is suggested that the remainder was torn in two to form the eastern and western continents. These then floated on the liquid surface like two large ice floes.

Pickering also suggested that Australia and Antarctica broke away from the area of the Indian Ocean. He plotted a tentative reconstruction of the lands across the Atlantic, and his diagram (Figure 17) is widely reproduced in reviews of early ideas on continental drift. Possibly the relative obscurity of Fisher's presentation results from his failure to use any map or illustration to catch the eye of the readers of *Nature*.

Later, in the same volume of the *Journal of Geology*, Pickering was taken to task by Joseph Barrell, professor of geology at Yale, for failing to check his references and give proper credit to Osmond Fisher and for maintaining a cavalier attitude toward the geological difficulties associated with earth fission. Barrell was one of the few eminent American scientists who favored the idea, widely held in Europe, that ocean basins have grown progressively larger by the foundering of slabs of continental crust.

The concept of earth fission as a source of the moon always has alarmed uniformitarians, and it long has been opposed by most geophysicists because it raises seemingly insoluble, dynamic problems. With respect to continental drift, earth fission is both unlikely and inconvenient as a triggering device because, from paleontological and structural evidence, the continents appear to have begun separating about 180 million years ago. In view of the comparatively orderly geologic record of the Mesozoic era, this date is much too recent to contemplate such violence as would accompany fission. And it is impossibly recent for the birth of the moon, which, we now know from radiometric dating of samples collected on the Apollo missions, has crustal rocks that crystallized more than 4 billion years ago.

Sunken continents

A masterful synthesis of geology as it was perceived around the turn of the 20th century was prepared by Eduard Suess, of Vienna, in *Das Antlitz der Erde* (The Face of the Earth), a four-volume work that was published between 1885 and 1909. Suess believed in a cooling, contracting earth composed of three concentric shells for which he coined the terms nife (Ni-Fe) for the core, sima (Si-Mg) for the mantle, and sal (Si-Al) for the crust. He thought that the sal originally covered the entire earth but that shrinkage of the interior caused immense slabs of it to founder into the sima, thus creating ocean basins. Reviewing the broad architecture of the earth, he noted that each continent has an extensive terrain of ancient crystalline rocks (now loosely called shields) that have remained stable since the Precambrian as shown by overlapping beds of flat-lying, early Paleozoic sediments. Suess also confirmed an observation made by the Austrian paleontologist Melchior Neumayr that there are two kinds of continental margins—an Atlantic, or fractured, type and a Pacific, or folded, type. Along fractured coastlines geological structures are truncated abruptly, as though the ocean basin were younger than the continental platforms. This type of continental margin can be seen on both sides of the Atlantic and Arctic oceans, in the Indian Ocean as far east as the mouth of the Ganges, and along the west coast of Australia. The Pacific margins, in contrast, are bordered by relatively young, folded mountain ranges and chains of active volcanoes that clearly control the configuration of the coastline itself. Suess concluded that the Pacific is a very old ocean, which might well have been initiated by the ripping away of the moon, and that it is now diminishing in area as mountain ranges are folded over its margins by tangential pressure from the continents. The Atlantic and Indian oceans he saw as much younger features, some portions of which were opened as late as the Mesozoic and Tertiary by the sinking of continental areas.

The idea of sunken continents is as old as the time of Plato and his story of Atlantis, a lost land which Suess identified with the North Atlantic floor bounding Greenland. That concept passed through many versions involving collapsed caverns or seismic convulsions until it gained renewed respectability with the publication of Charles Darwin's *Voyage of the Beagle* in 1839. Darwin described the beautiful coral atolls of the Pacific as marking sites where former volcanic islands have subsided to great depths below sea level. The subsidence of certain continental areas to account for the distribution of land plants and animals was discussed by the English biologist Edward Forbes in 1846—the same year that J. D. Dana founded the doctrine of permanence of continents and ocean basins.

Further investigations into species distribution led to the invention of Lemuria, a hypothetical land spanning the Indian Ocean from Madagascar to India and Ceylon. Lemuria was first described and named in 1864 by the English ornithologist Philip Lutley Sclater. Seven years earlier, on the basis of bird distribution, Sclater had subdivided the world into six major faunal zones: the Palearctic, Ethiopian, Indian, and Australian in the Old World and the Nearctic and Neotropical in the New World. Sclater's zoogeography was so inherently sound that, with minor readjustments—such as the creation of an Oriental region by enlargement of his Indian zone and the creation of a Holarctic region by partial merging of his two Arctic zones—it still is in use today to describe the broad distribution of all forms of land and freshwater animal life. Interestingly enough, Sclater first outlined his faunal zones two years before the appearance, in 1859, of Charles Darwin's *The Origin of Species*, the book we now tend to regard as the first rational attack on the idea that each species is a special creation.

Sclater was led to postulate Lemuria by the enigma of the Malagasy fauna shared by Madagascar and the Mascarene Islands. Lying some 450 kilometers from Africa across the Mozambique channel, Madagascar lacks all but a few of the common African birds, animals, and plants. For example, Madagascar has no monkeys, apes, baboons, lions, leopards, hyenas, zebras, rhinoceros, elephants, giraffes, or antelopes; but of 65 species of mammals known to occur there in the mid-19th century, 34 were species of the lemurs who lent their name to Lemuria. Some of these lemurs (Figure 18) and several other members of the Malagasy fauna seemed to show stronger affinities with their kindred in India than with those in Africa. As they are more primitive than monkeys, the lemurs' strange distribution suggested an early Tertiary connection between Africa and India that was broken at both ends before the introduction of mammals higher than lemurs on the evolutionary tree.* The last vestiges of ancient Lemuria were believed to be Madagascar itself and the islands and atolls of the Indian Ocean.

New evidence from the fossil record discovered

*If Atlantis was invented by a philosopher and later adopted by a few scientists, the reverse was true of Lemuria. Sclater's small Indian Ocean subcontinent was expropriated in the 1870s by Madame Helena Blavatsky who, in the arcane service of theosophy, expanded it over much of the southern hemisphere. Lewis Spence, a student of the occult, later removed Lemuria to the Pacific Ocean to explain the diffusion of Polynesian artifacts. In that location it competed for space with Mu, a Pacific continent invented in 1864 by the Abbé Charles Etienne Brasseur to account for mystic symbols in Asia and Central America. Geological sunken continents differ from the mythical ones in that the latter all foundered after the development of high civilizations; the former foundered before man arrived on the scene at all.

during the 1870s and 1880s seemed to confirm the need for former land links between India, Africa, and other southern continents. One of the most compelling examples was that of *Glossopteris*, a beautifully preserved fern found in coal beds of Permocarboniferous age in India, South Africa, South America, and Australia. By 1885 the distribution of fossils was sufficiently well known to prompt Melchior Neumayr to construct the first paleogeographic map (Figure 19). This map depicted the world of the Jurassic with three immense continental areas and numerous large islands.

Of special interest in the context of future debates over continental drift would be Neumayr's Brasilian-Ethiopian continent, to which Lemuria (Indomadagascar) is attached as an appendage. Also of interest would be his large eastern continent linking China with Australia. Neumayr's map inspired countless other scientists to reconstruct the lands and seas of past geologic periods, an occupation that is continuing apace. Unfortunately, Eduard Suess did not use any maps of this genre in his four-volume work. He followed the more conservative practice of describing his supercontinents with reference to a standard Mercator projection of the present world map. His own view of the late Paleozoic continents varied somewhat during the 20 years and more he worked on his book. In Volume 2 (page 254 of Sollas' English translation) Suess proposed that there were two large Paleozoic continents: Atlantis, centered on the North Atlantic Ocean, and Gondwána-Land in the south, the two being separated by an east-west seaway, the Tethys. In Volume 4, however, he described four upper Carboniferous continental areas that, despite the sinking of some fragments, have served as asylums for life. These were Laurentia and Angara Land in the northern hemisphere and Gondwána-Land and Antarctis in the southern. Laurentia consisted of the heartland of North America from Texas to the Arctic and eastward to Greenland. Angara Land included eastern Siberia and northern China. Gondwána-Land (literal translation, "Gond land land," meaning land of the Gonds) was named for the type locality in India of a series of Permocarboniferous sediments with coal seams containing *Glossopteris*. At its greatest extent, Suess' Gondwána-Land spanned 150° of longitude from Argentina to peninsular India, embracing part of the South Atlantic Ocean, Africa from the Cape moun-

Figure 18. The Malagasy fauna. As the common types of lemurs were well known in British zoological gardens, Alfred Russel Wallace chose to represent them by the "nocturnal and extraordinary aye-aye (*Chiromys madagascariensis*)." Also shown here are river hogs (*Potamochoerus edwardsii*), which he described as late immigrants related to African species, and three birds: *Euryceros prevosti* (upper left), of disputed identity; *Vanga curvirostris* (upper right), a peculiar Madagascar shrike; and *Leptosoma discolor* (center), a bird intermediate between the cuckoos and rollers. (From Wallace, 1876, plate 6.)

Suess' description of the marginal creep of Asia, his deduction from stratigraphic and fossil evidence that India belonged with the southern continents, and his use of the name Gondwána-Land were later adopted by proponents of the continental drift hypothesis. Some even claimed him as an early advocate of the idea. In fact, however, his geological world view in general and his vast continent of Gondwána-Land in particular were unreconcilable with the hypothesis of continental drift.

Speculating on the future, Suess saw that global contraction should logically result in a uniform shortening of all the planet's radii. When the topographic differences between continents and ocean basins are finally smoothed out, will not the surface of the shrunken earth once again be overwhelmed by pan-thalassa, the universal ocean? Suess closed the final volume of his masterwork by recommending a geological version of "Carpe Diem":

> In the face of these open questions let us rejoice in the sunshine, the starry firmament and all the manifold diversity of the Face of our Earth . . . recognizing, at the same time, to how great a degree life is controlled by the nature of the planet and its fortunes.

Suboceanic spread

To Eduard Suess, the map plan of Asia suggested a high dome of sal with borders overthrusting and crumpling against the rigid plates of India, Africa, and the Pacific floor. After extensive geological field work in Asia, Bailey Willis came to quite the opposite conclusion. In Volume 2 of *Research in China*, published in 1907, Willis described Asia as a mosaic of crustal units, formerly separated by mediterranean seas, that have been forced together by repeated thrusts from the direction of the Indian and Pacific oceans. All of the mediterraneans except the southern Tethys finally were closed and Asia was welded into one great patchwork in the "unequaled" Permo-Mesozoic diastrophism. Finally, in the Tertiary, further thrusting from the oceans caused the folding of the Himalayas and the uplift of the Tibetan plateau. Willis wrote:

> The theory is here entertained that these processes are due to what may be called *suboceanic spread*, i.e., to the expansion of suboceanic masses . . . say 100 miles deep, at the expense of subcontinental masses . . . occasioned primarily by molecular or mass changes under varying conditions of temperature and pressure. The general result is plastic flow in rigid and solid rock masses and it is held that in the great suboceanic regions such flow is a persistent condition, to which we may ascribe those accumulated stresses that have sufficed to produce the recurrent pronounced effects of diastrophism.

Although he believed that the ocean floors slowly spread beneath the continental margins, Willis saw the ocean basins as permanent features of the crust. Of the Pacific and Indian oceans he wrote:

> Their vast depressions have been oceanic basins throughout known history, however broadly their margins and connections may have varied. We may state this unequivocally, for, apart from other reasoning, there is no conceivable storage place for the enormous volume of their waters in other parts of the world had their reservoirs been drained at any time during the existence of the known continents.

Willis' own observations in the field thus led him to ideas that were germane to two concepts: continental accretion and sea-floor spreading. Despite his belief in a certain ocean-floor mobility, however, he never favored large-scale displacement of crustal blocks, and he was to become one of the most determined opponents of continental drift.

Indications of a core

Suess postulated a layered earth with a nickel-iron core, a magnesian silicate mantle, and a salic crust before any physical measurements had proved the existence of such structure in the planet. Speculations on the presence of a core in the earth date back at least to the 17th century and to Edmond Halley, who concluded that a metallic core rotating at a rate different from that of an adjacent shell gives rise to the earth's magnetic field.

The probability that the core is very large and consists of iron was proposed in 1897 by Ernst Weichert, of Göttingen, from calculations relating the earth's mean and surface densities to the abundances of the chemical elements. The densities of crustal rocks rarely exceed 3.3 g/cm^3, but the mean density of the planet is 5.5 g/cm^3. Clearly, some very heavy material in the interior is compensating for the light surficial rocks, and metallic iron, with a density of 7.9 g/cm^3, is the most abundant of the heavy elements.

The first seismic indications of the core were reported by Richard D. Oldham in England in 1906: "The modern seismograph has given to geology a new instrument of research . . . enabling us to see into, and determine the physical constitution of, the interior of the earth at depths removed from any other possible means of research." Oldham described the records of distant earthquakes as exhibiting three distinct phases. Two of these phases represent different forms of wave motion that have been propagated through the body of the earth, while the third travels along the surface. Oldham noted that the two types of body waves—which he called first-phase and second-phase waves—underwent a marked reduction in rate of transmission at a depth of about three-fifths of the earth's radius. At the same depth the second-phase waves are severely refracted, or possibly extinguished altogether, as they fail to emerge at the earth's surface anywhere within a zone ringing the epicenter at a distance of 120° to 130°. Oldham concluded that the waves must enter a central core of very different physical properties "if not also differing in chemical constitution."

The body waves described by Oldham are now called primary and secondary, or P- and S-waves. The P-waves vibrate back and forth in the direction of propagation, in the manner of sound waves, and pass through solids, liquids, and gases. The S-waves, sometimes called shake or shear waves, vibrate at right angles to the direction of propagation and are transmitted only in solids. The failure of S-waves to pass through the core is now taken as evidence that the core is liquid (Figure 21).

In 1914 Beno Gutenberg, in Germany, recalculated the depth of the core-mantle boundary as 2,900 kilometers, or only about two-fifths of the earth's radius. In 1936 Inge Lehmann, of Denmark, discovered that within the liquid core primary waves increase again in velocity at a depth of about 5,000 kilometers from the earth's surface. This is generally interpreted as meaning that the liquid core has a solid inner nucleus. Indeed, it should be solid; it is under a pressure of nearly 3 million bars.

The material of the core is widely believed to be

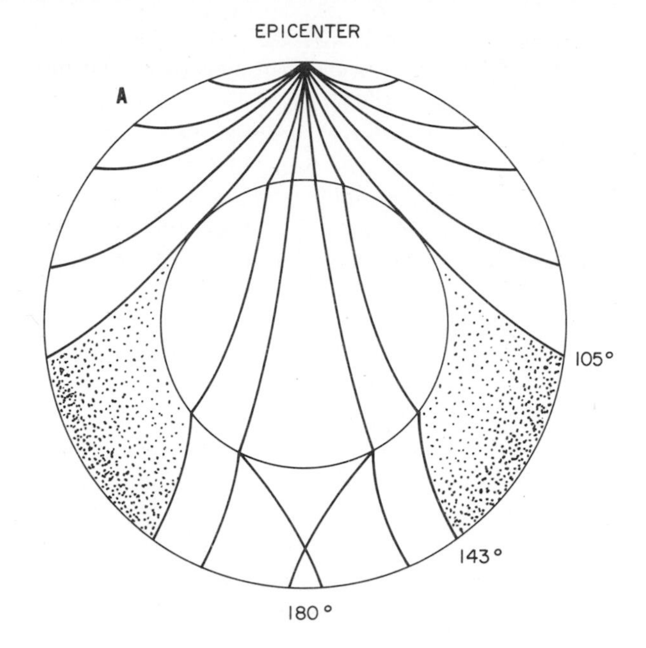

Figure 21. Two sketches of the earthquake-free shadow zone. **A:** Cross section showing paths of P-waves and S-waves through the mantle. S-waves fail to pass through the core. P-waves entering the core are severely refracted and emerge at distances of 143° to 180° from the epicenter. **B:** The shadow zone of an earthquake in the western Aleutian Islands.

nickel-iron, partly because of the core's calculated density and seismic properties and partly because of the sporadic fall upon the earth of nickel-iron meteorites. As we lack specimens of the earth's core, we find it convenient to assume that our core resembles these metallic specimens from space, an idea first proposed by Auguste Daubrée in Paris in 1868. We conclude, in turn, that the nickel-iron meteorites are fragments from the core of some other body or bodies in the solar system. According to physicists, such a conclusion is not circular reasoning but a self-consistent model in which all assumptions are equally probable.

"Probable" the assumptions may be, but alternative possibilities have been proposed. Harold C. Urey, the Nobelist in chemistry, has suggested that iron meteorites represent small pools of metal that formed in partially differentiated primitive bodies having the bulk composition of chondritic meteorites. John T. Wasson, of the University of California at Los Angeles, argues that at least one group of iron meteorites formed by accretion in the planetary nebula. Indeed, one of the newest theories of the earth, proposed in 1972 by Sidney P. Clarke, Jr., Karl K. Turekian, and Lawrence Grossman, of Yale University, assumes that the earth's core accreted directly from the planetary nebula at an early stage when the gas cloud cooled to about 1400° Kelvin and iron and nickel were the first abundant phases to condense as solid particles. After these particles coalesced, the silicates of the lower mantle accreted upon the metallic nucleus. As cooling continued and the nature of the gas changed, the upper mantle and outer crustal layer accumulated, the latter being a low-temperature veneer rich in oxides, volatile elements, organic compounds, and rare gases. Thus, the planet grew as a gravitationally stable body that has never heated to the point where near-surface materials have undergone convective interchange with the core. A variation on this theory, based on a study of a rare class of meteorites called Type III carbonaceous chondrites, suggests that the first condensates were high-temperature minerals rich in calcium, aluminum, titanium, uranium, and thorium, and these minerals formed a protocore that was later enclosed by a shell of metallic nickel-iron. Radioactive heating of the protocore led to melting and an overturn with sinking of the metal and floating of the lighter materials before the outer layers were accreted on the growing planet. All versions of this general model reject the time-honored geochemical picture of the planet as the product of a giant smelting operation in which the entire mass was melted and differentiated, with heavy metal sinking to the center while a light, siliceous slag floated to the surface.

Shock wave experiments in metals show that the density of the core is about 8 percent lower than that extrapolated for pure iron or nickel-iron at core pressures. To explain this low density, A. E. Ringwood, of the Australian National University, suggests that the core contains metallic silicon in the following proportions: iron, 70-80 percent; silicon, 10-20 percent; nickel, 10 percent. As an alternative, D. L. Anderson of the California Institute of Technology (who favors the second version of planetary accretion outlined above) argues for an older idea that the low density of the core reflects an admixture of sulfur with the iron and nickel. Geophysical measurements cannot discriminate between these two suggested compositions.

The mantle

Three years after Oldham's discovery of seismic evidence for the core, Andreiji Mohorovičić, studying the earthquake of October 8, 1909, in the Kulpa Valley of Croatia, found evidence of a sudden increase in the velocity of primary waves at a depth of some tens of kilometers. It was widely assumed that he had detected the interface between the sal and the sima, or, in structural terms, the crust and the mantle. The discontinuity was named for Mohorovičić in 1940 after two comprehensive tables of earthquake-wave travel times showed it to be a major boundary of worldwide extent where P-waves increase in velocity from 7.2 to 8.1 kilometers per second.

In 1923, V. Conrad, studying the earthquake of November 28 at Tauern, Austria, found evidence of a shallower discontinuity within the crust itself. That discontinuity, confirmed by other seismologists, was named for Conrad. The Conrad discontinuity, it was widely believed, marks the interface between an upper, granitic layer and a deeper basaltic layer of the crust.

What do these discontinuities mean in terms of rock composition? That question has preoccupied many geophysicists and petrologists in the decades since the discovery of the discontinuities. Terms such as "sima" and "sal" (later changed to "sial" by Alfred Wegener) denote a chemical contrast and have generally implied the melting and separation of lighter from heavier materials. In 1914, however, L. L. Fermor, the English geologist, pointed out that the Mohorovičić discontinuity could, alternatively, mark a phase contrast where familiar rocks of the crust transform under high pressures to denser forms of the same chemical composition. Argument continues to this day as to whether the Mohorovičić discontinuity represents a chemical or a phase change, with a probable majority of scientists favoring the former. In either case the main materials of the upper mantle are believed to be one of two rock types: peridotite, which consists mainly of olivine with minor pyroxene; or eclogite, which consists of pyroxene and garnet. Both are silicate rocks of uncommonly high density that are relatively rare in the crust, where their mode of occurrence suggests derivation from great depth.

Peridotite melts over a range of temperatures and yields magmas of basaltic or andesitic composition that are more siliceous than the parent peridotite. As these light magmas rise through zones of weakness to form intrusive or extrusive crustal rocks, a heavier residue, depleted in silica, remains in the mantle. Under other circumstances peridotite can incorporate water at low temperatures and transform, without melting, to serpentinite, a fibrous green rock commonly found in folded mountain ranges.

Eclogite has substantially the same chemical composition as basalt. Experimental evidence shows that a solid-state transformation from eclogite to basalt or basalt to eclogite can occur under a range of temperatures and pressures approximating those near the Mohorovičić discontinuity.

Although it was long believed to be a homogeneous domain, the mantle is now known to consist of two main layers (the upper and lower mantles) plus several less-well-defined regions of differing densities and temperatures. In a general way, the Mohorovičić discontinuity mirrors crustal topography in the manner implied by Airy, occurring at depths of about 30 kilometers beneath the continental platforms, 60 kilometers beneath mountain ranges, and only 5 kilometers beneath the ocean floors. Density contrasts in different portions of the crust also are important enough to have persuaded the Finnish geodesist W. A. Heiskanen, in 1924, that isostatic

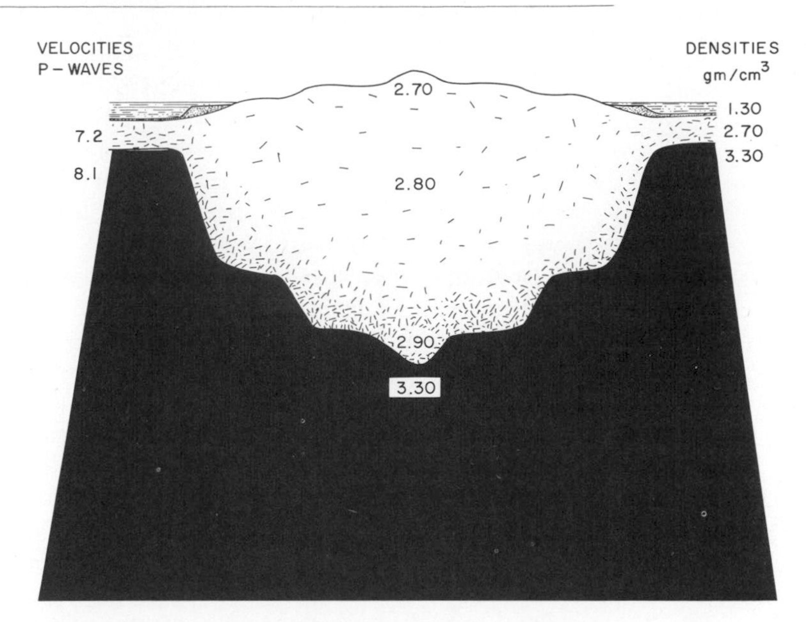

Figure 22. An idealized representation of the Mohorovičić discontinuity.

balance is achieved by a combination of principles in the proportions 63 percent Airy/37 percent Pratt. A simplified representation of the Mohorovičić discontinuity is shown in Figure 22.

Measurements made within the past decade indicate that the Mohorovičić discontinuity is not always a sharp boundary, nor is it always found at the predicted level, particularly under disturbed zones of the crust. All exceptions aside, however, the discontinuity retains its importance in defining the interface between the crust and the upper mantle.

Radioactive heat

The contraction hypothesis, so fundamental to Suessian geology, began with the assumption that the earth is still losing the heat it inherited when it formed from the primordial sun or a gaseous solar nebula. William Thompson (Lord Kelvin), in 1846, set himself the task of calculating the age of the earth from its cooling rate. Given the composition and thermal conductivity of crustal rocks and the measured increase in temperature with depth beneath the surface, Kelvin calculated that the molten globe, cooling by conduction from the outside inward, had acquired its solid crust between 20 and 400 million years ago—with the most probable time being about 100 million years ago. Kelvin made this pronouncement in 1862 when geologists had finally adopted the idea of an indefinitely long time scale. Charles Darwin, in *The Origin of Species* (1859), had speculated that more than 300 million years have elapsed since the beginning of the Tertiary and that the age of the earth itself is incomprehensibly vast. Other scientists agreed. T. H. Huxley, for example, regarded the age of the earth as utterly beyond the power of human understanding and scientific investigation.

To Kelvin, who believed that the solar system had a definite beginning and is undergoing chemical evolution at a measurable rate, the prevailing attitude toward geologic time was a scandal. "Led by Hutton and Playfair, Lyell taught the doctrine of eternity and uniformity in geology," wrote Kelvin, and he added that the results with respect to time were as disastrous as they would be if historians claimed that they could not tell whether the Battle of Hastings took place 800 years ago, 800 thousand years ago, or 800 million years ago. For their part, Kelvin's contemporaries found his "short" time span shocking. Said the geologist Andrew Ramsay to Kelvin: "I am as incapable of estimating and understanding the reasons which you physicists have for limiting geological time as you are incapable of understanding the geological reasons for our unlimited estimates." Kelvin, who felt not at all incapable of understanding geological problems, replied: "You can understand physicists' reasoning perfectly if you give your mind to it." And the two scientists agreed to "temporarily" disagree.

Far from relenting, Kelvin refined his calculations and reduced his upper limit by an order of magnitude. In 1897 he pronounced the earth only 20 to 40 million years old, "and probably much nearer 20 than 40." The inference was clear that geologic change occurred at immensely faster rates in earlier eras than those observed today. Catastrophism was receiving support from one of the pioneers of geophysics.

But one year earlier, in 1896, Henri Becquerel, in searching for phenomena comparable to x-radiation, discovered that uranium salts expose photographic film; and in 1903 Marie Curie found that the radium she isolated from pitchblende (UO_2) produced heat as well as radiation by the spontaneous decay process called radioactivity. By 1906 R. J. Strutt (Lord Rayleigh) and others had detected radioactivity in common rocks, soils, waters, and natural gases from all over the globe. The earth, therefore, was found to contain a hitherto unknown heat source—one that is ubiquitous and operates continuously. Suddenly the possibility arose that the planet is not cooling at all. Possibly it is even heating up! In any case, Kelvin's geochronology was exploded. His belief in a finite age of the earth was right and his calculations were sound, but due to a natural phenomenon that was unknown at the time his basic assumption was wrong.

Radioactivity was a wholly new factor in earth science. It not only upset all previous ideas of the earth's thermal history but it also held promise as an atomic clock for measuring in absolute terms the duration of geologic time. Several scientists appreciated the possibilities very quickly. By 1907 Bertram Boltwood of Yale had begun experiments in the dating of minerals on the assumption that lead is a stable product of the decay of uranium. In 1909 John Joly, in Ireland, published a book titled *Radioactivity in Geology*, and in 1913 Arthur Holmes, in London, published *The Age of the Earth*, the first of his many works on the subject. Holmes dated the oldest rocks of the earth's crust as about 1,600 million years old.

The year 1909 also saw publication of the fourth and final volume of *The Face of the Earth* by Eduard Suess, who, in his lifework, had encompassed global geology and built a theory on the assumptions of a cooling, contracting planet. Many of his observations and descriptions proved to be of lasting value. His broad interpretations did not. Indeed, how could they? Radioactivity does not appear in his index.

Continental Drift

". . . while the two continents on opposite sides have crept away in nearly parallel and opposite directions"

A radical departure from Suessian geology was proposed by Frank B. Taylor, an American geologist, at just about the time Suess published the fourth and final volume of *The Face of the Earth* in 1909. On December 29, 1908, Taylor presented a paper to the Geological Society of America advocating horizontal crustal movements on a grand scale. His ideas were published first in a privately printed pamphlet and later in the *Bulletin of the Geological Society of America*, issue of July 1910.

Taylor regarded Suess as having the most comprehensive knowledge and best grasp of the problems of global geology of any living scientist. Yet, Taylor could not accept thermal contraction as a force of any importance in determining the form and distribution of the continents. Impressed by the linear pattern, world-encircling extent, and lobate forms of the Tertiary mountain ranges, Taylor concluded that two large crustal sheets, originally located over the north and south poles, have undergone a massive creeping motion toward the equator, pressing their forward margins into arcuate folds and leaving behind great rifts and rents such as the Arctic basin and the South Atlantic and Indian oceans. Comparing the crustal movement to the advance of continental ice sheets, he worked out a detailed account of the Tertiary and Recent history of each continent. Like Suess, he believed that Greenland is a large block that is wedged in place. Unlike Suess, Taylor postulated that North America has split off and moved southwest, away from Greenland.

Taylor cited evidence that the northward creep of the south polar continent began at a somewhat later date and proceeded with less force than was the case with the northern continent. Antarctica he saw as a crustal fragment that remained stranded over the pole while South America and Australia split off and moved equatorward.

The Mid-Atlantic Ridge was discussed in some detail by Taylor, who characterized it, very perceptively, as a submerged mountain range of a different type and origin from any other on earth. Taylor noted that the coastlines of Africa and South America seem to fit the trend of the ridge crest, and he suggested that the ridge is a kind of linear horst marking a line of rifting where the crust split open. In 1910 he wrote: "It is probably much nearer the truth to suppose that the mid-Atlantic ridge has remained unmoved while the two continents on opposite sides of it have crept away in nearly parallel and opposite directions." He suggested that Africa moved eastward to its present position in the Permian and that South America moved westward in the Tertiary and Recent.

As the cause of these prodigious crustal movements that involved thousands of miles of horizontal displacement, Taylor rejected cooling and contraction in forthright terms. In 1926 he wrote:

The Earth never was a molten globe. The whole idea is pure Cartesian fiction. And because the globe was never molten, it has not suffered contraction in consequence of cooling. Its internal temperature has been substantially constant for unnumbered ages. The interior of the Earth is hot, but so far as is known it has never been hotter and it is not molten or liquid, but solid and more rigid than steel.

Figure 23. The equatorward creep of the north polar continent as depicted by Frank B. Taylor. The Tertiary mountain ranges are represented by heavy arcs and loops; areas that remained stable during the Tertiary, by horizontal shading. According to Taylor, the front of Asia reaches from Alaska (A) to Damascus (C), just halfway around the world. (From Taylor, 1928, in *Theory of Continental Drift*; used with permission of The American Association of Petroleum Geologists.)

To Taylor, the only force likely to affect the crust so as to cause massive distortion and movement was tidal action. Having outlined his analysis of the direction and magnitude of the motions (as shown in Figure 23), he said that the crust has behaved as it would if two large, low-density continents had been located over the poles and the earth had suddenly changed from a perfect sphere to an oblate spheroid. Such a change would take place as a result of an increase in the earth's speed of rotation—with both the water of the oceans and the rocks of the crust being pulled into an equatorial bulge. For years Taylor resisted the temptation to suggest an ultimate causal factor for this change. Finally, in 1926, he pointed out that the Tertiary mountain belts dominate the geology of the globe today; that they probably are the greatest ranges ever known on earth; and that, geologically speaking, they arose in all parts of the earth simultaneously, suddenly, and recently. To account for them requires a general, relatively sudden, and permanent increase in the tidal forces causing crustal movements. Then he said that if we could postulate capture of the moon in the Cretaceous many problems would be solved, including not only the increase in speed of rotation and the oblateness of the earth's figure but also the displacement of continents and the great transgressions of Cretaceous oceans over low-lying lands throughout tropical and subtropical latitudes.

Because Alfred Wegener's hypothesis of continental drift was introduced soon afterwards, it is fruitless to speculate on the influence Taylor's ideas might have had. Taylor made no overall attempt to reassemble the continents in their former positions except to match the west coast of Greenland to North America and to describe the creep of Euro-Africa and the Americas away from the Mid-Atlantic Ridge. (This idea was adopted in the late 1960s by advocates of sea-floor spreading with no credit accorded to Taylor.) Taylor's model left unexplained all mountain ranges older than the Tertiary, and his postulate, however tentative, of lunar capture so late in earth history smacked of an ad hoc catastrophe for which there seemed to be too little geological and biological evidence. His papers are, nevertheless, a rich source of information and ideas, many of which were well in advance of his time. Surprisingly, these writings do not appear to have caused any perceptible stir in geological circles, either in the form of enthusiastic support or angry rebuttal. Possibly this is because Taylor, following Suess insofar as he could throughout his geological descriptions, inadvertently made his ideas seem less radical than they actually were.

The Atlantic rift

In 1911 Howard B. Baker, another American, published the first of a series of articles on continental displacement, which he ascribed to planetary perturbations in the solar system and the fissioning of the moon from the earth. His first article, entitled "Origin of the Moon," appeared in the *Detroit Free Press* of April 23, 1911. Several later papers on "Origin of Continental Forms" (parts 2-5) were published in the *Annals of the Michigan Academy of Science*. A privately printed summary of his ideas, *The Atlantic Rift and Its Meaning*, was issued in a limited edition in 1932. Baker conjectured that, in the late Miocene, the orbits of the earth and Venus approached in such a way that the moon was ripped out of the Pacific basin, taking with it most of the earth's ocean water. Water was replenished by being captured from another nearby planet which was shattered to the small bits now represented by the thousands of asteroids. Meanwhile, the Atlantic Ocean opened as a great rift in the earth's crust and the continental fragments moved very rapidly away to either side, toward the edge of the new Pacific hollow.

Baker's reconstruction of the Atlantic rift is shown in Figure 24. He backed his arguments for the matching of the shorelines with structural and stratigraphic evidence, although India, the type locality for the Gondwana sediments, is scarcely mentioned in his text. Today, geologists are in general agreement that the Miocene, which began only 25 million years ago, is an impossibly late date to consider the shattering of the planet, the emptying and refilling of the oceans, and the displacement of continents.

Such were the arguments for continental displacement by advocates preceding Alfred Wegener. There were other advocates, not included here, who proposed at least some aspects of the idea. Their number, however, was small, and they worked singly amid the legions of geologists who accepted as established fact the contraction hypothesis or the doctrine of permanence of continents and ocean basins. As we have seen, the company of drift proponents tended toward the eccentric, although Fisher, Pickering, and Taylor stand out as men who based their interpretations of earth science on the principles of physics. Every advocate associated continental drift with a catastrophe, chiefly the fissioning-off or the capture of the moon.

Wegener said that he conceived his idea of continental drift in 1910 while contemplating, on a

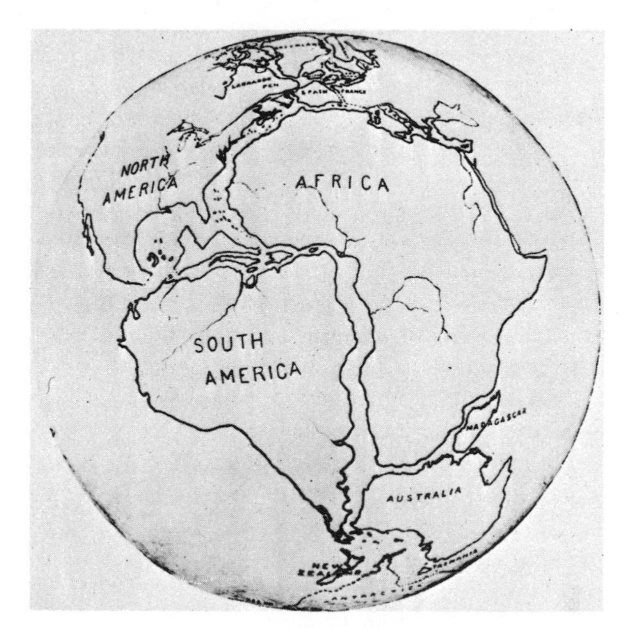

Figure 24. The Atlantic rift of Howard B. Baker, 1932. Features of special interest include the rotation of the British Isles, the matching of the mountains of Spain with those of Newfoundland, and the distribution of lands in the Caribbean region. (Used with permission of Dr. Burton P. Baker and Mrs. Caroline Baker Davis.)

world map, the congruence of the coastlines across the Atlantic Ocean. His claim of having developed a wholly original hypothesis has never been challenged. He gave his first lectures and published his first two papers in Germany in 1912. Unlike his predecessors, however, Wegener unleashed a storm.

Alfred Wegener: The Hypothesis

As we have seen, the striking parallelism of the coastlines on opposite sides of the Atlantic Ocean began to excite interest as soon as the first maps were compiled of the newly discovered western continents. Since the 16th century the apparent matching, particularly of South America and Africa, has been noted by scientists, laymen, and apparently also by large numbers of children, if we may judge from the multitude of adults who claim to have "noticed that matching" at a very early age.

"But now this similarity had been noticed by an expert geophysicist, a brilliant man of unbounding energy, who would spare no pain in following up the matter and gaining any facts from other fields of science that might seem to have a bearing on the question," wrote Professor Wladimir Köppen, who was known in Germany as the Grand Old Man of Meteorology and who became a scientific collaborator (and also father-in-law) of Alfred Wegener.

The congruence of the Atlantic coastlines intrigued Wegener at least as early as 1903 when he pointed it out to W. Wundt, a fellow student in Berlin. At that time, Wegener was working toward a doctoral degree in astronomy, which he received in 1905 after submission of a thesis entitled "The Alphonsine Tables for the Use of Modern Computers." The Alphonsine Tables of planetary motions were compiled about the year 1273 under the patronage of King Alfonso of Castile. They were in a sexagesimal numerical system but Wegener transcribed them into the decimal system for the convenience of modern computers, who, in 1905, were persons rather than machines. Later in the same year Wegener published an article on the general history and usage of the Alphonsine Tables, and that article, according to astronomers, was a contribution more worthwhile than his thesis. It seems likely that Wegener's main interest lay in the subject matter of the article, but, as doctoral degrees were not then granted for efforts in the history of science, he fulfilled his requirement by transcribing the Alphonsine Tables.

From his youth Wegener had cherished the hope of exploring northern Greenland, and while a student he systematically built up his endurance by long days of walking, skating, alpine climbing, and skiing. He also was fascinated by the new science of meteorology (the stratosphere was first discovered in 1900), and at the Royal Prussian Aeronautical Observatory he carried out experiments with kites and balloons. In 1906 he and his brother Kurt won an international free balloon contest by staying aloft for 52 hours, a duration that was 17 hours longer than the world record. Shortly thereafter Wegener's great dream was fulfilled when he left for Greenland as official meteorologist with the *Danmark* Expedition led by Mylius-Erichsen.

After returning from Greenland in 1908, Wegener took up residence at the University of Marburg where he was appointed "Privatdozent" in meteorology, practical astronomy, and cosmic physics. His lectures, which were very popular with students, were said to be remarkable for clarity, candor, and the lack of Prussian professorial style. Wegener readily admitted a distaste for treatises heavily larded with mathematical formulae. He preferred to make his points in language. In a letter to Köppen, written in 1913, he commented: "When one cannot follow the printed or written word one should not always put the blame on oneself. When logic is lacking, one can still usually fill out a few lines with formulae." This is a most interesting statement from one whose first book, published in 1911, was entitled *Thermodynamik der Atmosphäre.*

According to Wegener's own testimony, the first glimmerings of the concept of continental drift occurred to him about Christmastime in 1910 as he examined the world maps in the large Andrée *Allgemeiner Handatlas*, but he immediately dismissed the idea as improbable. In the autumn of 1911, however, he came "quite by accident" upon a compendium of references describing the faunal similarities of Paleozoic strata in Africa and Brazil. The compendium presented arguments for a former land bridge across the Atlantic. Wegener saw the problem differently; to him this additional evidence of matching between the two continents suggested, once again, the possibility of continental drift. With his interest rekindled, he sought out confirmatory data, and within four months he presented his hypothesis in two addresses. The first, entitled "The Geophysical Basis of the Evolution of the Large-Scale

Features of the Earth's Crust (Continents and Oceans)," was delivered to the Geological Association in Frankfurt-am-Main on January 6, 1912; the second, called "Horizontal Displacements of the Continents," was presented four nights later before the Society for the Advancement of Natural Science in Marburg. Before leaving shortly thereafter, on his second expedition to Greenland, Wegener—in view of the unquestioned dangers he faced—bequeathed his first manuscript on continental drift to *Petermanns Mitteilungen*, which published the three-part paper in its April and June issues of 1912. The paper was summarized in the *Geologische Rundschau* the same year.

Returning from Greenland in 1913, Wegener married Köppen's daughter, Else, whose photographs depict an enchanting young woman. The following year, on July 29, 1914, World War I opened with the Austrian bombardment of Belgrade, and on August 1 the Kaiser ordered general mobilization. Wegener was called into the army as an officer, serving, as his wife wrote later, from the first day to the last. According to Professor Benndorf, one of his colleagues, Wegener took this duty very hard because he harbored a profound sense of the futility of war.

Throughout the war, in which he was wounded twice, Wegener continued to work on his Greenland data, to carry out meteorological observations, and to develop his ideas on continents and oceans. His scientific efforts in that period never ceased. He published a book in 1915 and from two to seven articles in German scientific periodicals every year from 1914 to 1920. On April 3, 1916, he observed the fall of a bright meteor, and this turned his thoughts toward impact phenomena and the craters of the moon. One of his first projects after the war was a series of experiments in which he reproduced the forms of lunar craters, with and without central peaks, by impacting projectiles against groundmasses of various types, both viscous and pulverized. His scientific interests were wide-ranging indeed.

Wegener's book, *Die Entstehung der Kontinente und Ozeane* (The Origin of Continents and Oceans), published in 1915, was written during a sick leave. The first edition, a slim volume of 94 pages with no index, never enjoyed wide circulation. Today, copies of that edition are cataloged in only two libraries in the United States—the New York Public Library and the John Crerar Library in Chicago.

In 1919 Wegener left his position in Marburg to become Köppen's successor as director of the meteorological research department of the Marine Observatory in Hamburg. In the next year he published the second edition of his book, a thoroughly reorganized version incorporating much new material. This second edition excited immediate interest and controversy on the continent, but within the English-speaking world there was little awareness of his hypothesis until critical reviews were written in 1922 by the English geologist Philip Lake and by the American geologist Harry Fielding Reid. The book was widely read for the first time in 1924 when the third edition (1922) was translated into English, French, Spanish, Swedish, and Russian.

In brief outline, the main premises of Wegener's hypothesis as presented in the 1922 edition were as follows:

(1) The continents and ocean floors are fundamentally distinct. The continents are blocks of light granitic rock (sial), about 100 kilometers thick, that float isostatically in denser basaltic rock (sima), out of which they project only about 5 kilometers. The denser medium is exposed in the ocean floors.

(2) The sial no longer completely covers the entire earth; whether it ever did can be left undecided. In any case its area has grown smaller and its thickness has increased as a result of folding and thrusting during geologic time. It has also split into fragments. Today, continental rock covers only about one-third of the earth's surface.

(3) The continental blocks retain the approximate outlines they acquired during a breakup that began in the Mesozoic. If the younger, Tertiary folded mountains could be flattened out, the pieces could be reassembled into one large protocontinent partially flooded by shallow seas.

(4) The correctness of the proposed reconstruction—made by fitting the blocks along the edges of the continental shelves—is confirmed by the matching of truncated mountain ranges, sedimentary formations, basaltic dikes and flows, glacial tillites, and the distribution of fossil and living flora and fauna.

(5) The first signs of breakup are continental rift valleys, and these gradually widen to new oceans. The Tertiary mountain ranges resulted from the crumpling of the continental blocks along their forward margins as they were pressed together (India-Asia) or met resistance from the cooled sima of the Pacific floor (the American, Antarctic, and Australian cordilleras). Islands and island arcs are remnants sloughed off in the wake of drifting continents. The trenches of the western Pacific are tension fractures ripped open as Asia moved westward. Those of the eastern Pacific were caused by the sinking of the ocean floor at the margin of the oncoming continents.

(6) The pattern of late Paleozoic climates indicates that, in addition to continental drift, polar wandering has occurred. Fossil distributions show that there has been an apparent migration of the earth's surface relative to the poles for a distance of at least 4,000 kilometers since the Permian.

(7) The forces causing the drift of continents are all intrinsic to the earth's rotation system. They include the Eötvös, or pole-fleeing force, which impels continents equatorward, and the tidal attraction of the sun and moon, which exerts a drag on the crust, slowing its rate of rotation and so causing it to move westward with respect to the interior. These forces are very small but when applied for long periods of time they become effective enough to impel crustal blocks through the substratum, which opens ahead of them and closes behind them.

(8) Geodetic determinations of longitude and latitude, repeated at intervals from various stations, show that Greenland and certain other continents and islands are moving at measurable rates.

With premises 1 and 2, Wegener was rejecting the possibility, so basic to Suessian geology, that the ocean basins could have formed by the foundering of continental blocks. He compared the continents to tabular ice floes in seawater. They could split into fragments, however small, but none of them could sink.

Wegener accepted Suess' concept of three earth layers—the nife, the sima, and the sal. He used the latter term throughout the first edition of his book, but in the second edition he followed the suggestion of a friend, G. Pfeffer, and substituted the term "sial," thus abbreviating silica-alumina and ending the semantic confusion with the Latin word for salt. Unlike Suess, Wegener was not convinced that the sial ever constituted a continuous crustal layer, and he viewed all of the ocean floors as the exposed surface of the sima.

Two favored levels on the earth

To support his contention that the continents and ocean floors are fundamentally distinct, Wegener marshalled several types of evidence: the presence of two main topographic levels on the globe, magnetic indications that oceanic rocks are relatively enriched in iron, the greater seismic velocities recorded in oceanic crust, and the close approach to isostatic equilibrium of continents and ocean basins.

Rough and irregular as the land surface may seem to its inhabitants, geodetic surveys have shown that the continents are essentially vast platforms with an elevation of about 230 meters above sea level, and the ocean floors are predominantly abyssal plains lying about 4,700 meters below sea level. Mountain ranges and deep trenches give the earth a maximum relief of 19,881 meters from the top of Mount Everest (8,848 meters) to the bottom of the Marianas Trench (−11,033 meters), but such dramatic features are minor wrinkles on the two main levels that are separated by a vertical distance of approximately 5 kilometers (Figure 25). This distribution of elevations appeared highly significant to Alfred Wegener:

> In the whole of geophysics there is scarcely another law of such clearness and certainty as this one, which states that there are two favored levels on the earth, which occur alternately side by side and which are represented by the continents and the floors of the ocean. Therefore it is very remarkable that for this law, which has been well known for at least 50 years, no explanation has ever been sought . . . it must be concluded that there are already two undisturbed original levels, and from this the step seems inevitable that in the continents and the floors of the oceans we have two different layers of the body of the earth which—expressed in somewhat exaggerated form—act as water does between great sheets of ice. This step seems so easy and obvious that the next generation will certainly wonder that we should have hesitated such a long time over taking it.

[In fact, however, the next generation also hesitated. A widespread belief in sialic floors under the Atlantic and Indian oceans persisted well into the 1950s.]

In defense of his conclusion that the two topographic levels represent distinct earth layers, Wegener argued that the deformation of a single layer would result in a Gaussian distribution of elevations as shown by the dotted line in Figure 26.* To this topographic argument Wegener added that investigators in geomagnetic research were of the opinion that the ocean floors are more strongly magnetizable and therefore probably more iron-rich than continental rock. He also cited seismic evidence that the

*Wegener cited the figures of +100 meters and -4,700 meters as representing the two favored levels. On a modern histogram these levels, which are averages for continental and ocean floor platforms exclusive of mountains and deeps, would be about +230 meters and -4,700 meters.

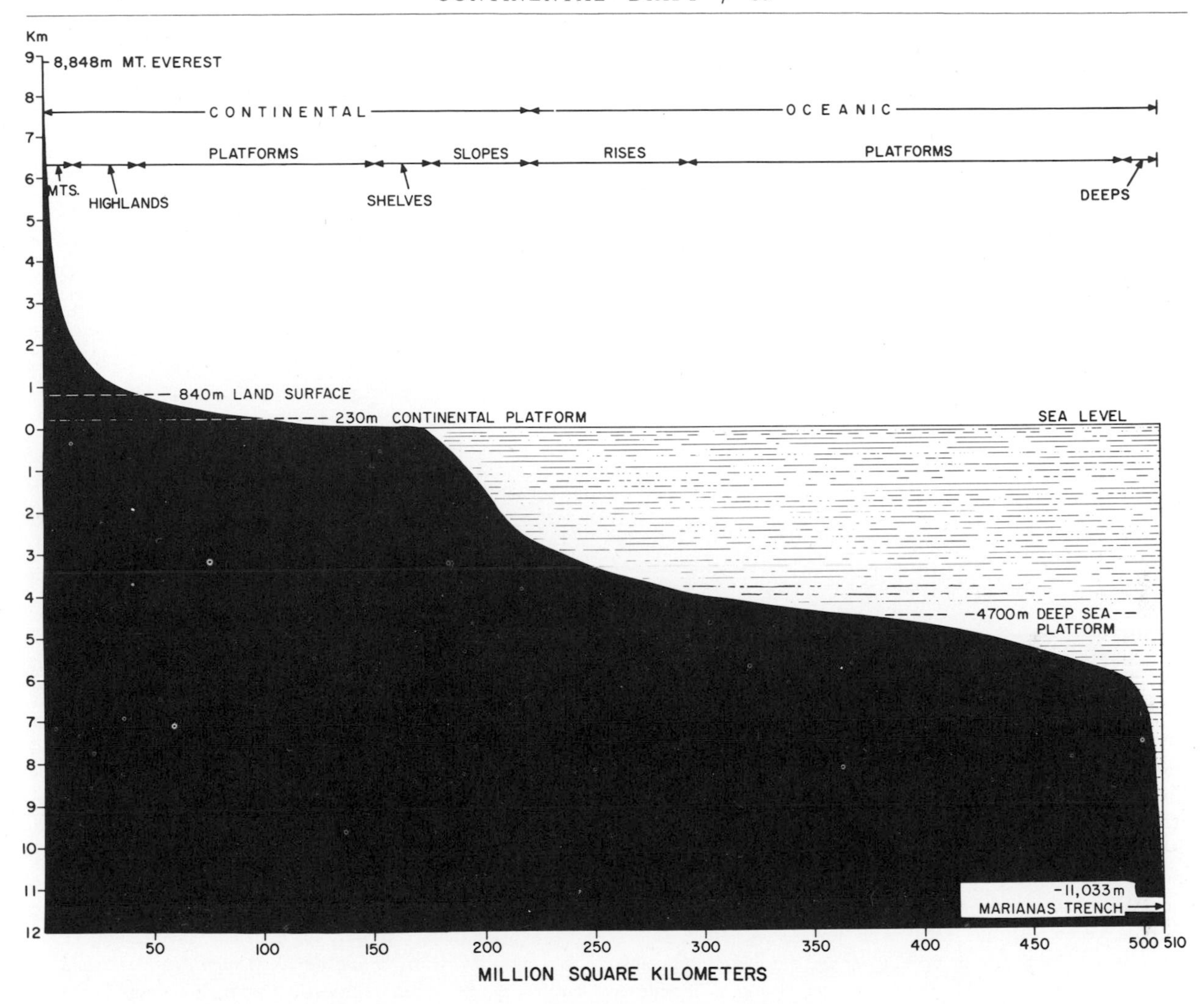

Figure 25. Hypsometric curve showing the distribution of elevations on the earth's surface.

velocities of surface waves over oceanic areas are, on the average, about 0.1 km/sec higher than over continents. The lower value is for the continental sial; the higher one matches the predicted velocity for basaltic rock, which he proposed as the material of the ocean floors.

Fundamental to Wegener's hypothesis was the assumption that the earth and its large surface features (continents and oceans) are in a state of nearly perfect isostatic equilibrium. Wegener's floating iceberg model combined Airy's concept of deep roots balancing topographic highlands with Pratt's idea of density differences between crustal provinces. Wegener's own sketches of the two concepts of isostasy are shown in Figure 27. Referring to the evidence for the postglacial uplift of Scandinavia, Wegener reasoned that if pieces of the sialic crust can ride passively upward and downward they can also float sideways provided they are subject to sufficient force. The force of gravity is operative in the case of vertical motions; Wegener's problem was to identify forces of sufficient magnitude acting in a direction tangential to the earth's surface.

Unlike his predecessors Fisher, Pickering, and Taylor, Wegener avoided calling upon a catastrophic event such as earth fission or capture of the moon to initiate continental drift. He concluded that the forces that displace continents are the same ones that

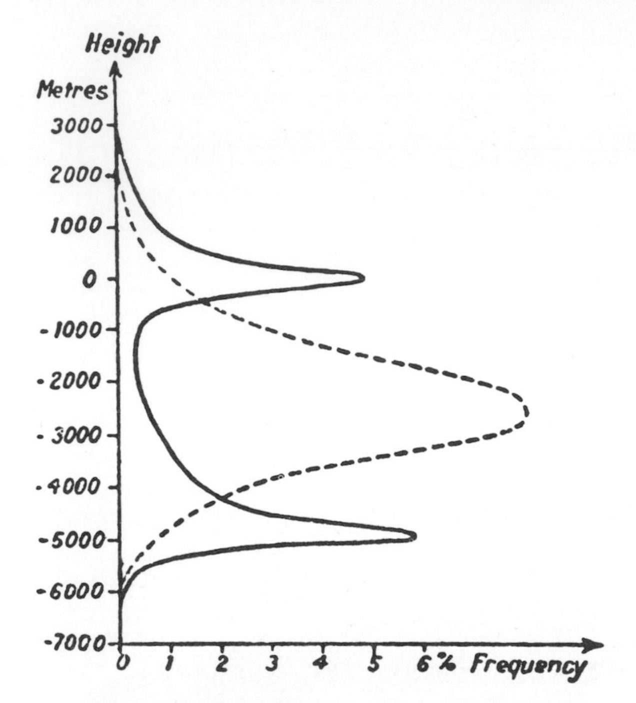

Figure 26. The distribution of the earth's elevations. The solid line represents Alfred Wegener's double-peaked curve of surface elevations. In Wegener's view the two peaks represent two fundamental levels, the surfaces of the sial and the sima. The dashed line is the Gaussian distribution which Wegener would expect if the earth had only one surface level that has been deformed to create continents and ocean basins. (The diagram was first used in *Die Entstehung der Kontinente und Ozeane*, second edition, 1920. This English version is from page 30 of *The Origin of Continents and Oceans*, translated by J. G. A. Skerl, 1924; used with permission of Friedr. Vieweg & Sohn, Braunschweig, and Methuen & Co. Ltd., London.)

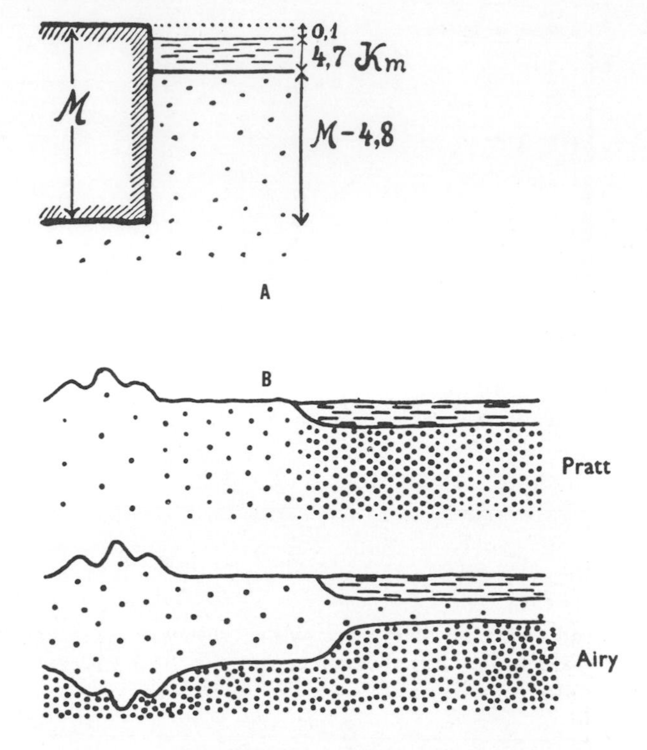

Figure 27. Two sketches by Wegener to illustrate the principles of isostasy. A: Block of sial (M), 100 kilometers thick, floats to a depth of 95.2 kilometers (M-4.8) in the sima. B: Illustration of Pratt's idea that continents and oceans are underlain by materials of different densities and of Airy's concept that they are underlain by varying thicknesses of light material "afloat" in a denser substratum. (From Wegener, *Die Entstehung der Kontinente und Ozeane*. A, from page 26 of first edition, 1915; B, from page 41, fourth edition, 1929. Used with permission of Friedr. Vieweg & Sohn, Braunschweig.)

produce great folded mountain ranges, and he believed these forces derive from the earth's rotation system. The forces he postulated were the following two: the Eötvös, or pole-fleeing force (*Polfluchtkraft*), and lunar-solar tidal drag on the viscous earth.

The displacement forces

The Eötvös force is centrifugal force applied to a body floating in isostatic equilibrium. It arises from the fact that Archimedes' principle of buoyancy is not quite sufficient—except at the poles and the equator—to describe the motion of floating bodies on a rotating spheroid. At all latitudes between 0° and 90° the flattening of the spheroid is such that a "radius" drawn normal to the surface does not pass exactly through the center of the spheroid. The amount of flattening decreases with depth, however, so that a series of normals drawn from deeper and deeper levels describes a curve that is concave toward the pole (Figure 28). Gravity, acting at the center of gravity in the floating body, pulls the body perpendicularly downward from the earth's surface; buoyancy, acting at the center of gravity of the displaced

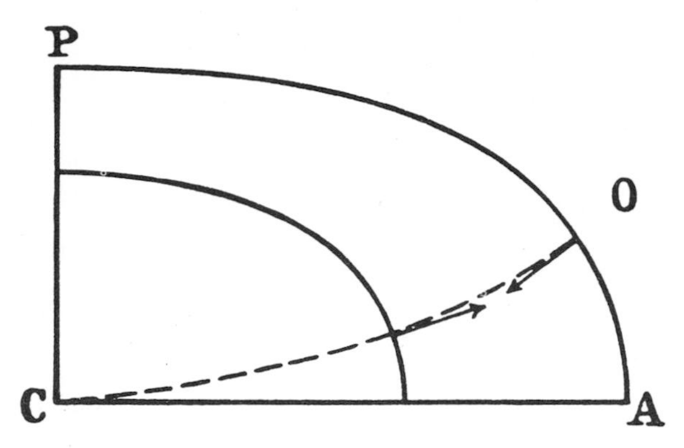

Figure 28. The Eötvös force as depicted by Wegener. PCA is one quadrant of an oblate spheroid with two "concentric" layers. A landmass floating in isostatic equilibrium at O will be impelled toward the equator (A) because the pull of gravity (arrow pointing inward at 90° from the surface) is not directly opposed by the buoyancy of the displaced medium (arrow pointing outward at 90° from deeper level). The dashed line is the line of force of the gravitational field. The floating mass will come to rest only at the equator where the two arrows point directly toward each other. (From Wegener, *Die Entstehung der Kontinente und Ozeane*, page 133, third edition, 1922; used with permission of Friedr. Vieweg & Sohn, Braunschweig.)

medium, impels the body perpendicularly upward from a deeper surface. But as these surfaces are not quite parallel, the forces are not exactly opposed and so a small resultant tends to move the body equatorward. Although most geophysicists agree that the Eötvös force exists, few, if any, believe with Wegener that it has the magnitude to move continents and fold mountain ranges.

The tidal forces of the sun and moon cause a frictional drag on the earth's crust and its ocean waters that is slowing the planet's rate of rotation. Wegener concluded that on an eastward-rotating earth this drag would be most effective on the high, sialic continental blocks and would tend to make them lag behind the more rapidly rotating sima. This lagging behind in a race to the east would give the continents an apparent westward motion. Thus, along with an equatorward movement Wegener pictured the continental fragments as lagging at different rates and so appearing to drift westward for different distances. Today, the existence of this component of the tidal force is in some dispute, but, as with the Eötvös force, its magnitude is not in dispute. If it exists, it is too small to move continents.

Wegener knew that both forces were vanishingly small but he believed that they would become effective when continuously applied for long periods of time. He pictured the light blocks as gently pushing their way through the sima, which, millimeter by millimeter, opens before them and closes behind them, always retaining an apparent solidity. The concept of sialic blocks riding passively along in currents of moving sima was not a part of Wegener's model until 1929 when he revised his book for the fourth time. By then he commented favorably on the ideas of Otto Ampferer, John Joly, and others that thermally driven convection currents occur in the mantle and cause the deformation or displacement of the crust.

The westward drift of Greenland

For proof of continental migration Wegener pointed to the geodetic measurements of longitude made in Greenland, beginning with those of Edward Sabine in 1823. By observations on the moon Sabine determined the longitude of a site at the south end of an island (later named Sabine Island). In 1870 a measurement was made by Börgen and Copeland, on the *Germania* Expedition, at a site a few hundred meters east of Sabine's station. In 1907 J. P. Koch measured the longitude of a site farther north but tied to Sabine's location by triangulation. All three determinations were based on lunar observations, a method known to be inaccurate, but Koch and Wegener believed that the measured differences were significantly greater than the mean errors, which were calculated at about 124 meters. From the three sets of measurements Wegener calculated that Greenland had moved westward relative to the Greenwich meridian for a distance of 420 meters (or 9 meters per year) between 1823 and 1870 and for a distance of 1,190 meters (or 32 meters per year) between 1870 and 1907.

The substitution of radio time transmissions for lunar determinations of longitude was begun in 1922 by P. F. Jensen of the Geodetic Institute in Copenhagen. A permanent station at Kornok, in western Greenland, was established where measurements could be repeated at five-year intervals. The second determination at Kornok, made in 1927, indicated a westward drift of 0.9 seconds of time in five years, or 36 meters per year. In 1929 Wegener wrote: "*The result is therefore proof of a displacement of Greenland that is still in progress. . . .*" (Italics are Wegener's.)

Wegener also cited figures for changes in the longitude of several other continents and islands and for some decreases in latitude. His faith in these geodetic determinations persuaded him that his continental drift theory was fully confirmed. Discussion, he said, should cease to turn upon the basic soundness of the theory and should center upon the quantitative aspects of its individual assertions.

None of these inferred changes in longitude or latitude was confirmed by the more precise measurements of succeeding years. Geodetic techniques of the accuracy required to measure the displacement of crustal units on a continental scale (as distinct from relative slip along specific fault planes) are not yet in operation. Two possibilities promise well for the future: measurements taken by means of very-long-baseline interferometry or by laser tracking of earth-orbiting satellites. The laser method may be refined so as to involve an uncertainty of only about ±2 centimeters in positioning of each tracking station. Even so, measurements from a global network—in which each station can be related to all others and to the center of mass of the earth—must be repeated over periods of years or decades to clearly detect and measure the migration of large crustal units if such migration is occurring at the presently inferred rates of up to 15 centimeters per year.

The jigsaw puzzle

The concept of continental drift first occurred to Alfred Wegener as he contemplated the apparent fit of the coastlines of the Atlantic Ocean. To test this fit he chose the edge of the continental shelf as the boundary between the sial and sima. After using tracing paper slashed to fit over a globe 50 centimeters in diameter, he flattened out the young, Tertiary, folded mountain ranges according to what he thought was the best available estimate of the amount of crustal shortening in each. After fitting northeastern Brazil into the Gulf of Guinea, he found that by suitable translations and rotations he could fit together all of the Atlantic continents and islands. His earliest predrift reconstruction (published in the first edition of his book and shown in our Figure 29A) includes only the Atlantic continents, each of which is notably lengthened and broadened. As every adept at jigsaw puzzles knows, the fit of the pieces is not enough; the design must continue across the borders. Wegener sketched-in two continuous transatlantic Paleozoic structural features: the Appalachian mountain chain which he extended to the British Isles and northern Europe, and the Sierra de la Ventana of Buenos Aires which he matched with the Cape mountains of South Africa.

In his book's second edition (1920) Wegener transferred his predrift world-continent to a globe (Figure 29B) where he found places for all of the continental blocks and indicated their positions with respect to the Permocarboniferous north and south poles. He added several structural features to his earlier map of the Atlantic continents and updated that map to illustrate the world of the Eocene after the initiation of drift (Figure 30). Both of these maps were dropped from his book's third edition (1922), in which he published, for the first time, his now famous globes depicting the fragmentation of a massive Carboniferous protocontinent (Figure 31). In both the second and third editions Wegener, in a rather offhand manner, referred to such a continent as a "Pangäa." Although he never used this term as a specific proper name or printed it on a map, his

Figure 29 (facing page). Two of Alfred Wegener's predrift reconstructions. **A:** The Atlantic continents as proposed in 1915. Heavy lines outline the continental margins. Dashed lines represent Paleozoic mountain ranges extending from North America to Europe and South America to the Cape of Good Hope. (From Wegener, *Die Entstehung der Kontinente und Ozeane*, figure 17, 1915; used with permission of Friedr. Vieweg & Sohn, Braunschweig.) **B:** The world's continents in the Permocarboniferous as proposed in 1920. The glaciated lands are grouped at the south pole while the ancestral Pacific Ocean surrounds the north pole. (From Wegener, *Die Entstehung der Kontinente und Ozeane*, page 61, second edition, 1920; used with permission of Friedr. Vieweg & Sohn, Braunschweig.)

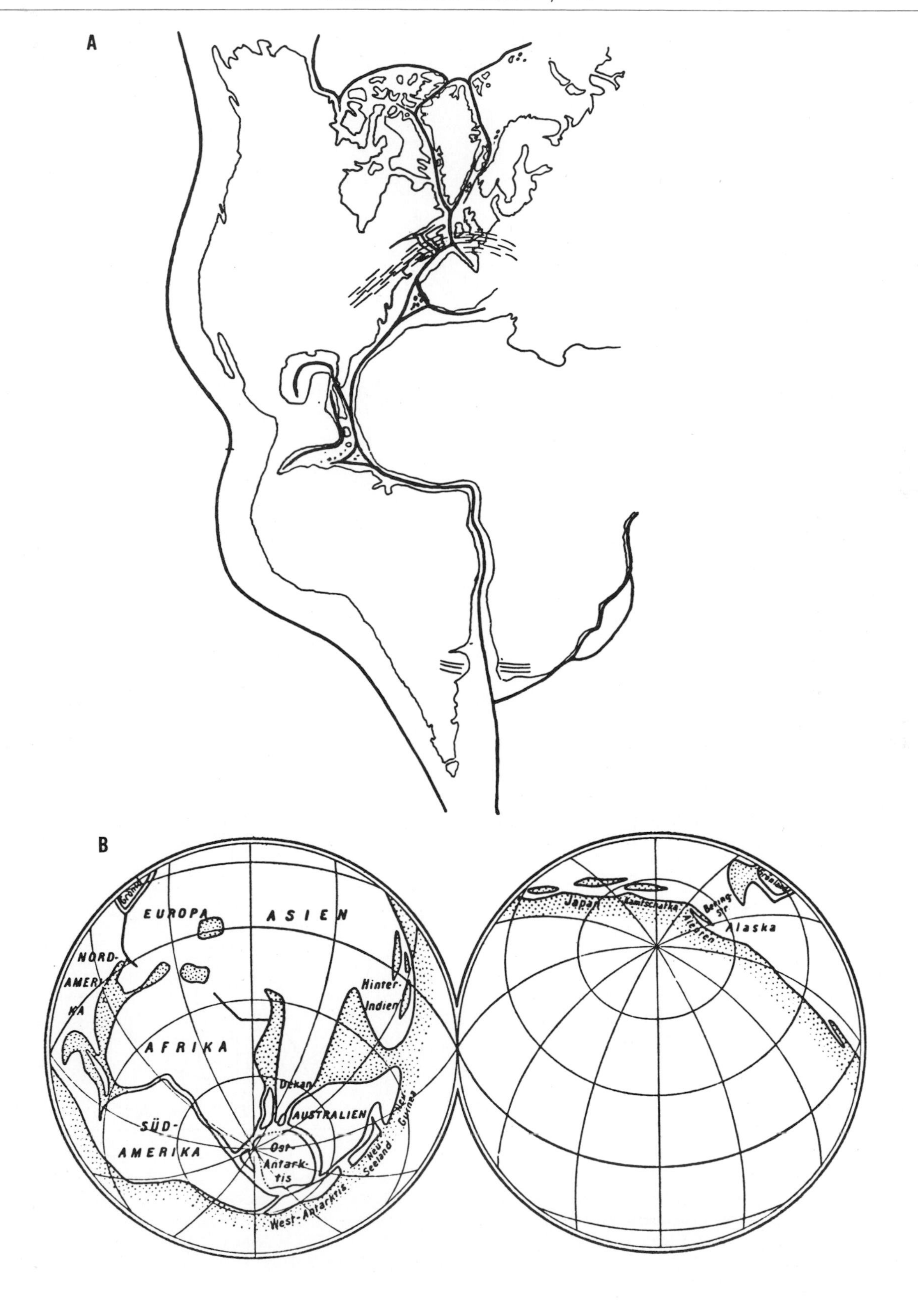
A
B
EUROPA
ASIEN
NORD-
AMERI-
KA
Hinter-
Indien
AFRIKA
Dekan
AUSTRALIEN
Neu-
Guinea
SÜD-
AMERIKA
Ost-
Antark-
tis
Neu-
Seeland
West-Antarktis
Japan
Kamtschatka
Bering-
Str.
Aleuten
Alaska

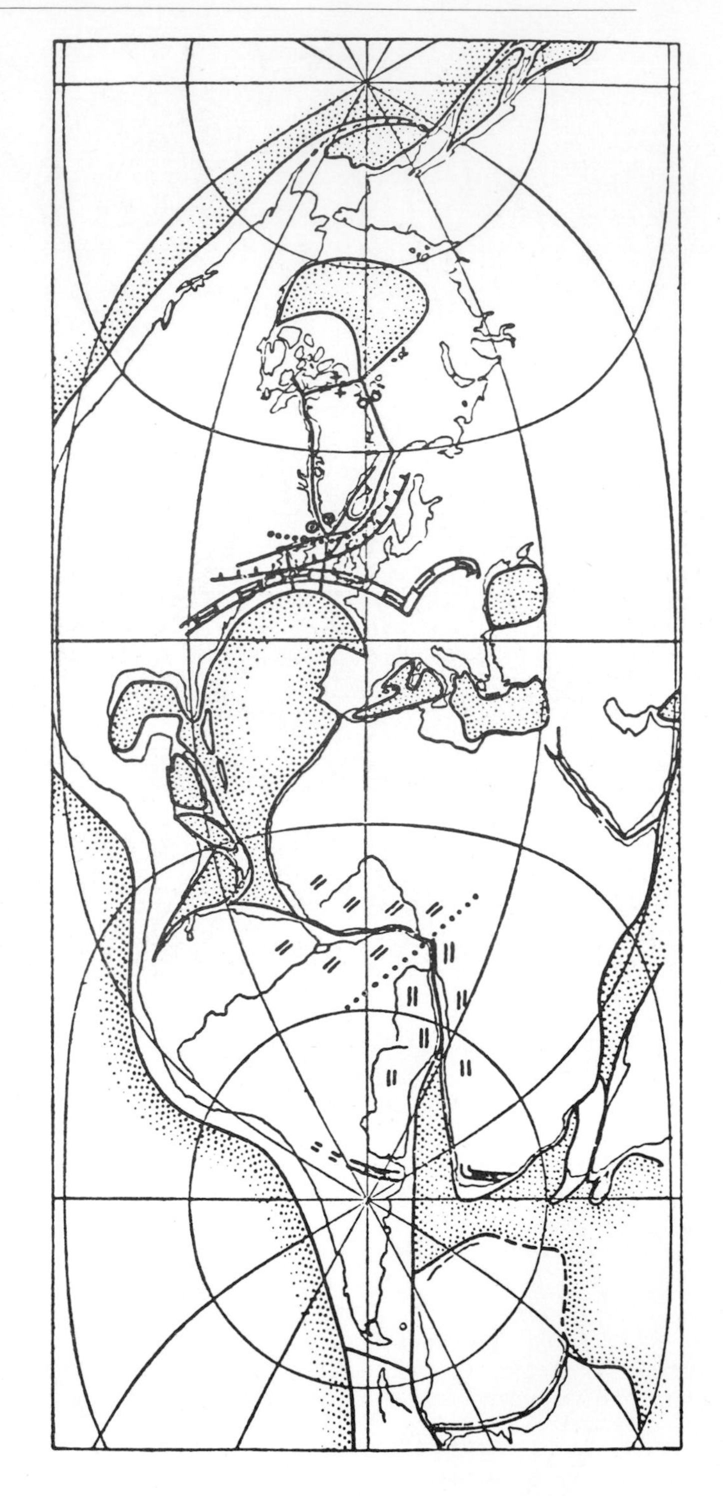

Figure 30. The Atlantic continents in the Eocene as drawn by Wegener in 1920. Matching structural features include the Caledonian mountain chain, extending from Newfoundland to Norway; the Appalachian-Hercynian chain, from New England to Old England and central Europe; the lineations in Precambrian gneisses of Brazil and West Africa, which show an abrupt change in strike at the dotted line; and the mountains of Buenos Aires Province and the Cape of Good Hope. Note how the south pole has shifted from its position in the Carboniferous shown in Figure 29B. (From Wegener, *Die Entstehung der Kontinente und Ozeane*, page 67, second edition, 1920; used with permission of Friedr. Vieweg & Sohn, Braunschweig.)

readers quickly titled his protocontinent Pangaea.* Wegener listed eight geological features—now truncated by the Atlantic coastlines—that "matched perfectly" on his new reconstruction. In summary, his description of these features was as follows:

(1) The Zwartberg folded range of Permian age which strikes east-west in South Africa comes "exactly in line" with the Sierras of Buenos Aires Province. Both ranges have a similar structure, age, and fossiliferous stratigraphic section.

(2) The lineations of the gneissic plateaus of Africa and South America have a matching change in strike and matching suites of magmatic rocks that include uncommon varieties of alkaline intrusives and kimberlite pipes. Diamond fields, for example, occur in both South Africa and Brazil.

(3) The Permocarboniferous system of Karoo sediments in South Africa, with a basal bed of glacial tillite overlain by 30,000 feet of continental sediments and capped by a thick layer of plateau basalts, is matched by the Santa Caterina Series in Brazil. Both contain *Glossopteris* flora and *Mesosaurus* fossils.

(4) The Carboniferous Armorican folded ranges of Brittany and southwestern Ireland continue in Nova Scotia and southeastern Newfoundland, bringing the coal fields of Belgium and the British Isles in line with those of the Appalachians.

(5) The Caledonian folded mountains of Scotland and Ireland continue in Newfoundland.

(6) The gneissic mountain systems of Algonkian age occurring in the Hebrides and northeastern Scotland strike northeast-southwest and continue in Labrador where they strike east-west.

(7) The terminal moraines of the Pleistocene ice sheets in Europe "unite without a break" with those of North America although they are now separated by 2,500 kilometers of

*Pangäa, or Pangaea, comes from Greek roots meaning "all-earth" or "all land." Gaea was the earth goddess who lent her name to geology, geochemistry, geophysics, and also to geomorphology, the science of land-forms. Land belongs to Gaea wherever it occurs, on earth, moon, or planets, despite much wit in recent periodicals decrying such terms as lunar and Martian geology.

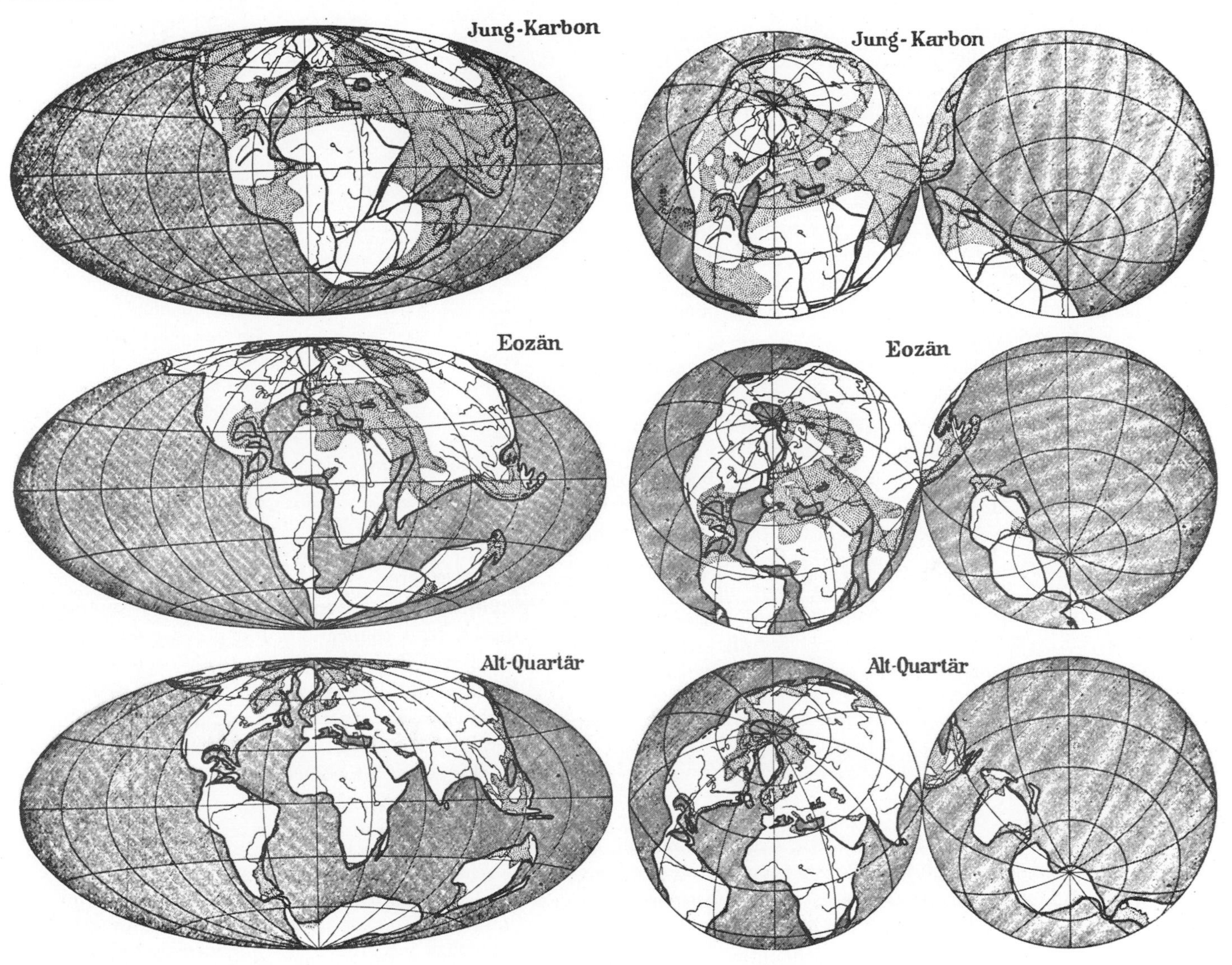

Figure 31. The world-continent as it appeared in the late Carboniferous, Eocene, and Early Quaternary according to Wegener, who first used these maps in the third German edition (1922) of his book. Stippling marks shallow seas. Africa is depicted as fixed with respect to its present latitudes and longitudes, so these diagrams show continental drift without polar wandering. (From Wegener, *Die Entstehung der Kontinente und Ozeane*, pages 4, 5; used with permission of Friedr. Vieweg & Sohn, Braunschweig.)

ocean and, on present maps, the American ones are 4.5° of latitude south of the European.

(8) Thick occurrences of the Devonian Old Red Sandstone form a band extending from the Baltic states, southern Norway, and the British Isles to Greenland and North America.

Wegener found no evidence of a match between Spain or North Africa and the Antillean region of the Americas. A wedge of deep ocean appeared there in his 1922 reconstruction of Pangaea. These lands could not, he said, have been in direct contact because between them lie the Azores, which he presumed were the uppermost peaks of a great submarine massif partially covered with continental sediments. He regarded the Azores, Madeira, and numerous other islands as pieces sloughed off in the wake of drifting continents in the manner of calf ice from icebergs. As another process contributing to islands and ocean ridges, he cited the fusion of sial at depth and its smearing outward and floating upward behind drifting continents. He suggested such an origin for the Abrolhos Bank off Brazil and the Seychelles of the Indian Ocean.

Wegener concluded that the South Atlantic began splitting open in the late Mesozoic and slowly widened, while the far northern Atlantic remained closed until after the melting of the Pleistocene ice caps. North America subsequently moved westward and also southward, offsetting the terminal moraines by 4.5° of latitude.

Looking at the Atlantic rift, Wegener commented that it was like examining a torn newspaper, then fitting together the ragged edges and finding that the lines of type continue unbroken across the seam. He said that if two segments of even one geological feature—like one line of print—were found to match on his map the odds would be 10 to 1 in favor of his theory. The matching of six independent controls raised the odds to 10^6, or one million to 1 in his favor. Wegener said that even if this was exaggerating the odds somewhat, he believed that the case for continental drift would still remain unassailable.

Gondwanaland

Wegener conceded that the close matching of geological structures is less striking for other oceans than it is for the Atlantic. On the other hand, an examination of paleontological evidence and the distribution of Permocarboniferous glacial tillites brought him to the problem that Suess had faced when he reconstructed the vast southern continent of Gondwána-Land.* Wegener solved the geological problem by a combination of two phenomena: continental drift and polar wandering.

According to Wegener, by the 1920s Suess' late paleozoic continents had evolved to the configuration shown in Figure 32. All of the segments now occupied by oceans, with the exception of the Pacific and remnants of the Tethys, were presumed to have foundered since the Permian.

*Wegener, dropping the accent and hyphen, made the name Gondwanaland.

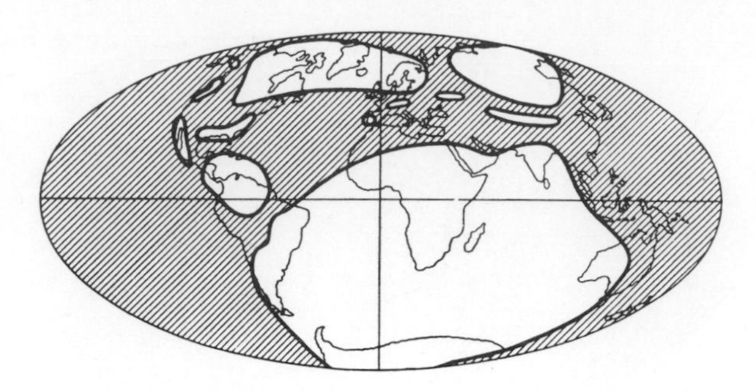

Figure 32. Wegener's sketch of the distribution of land in the Carboniferous "according to the usual conception." The northern continents approximate the Laurentia and Angara-Land of Suess, but Gondwanaland has expanded to include Suess' Antarctis. (From Wegener, *Die Entstehung der Kontinente und Ozeane*, page 8, fourth edition, 1929; used with permission of Friedr. Vieweg & Sohn, Braunschweig.)

To Wegener the concept of sunken blocks of sial was a geophysical impossibility—low-density continents could not sink into a heavier substratum. Massive land connections also posed a severe hydrological problem: to fill up large stretches of the Atlantic and Indian oceans with land would raise worldwide sea level so drastically that, far from providing bridges between continents, dry land would be obliterated altogether. (Unless there were, coincidentally, much less ocean water or the Pacific basin were vastly deeper.) Wegener concluded:

> Where the ocean basins are involved, it is not a question whether drift theory or the theory of sunken continents is to be preferred, because the latter idea just does not come into the picture. It is simply a matter of choosing between drift theory and the theory of the permanence of ocean basins.

The number of biological as well as geological facts favoring drift over permanence was, he said, legion. Most persuasive of all was the occurrence of the Gondwana-type sediments on seven landmasses that now are widely separated by ocean basins.

The Gondwana Series, in its Indian type-locality, begins with a Late Carboniferous to Permian glacial tillite which lies directly upon the eroded basement of ancient Precambrian metamorphics. Associated with the tillite are coal seams containing beautifully preserved leaf imprints of *Glossopteris* (see Figure 33), a genus of seed-bearing plants embracing over 40 species. Overlying the tillites is a succession of

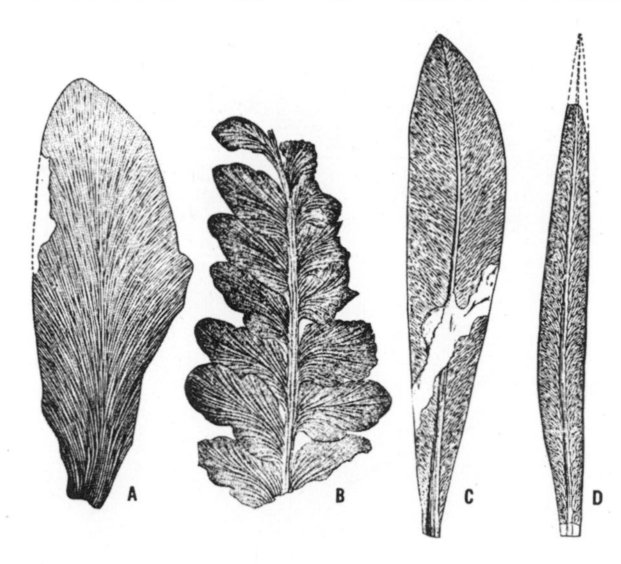

Figure 33. *Gangamopteris* leaf (a) with midrib, and three *Glossopteris* leaves (b-d) with midribs. *Glossopteris* is the more famous name but both genera occur in the Gondwana coal measures. (From E. W. Berry, 1920, plate 5.)

detrital sediments deposited in a fluvatile rather than a marine environment. These sediments include sandstones, shales, and clays interbedded with seams of coal. Some exposures of the series are capped by the Jurassic Rajmahal basalts, a succession of flows 3,200 meters thick. Where erosion has removed the horizontal lavas it has revealed an abundance of feeder dikes cross-cutting the clastic sediments. The Gondwana Series as a whole attains a thickness of at least 7,000 meters and ranges in age from Late Pennsylvanian to Early Cretaceous.

This unmistakable series, beginning with a glacial tillite associated with *Glossopteris* coal flora and often ending with flood basalts, is duplicated, with local variations, in South Africa, Madagascar, South America, the Falkland Islands, Australia, and Antarctica. In Africa, as the Karoo Series, it occurs in large discontinuous outcrops from the equator to the Cape of Good Hope and achieves a maximum thickness of 15,000 meters. In South America, the equivalent Santa Caterina Series is exposed in Brazil, Paraguay, Uruguay, and Argentina. Here the series is capped by the Paraná basalts, probably the greatest outpouring of lava on any one continent, covering well over a million square kilometers and varying in thickness from 100 to 1,000 meters.

On the Falkland Islands, the Gondwana-type deposits are of special interest because the glacial tillite contains numerous erratic boulders of igneous and metamorphic rocks that are entirely foreign to the islands. (Wegener was convinced that specimens of South African bedrock lay in the tillites of Brazil and that the diamonds of Brazil, South Africa, and India might have a common source, but neither idea has been substantiated.)

Marine beds are not absent from the Gondwana deposits. A few thin limestones are interbedded with the continental sediments, most conspicuously in Madagascar and Australia. All variations aside, the Gondwana-type rocks demonstrate clearly that six landmasses in the southern hemisphere and one landmass in the northern had strikingly similar geological histories from the Carboniferous to the Jurassic. Nothing resembling the Gondwana Series occurs in the northern hemisphere outside India.

Eduard Suess had visualized the present exposures of the Gondwana strata as fragments that remained elevated after the post-Cretaceous downfaulting of the intervening oceans. His reconstructions implied that the ice sheets, detrital sediments, and basalt flows had once been continuous over large stretches of a landmass covering nearly a third of the globe. Wegener saw the same exposures as fragments that have been displaced horizontally. By fitting them together again he recreated southern Pangaea, where the ice sheets, sediments, and basalts spread over a much smaller—albeit very impressive—area.

The biological evidences

Gondwanaland, Lemuria, and most of the other sunken continents and land bridges were originally postulated for biological reasons—to account for the distribution of flora or fauna. Geological considerations were secondary. Wegener did not believe in sunken land bridges so he used the arguments in their favor as evidence for continental drift. He was especially pleased with the results of a survey by the paleontologist Theodor Arldt that was published in 1917. Arldt searched the writings of 20 paleontologists and tabulated their opinions with respect to the existence or nonexistence in each geological period of four specific land bridges. A resounding majority of the paleontologists favored bridges between the "Gondwana" continents throughout the Paleozoic. A majority also favored a bridge between Europe and North America from the Silurian to the Permian, but opinions were divided about its existing as early as the Cambrian and as late as the Mesozoic. Virtually all of these bridges were believed broken in the early Tertiary.

Among the organisms listed by Wegener as supporting his hypothesis we will mention but a few. Along with the *Glossopteris* flora, the basal Gondwana beds of South Africa and Brazil contain the fossilized remains of *Mesosaurus*, a small Permian reptile with a long, slender head and delicate, needle-like teeth that suited him for scavenging minute crustaceans from brackish coastal shallows. No paleontologist has supposed that *Mesosaurus* (Figure 34) was a sea-going creature capable of swimming the present Atlantic, yet his skeletons, well preserved in mudstones, lie on both sides of that ocean.

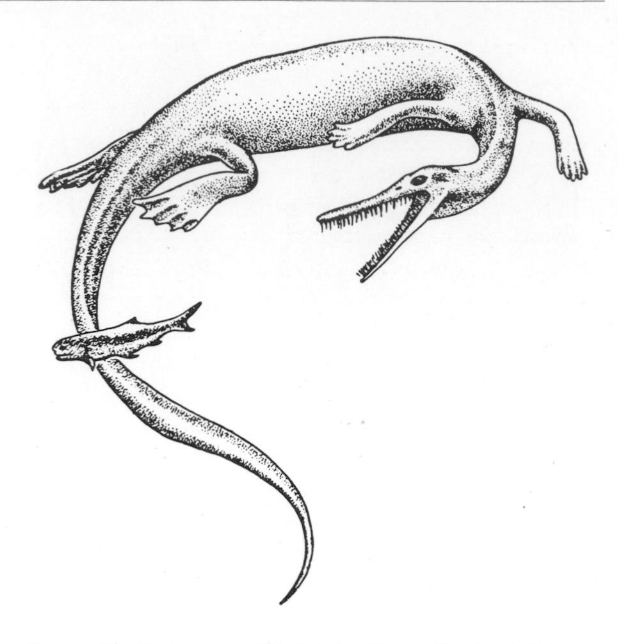

Figure 34. *Mesosaurus*. (Sketch by Henry Fairfield Osborn.)

Wegener maintained that a contemporary situation obtains with respect to the manatee, whose only haunts are the mouths of tropical rivers of South America and western Africa. Numerous freshwater perches swim the lakes and streams of Eurasia and eastern North America; the garden snail is found on leafy plants from southern Germany to the British Isles, Iceland, Greenland, Canada, and the eastern (but not the western) United States; and the earthworms of Europe and Siberia also occur in the eastern (but not western) United States.

Wegener also discussed the unique floral and faunal assemblages of Madagascar and Australia. We have already seen how the Malagasy fauna led to the invention of Lemuria, a continent which was tentatively approved in 1876 by the biologist Alfred Russel Wallace. In 1880, however, Wallace published *Island Life*, an epoch-making book in which he rejected Lemuria altogether and expressed a wish that that hypothetical land could be erased from memory. Wallace had become an advocate of the permanence of continents and ocean basins and he believed that the opening and closing of land bridges due to minor fluctations along continental margins and in island groups could account for all species distributions, however unexpected. (And some of them were unexpected indeed: Wallace pointed out that the lemurs of Madagascar were a minor problem compared with that island's colubrine snakes, which had no overseas relatives except for two genera found in America!)

Mere distance is one of the least important factors determining species distributions, according to Wallace, who drew the now famous Wallace Line through the Malay Archipelago between the islands inhabited by the oriental (Indian) and by the Australian fauna. This line passes through the 38-kilometer strait between the islands of Bali and Lambok which, Wallace discovered, have assemblages of birds and quadrupeds that differ from each other far more than do those of England and Japan. Problems such as this one could hardly be solved by any appeal to sunken continents, but to Alfred Wegener it seemed manifest that they could be solved by an appeal to continental drift. He cited Wallace himself to the effect that the Australian fauna includes three stocks introduced at different periods from different sources. The oldest is a pre-Jurassic warmth-loving fauna sharing affinities with those of India, Madagascar, and Africa. The next oldest is characterized by the curious marsupials that are also found in the Moluccas, some Pacific Islands, and South America. Even their parasites, the flatworms, are, Wegener said, identical in hosts inhabiting these far-flung areas. Strangely enough, marsupials do not occur in the Sunda Islands, next door to Australia. The youngest Australian stock is represented by mammals such as rats, bats, and the dingo, introduced from Asia since the Pleistocene.

To explain such distributions, Wegener represented Africa, Madagascar, India, Australia, Antarctica, and South America as adjacent parts of southern Pangaea until the end of the Triassic (see Figure 31). He supposed that Australia split away from India in the early Triassic, thus losing contact with the oriental fauna and flora but maintaining contact with Antarctica and South America for easy interchange of marsupials until the Eocene. At that time Australia floated free and developed its unique insular species until it approached southeast Asia in the Pleistocene and began to receive island-hopping immigrants. Meanwhile, in the upper Cretaceous, India broke away from Madagascar and moved rapidly northward, while its northern margin was compressed into great folds which thickened the crust and resulted in the uplift of the Himalayas and the Tibetan plateaus. Wegener described the east-west Eurasian mountain chains as due to a great "Lemurian compression" which occurred in the Tertiary as northern and southern lands moved equatorward. The compression cleared the Indian Ocean of land without recourse to a sunken continent of Lemuria. Finally, Madagascar split away from Africa in the Pleistocene before any of the common African mammals or birds could enter.

All of the biological evidence cited by Wegener was gleaned from the literature. Since he was neither a biologist nor a paleontologist he depended on other authorities for his information. His critics made much of this point and we shall be hearing from them shortly.

Polar wandering

Having proposed the drift of continental blocks and reconstructed Pangaea on the basis of the geometry, geology, and biota of the continents, Wegener postulated polar wandering on paleoclimatic evidence. Polar wandering, in Wegener's day, was a concept derived from the apparent shifting of climatic zones. It had no association with the drift of the earth's magnetic field, an implication that became important thirty years later.

> Polar wandering [Wegener wrote, in 1929] is a geological idea . . . a rotation of the whole surface [of the globe] relative to the system of parallels. . . . To be effective, this rotation must obviously be about an axis which differs from that of the earth's axis of spin. . . . Superficial polar wandering in this sense can only be detected in the remote past by fossil evidence for climate. Geophysics cannot make any judgement about its reality or possibility.

That statement sounds very quaint today, when polar wandering is mainly a geophysical idea based on studies of paleomagnetism.

Alfred Wegener was by no means one of the first scientists to suggest a history of polar wandering. He listed 17 writers who had discussed it between 1870 and 1918, and he remarked that the number had snowballed since then.

Wegener took special pains to separate what he called superficial polar wandering from astronomical polar wandering. He pointed out that superficial polar wandering could be accomplished either by a rotation of the earth's crust as a whole over the interior or by an internal shift of the mass of the earth with respect to the rotation axis. He was persuaded that any rotation of the crust as a whole—as though the skin of an orange could be loosened and moved over the pulp—takes place only in a westerly direction, around the earth's spin axis, and such motion does not constitute polar wandering. Nor, by Wegener's definition, does crustal creep from the poles toward the equator constitute polar wandering. Wegener could find no theoretical basis for believing that the crust can rotate about any axis different from the spin axis. He therefore concluded that polar wandering results not from movement of the crust over the interior but from internal axial displacement; for example, a shifting of the earth's mass along the entire length of the spin axis. In describing such a shift he wrote:

> Let us assume that the inertial pole . . . has been displaced by a small amount x as a result of geological processes. The pole of rotation must follow suit. The earth now rotates about an axis which is slightly different from the previous one. It must follow that the equatorial bulge reorients itself.

Wegener believed that clear evidence of such motion, operating in small increments, can be read in the record of changing climatic zones and also in the history of marine transgressions and regressions on the continents. From the paleoclimatic record, Köppen and Wegener traced the course of polar wandering since the Carboniferous (as shown in Figure 35B, which is taken from their book *Die Klimate der geologischen Vorzeit*, published in 1924). They show the Carboniferous (*Karbon*) south pole in the midst of Gondwanaland and the north pole in the Pacific. This orientation explained why southern Pangaea supported continental ice caps while northern Pangaea remained ice-free and developed, in the rain belt, a series of luxuriant coal swamps that extended from North America to Europe. Wegener distinguished between two types of coal: a lush subtropical variety derived from tree ferns and other vegetation without annual rings, and a polar variety derived from *Glossopteris* and *Gangamopteris* flora, some specimens of which do show annual rings indicative of fluctuations in humidity or in availability of sunlight.

The history of marine transgressions also bears witness, according to Wegener, of polar wandering. For example, most of Europe was dry land from the Carboniferous to the Jurassic when waters began flooding the coastal areas and spreading inland until the area became one vast marshy seascape in the Cretaceous. Withdrawal began in the Eocene, and Europe has been dry land since the early Tertiary. Wegener ascribed such changes to the fact that the ocean conforms immediately to a redistribution of the equatorial bulge, whereas the rocky mass of the earth adjusts more slowly. He pictured what amounted to a slight dip in sea level (regression) preceding, and a slight bulge in sea level (transgression) following, each pole of the earth in its path of wandering. Thus, the spread of shallow seas over a continental landscape signals the retreat of a pole away from that area. The pattern of polar wandering that Wegener deduced from these marine cycles correlated satisfactorily with that indicated by the shifting of climatic zones.

Wegener also inquired into the possibility of astronomical polar wandering—the displacement of the earth's axis in space with respect to the stars. At present the earth's spin axis is tilted about 23.5° from normal to the ecliptic, the plane in which the earth and other planets orbit the sun. Throughout historic time the north pole of the earth has been approaching its present position nearly under Polaris, the north star. The axis, however, is undergoing a grand precession—so slow as to be almost imperceptible—which causes each pole to describe a full circle every 25,800 years. As it wheels about this circle, the north pole points toward a succession of constellations—including Cepheus, Lyra, and Draco—before returning to the vicinity of Polaris in Ursa Minor. In the year 14,000 the north pole will achieve its maximum deviation of 47° from Polaris and will point toward the bright star Vega; in the year 23,000 it will lie under α-Draconis, as it did in 2700 B.C. when Chinese astronomers regarded that star as "polaris." The angle of tilt actually fluctuates very slightly above and below 23.5° over a period of 41,000 years, and the earth's axis also undergoes a minor wobble. However, the dynamic system as a whole is so remarkably stable that few scientists believe in the extraordinary changes in the angle of tilt that would be consonant with astronomical polar wandering. (We remember that Buffon speculated on a tilting of the earth's axis to explain the former range of elephants in the arctic but rejected the idea as untenable. For it he substituted the much more original but absolutely untenable idea that northern latitudes were warm because of a high rate of terrestrial heat flow which diminished with the thickening of the crust, forcing warmth-loving animals to migrate equatorward.) Wegener, in contrast, did not reject the idea of astronomical polar wandering. He thought it quite possible that, over geologic time, the tilt of the earth's axis has varied considerably from its mean value. To support his argument he pointed to long periods when the climates of the entire earth appear to have been more severely zoned than they are at present (an indication of increased tilt) and to other periods when the climate was so uniformily mild that no ice caps formed even on lands surrounding the poles (indicative of decreased tilt). Wegener admitted that the idea was highly speculative, but it was his opinion, as an astronomer, that pronounced oscillations in the angle of tilt are likely.

Recapitulation

Wegener's interpretation of post-Paleozoic earth history may be summarized briefly. The large protocontinet, Pangaea, which was always partially flooded by shallow-shelf seas, began rifting into fragments that

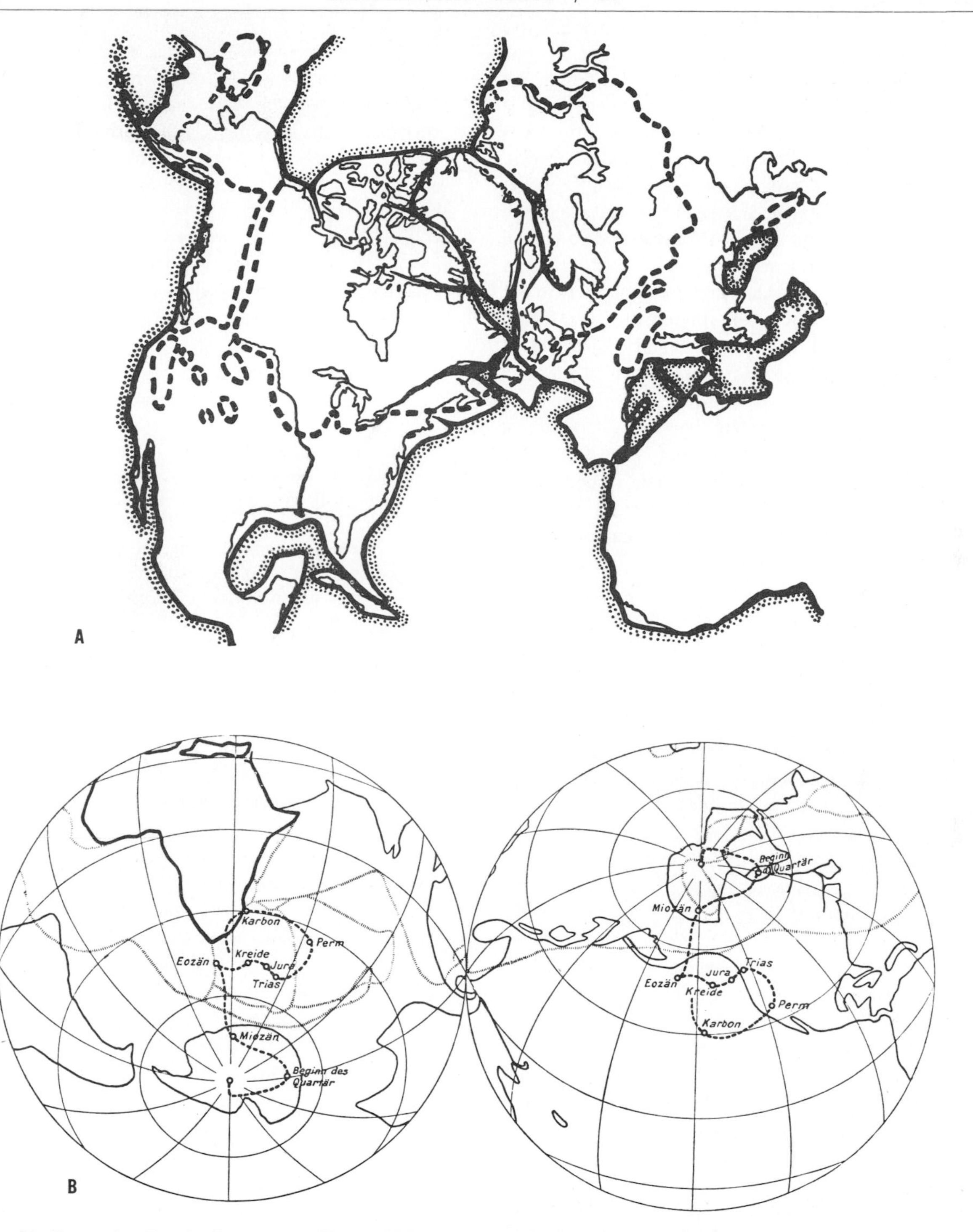

Figure 35. Two paleoclimatic diagrams. A: Wegener's illustration of the boundaries of the Pleistocene ice caps (dashed lines) before separation of the northern continents. (From Wegener, *Die Enstehung der Kontinente und Ozeane*, page 89, first edition, 1915; used with permission of Friedr. Vieweg & Sohn, Braunschweig.) The terminal moraines actually occur somewhat farther south along Cape Cod, Long Island, and the Ohio River. B: Concurrent continental drift and polar wandering from the Carboniferous to the present according to Köppen and Wegener. (From Köppen and Wegener, *Die Klimate der geologischen Vorzeit*, page 155, 1924; used with permission of Gebrüder Borntraeger, Stuttgart.)

tended to move equatorward and westward through the ocean floors. The Atlantic began opening in the Jurassic as a great rent, hinged at the north; and a long strip of continental crust, including the parts that would become North America, South America, Antarctica, and Australia, moved westward (clockwise as viewed from the north). The resistance of the cooled oceanic sima of the ancestral Pacific caused the folding of cordilleras along the forward margin of the entire strip from Alaska to Australia. Where the continental links were weak, the resisting sima forced thin strands of sial into arcuate island chains such as those of the Antilles and South Sandwich Islands; and it bent eastward the extremities of Greenland, North America, South America, and the Palmer Peninsula of Antarctica.

A massive westward shift of Asia ripped gaping tension fractures (trenches) in the western margin of the Pacific and left stranded long, arcuate strips of the continental coastline (island arcs). India split off from Australia, and then from Madagascar, and moved swiftly equatorward, crumpling the sedimentary layers of its northern reaches into great mountain ranges. Australia, splitting first from India and later from Africa and Antarctica, underwent an abrupt counterclockwise turn in its direction of drifting and so eventually began to approach southeast Asia. As it turned, its newest cordilleras were sheared off and left stranded to become New Zealand. A vast compression (which Wegener called the Lemurian compression) between "northern" and "southern" landmasses crumpled the strata that form the Tertiary mountain ranges stretching from Morocco to Manchuria.

These motions can be traced, in a very sketchy fashion, on Wegener's series of maps shown in Figure 31. The picture is incomplete, however, because the maps show continental drift without polar wandering. Africa, for example, remains stationary because it is drawn as a fixed reference relative to our present system of latitudes and longitudes. To follow Wegener's descriptions, one must imagine continental drift simultaneously with a shift of the crust relative to Wegener's Permocarboniferous equator and poles. And a very nimble imagination is required as one tries, for example, to justify Wegener's argument, by reference to his figures, that Australia has drifted westward. Finally, Wegener reckoned that the actuality of continental drift was confirmed by a westward flight of Greenland as determined by geodetic measurements.

Wegener knew that the problem of a causal mechanism of sufficient force was far from solved. In the 1929 edition of his book he wrote:

> The Newton of the drift theory has not yet appeared. His absence need cause no anxiety; the theory is still young and still often treated with suspicion. . . . Continental drift, faults and compressions, earthquakes, volcanicity, transgression cycles and polar wandering are undoubtedly connected causally on a grand scale. . . . However, what is cause and what effect only the future will unveil.

Wegener's geological world view was highly dramatic, undeniably imaginative, and all-embracing. No other theorist had, in one broad sweep, posed such a challenge or a threat to geologists, geodesists, geophysicists, paleontologists, botanists, zoologists, and climatologists. And Wegener proposed his radical new model in the first quarter of the 20th century, when many investigators believed that earth science was established, at last, on sound physical principles.

Once scientists grasped the full import of what Wegener proposed, the reaction was swift and devastating. Overawed by the task of trying to disprove the theory as a whole—which seemed in any case to be too dependent upon Wegener's own frame of reference—most critics set out to demolish the details, many of which proved all too vulnerable to attack.

The Reaction

Disciples of the permanence of continents and ocean basins had never taken kindly to the Suessian tradition of sunken continents.

"The trend of a mountain range, or the convenience of a running bird, or of a marsupial afraid to wet its feet, seems sufficient warrant for hoisting up any sea-bottom to connect continent with continent," said A. P. Coleman in a presidential address to the Geological Society of America on December 29, 1915. "A Gondwana Land arises in place of an Indian Ocean and sweeps across to South America so that a spore-bearing plant can follow up an ice age; or an Atlantis ties New England to Old England to help out the migrations of a shallow-water fauna."

However disapproving they might be of Suess' conceptions, such critics found the ideas of Alfred Wegener eminently less agreeable. Many American readers were first made aware of Wegener's book when the second (1920) edition was reviewed by Harry Fielding Reid in the October 1922 issue of *The Geographical Review*. In less than three pages of terse criticism Reid disposed of Wegener's book and two articles by Köppen. Both continental drift and polar wandering were temporarily laid to rest before causing a stir in America.

". . . it is not easy to avoid bias"

In England the reaction was more spirited. In its issue of August 1922 *The Geological Magazine* carried a critical review by the geologist Philip Lake. A month later, Wegener's hypothesis was the subject of lively debate at a meeting of the British Association for the Advancement of Science. In January 1923 Lake presented his views orally to the Royal Geographical Society. The following outline includes excerpts from both of Lake's reviews.

Lake opened fire:

> In examining ideas so novel as those of Wegener it is not easy to avoid bias. A moving continent is as strange to us as a moving earth was to our ancestors, and we may be as prejudiced as they were. On the other hand, if continents have moved many former difficulties disappear, and we may be tempted to forget the difficulties of the theory itself and the imperfection of the evidence. . . . Wegener himself does not assist his reader to form an impartial judgment. Whatever his own attitude may have been originally, in his book he is not seeking truth; he is advocating a cause, and is blind to every fact and argument that tells against it. Nevertheless, he is a skillful advocate and presents an interesting case.

Having made this concession, Lake set out to demolish the "interesting case" by attacking Wegener partly on the substance of his arguments and partly on what he perceived as a certain intellectual wooliness in Wegener's presentation. Wegener, for example, based his argument that the continents and ocean floors represent two distinct earth layers (which he equated with the sial and the sima) on the double-peaked curve of altitude frequencies (see Figure 26). The tectonic deformation of a single surface would not, according to Wegener, produce two maxima unless physical causes were present which would give preference to two particular altitudes. "Since this is not the case," wrote Wegener, "the frequency should simply be controlled by the Law of Errors of Gauss . . . ." Lake retorted that Wegener should have said "Since, so far as I know, this is not the case," and he added that Wegener's reasoning leads us

> to the remarkable principle: If we do not know the law, the law must be the law of errors. . . . And so the actual heights [of the earth's surface] must be influenced by the extent of our knowledge. But Wegener's world is not an ordinary one. In his diagram he draws a broken line, and this, he says, follows approximately the course of the law of errors, according to which the frequencies would be regulated if one level only had been involved in the subsequent movements. The frequencies in the diagram are expressed as percentages, and in the ordinary world would add up to 100. Wegener's line gives a total of about 200. . . . A mere inspection of the diagram is enough to show that the two curves cannot both be correct, for the areas contained between them and the vertical axis should be equal.

Lake cited an analysis by G. V. Douglas and A. V. Douglas showing that undulations of a single crustal level would, in fact, "necessarily" produce a double-peaked hypsometric curve similar in all essential respects to the one invested by Wegener with such geophysical importance.

With reference to Wegener's geophysics, Lake pointed out that the concept of lighter crustal masses balanced isostatically in a heavier substratum was very generally conceded but that Wegener imagined these masses as moving laterally, a very different matter. "There is the force of gravity to press them downwards into the sima," Lake said, "but there is no known force comparable in magnitude to move them sideways." Granted, Osmond Fisher and W. H. Pickering had suggested the movement of crustal fragments toward the cavity after the moon was thrown off from the region of the Pacific, but this suggested occurrence belonged to a very early period of earth history. Neither scientist imagined that such movements began as late as the Cretaceous and are still taking place.

On the other hand, Lake was certain—Wegener's objection notwithstanding—that large areas of sal (Lake could see no advantage to the term "sial") have, in fact, sunk beneath the sea and risen again many times. Some of these areas, Lake said, have achieved great depths, and all have gone below the 100-fathom line which, as the edge of the continental shelves, Wegener chose as the sial-sima boundary. Lake was convinced that if this boundary occurs at all, which he doubted, it should be found at the foot rather than at the top of the continental shelves. Wegener had defended his choice by arguing that the continental masses had bulged somewhat under their own weight, pushing the bottom of the continental slope outward. Lake asked why the top of the slope was not also deformed by this bulging.

Wegener's fitting together of the continental margins became a source of great amusement to his critics, beginning with Lake. A glance at his reconstruction of Pangaea shows that each continent has been stretched, broadened, or generally distorted, a result he arrived at by flattening out the Tertiary mountain ranges and performing various rotating and hinging motions with what he called "a certain freedom." Said Lake:

> It is easy to fit the pieces of a puzzle together if you distort their shapes, but when you have done so, your success is no proof that you have placed them in their original positions. It is not even a proof that the pieces belong to the same puzzle, or that all of the pieces are present.

Lake also stated that Wegener

> does not think that the flattening out of the Alpine folds would obliterate the Mediterranean, but he imagines that the unfolding of the Himalayas would produce an elongated Indian Peninsula extending through nearly sixty degrees of the earth's surface. He does not enter into details.

According to Lake, Wegener's matching of geological features was equally questionable. As an example, he said that the strike of the Algonkian gneisses of the Hebrides and northern Scotland is not northeast to southwest, as stated by Wegener, but is, according to the geological survey of Scotland, west-northwest to east-southeast; therefore, if Wegener's direction does fit with Labrador the real direction obviously does not. Lake added several more observations: The Caledonian folds of Scotland and Ireland have no counterpart in North America—unless one accepts the suggestion of the French geologist Termier that the Newfoundland folds are a delayed ramification of a considerably later date. The Armorican and Appalachian folds do correspond in age, and Wegener has bent North America in such a way that they fall into line. Such a coincidence proves nothing. The folded gneisses of Africa and of South America are not mapped well enough to bear out Wegener's claim that they show a matching change of strike. In West Africa, for example, a predominant northeast-to-southwest strike is not obvious on Lemoine's map, which was reproduced by Wegener. Wegener admits the map does not illustrate his point very well, but says it was drawn for other purposes. "No doubt he is right," said Lake. "Lemoine was collecting facts and not supporting hypotheses."

In South Africa, Lake said, the folded Zwartberg range actually bends north before it reaches the Atlantic coast. Wegener assumed that the north-trending branch was simply a local spur (an assumption not shared, according to Lake, by many South African geologists), so he extended the main range westward to join with the Sierras of the same age in Buenos Aires Province. These, however, have not been examined closely.

Lake pointed out that without distortions the continental margins across the Atlantic do not really match at all. If Newfoundland is joined to Ireland, the Zwartberg range cannot be brought within 1,200 miles of the Sierras in Buenos Aires. Elsewhere on the globe one could postulate matches that are much more remarkable. For example, Australia, without New Guinea attached, fits handsomely into the Arabian Sea, but this, said Lake, is not part of Wegener's scheme.

Lake conceded that the matching of the Gondwana system from continent to continent together with the assumption of polar wandering helps resolve many difficulties related to the Permocarboniferous glaciation and distribution of *Glossopteris*. But he added that the statement is incomplete. *Glossopteris* also is found in Kashmir, Afghanistan, northeastern

Persia, Tonquin, and northern Russia. Furthermore, numerous glacial or pseudoglacial beds of Permocarboniferous age lie in Siberia and North America.

Lake expressed the gravest doubts about the very late post-Pleistocene date that Wegener assigned to the opening of the North Atlantic and about the geodetic measurements allegedly showing the continued rapid flight of Greenland from Europe.

In conclusion Lake commented that his own brief account should have made it clear that the "geological features of the two sides of the Atlantic do not unite in the way that Wegener imagines, and if the continental masses ever were continuous they were not fitted as Wegener has fitted them."

"Not for the first time, but for the first time boldly"

The discussion that followed Lake's presentation to the Royal Geographical Society provides a valuable insight on the geological attitudes of 1923. Mr. G. W. Lamplugh was the first to comment:

> It may seem surprising that we should seriously discuss a theory which is so vulnerable in almost every statement as this of Wegener's. . . . But the underlying idea that the continents may not be fixed has in its favor certain facts which give every geologist a predilection towards it in spite of Wegener's failure to prove it.

After citing some of these facts, Lamplugh concluded by saying:

> We are discussing his hypothesis seriously because we should like him to be right, and yet I am afraid we have to conclude, as Mr. Lake has done, that in essential points he is wrong. But the underlying idea may yet bear better fruit.

At that, Mr. R. D. Oldham, discoverer of the seismic evidence for the earth's core, arose to voice surprise at the unanimity with which Wegener's theory is regarded as a novel idea. He recalled that when he started as a geologist there was much evidence suggesting that the continents have not always occupied their present positions on the globe:

> But also I can remember very well that in those days it was unsafe for anyone to advocate an idea of that sort. The physicists, who before that had forced on us the notion of a fiery globe with a molten interior and thin crust on it, had gone round and insisted on a solid heated sphere, and they would allow us to appeal to nothing, as the cause for various structures and changes that we knew in geology, but the slow cooling and contraction of this solid globe, and any notion of the shifting of continents was incompatible with that theory. Those ideas held the ground so strongly that it was more than any man who valued his reputation for scientific sanity ought to venture on to advocate anything like this theory that Wegener has nowadays been able to put forward. . . .
>
> But there was one man, even then, who did quite formally propose and maintain something . . . almost identical with the Wegener hypothesis. That was Osmond Fisher, a man who in his time was a scientific Ishmaelite.

"Ishmaelite," an outsider—the word gives us the necessary clue to Fisher's lack of influence despite his textbook *Physics of the Earth's Crust* and his article in *Nature*, in 1882, which we have cited earlier. Oldham pointed out the many parallels between Fisher's and Wegener's ideas and he recommended the belated recognition of a fellow countryman. He also stated that the important question is not whether Wegener is right or wrong in his details but whether the doctrine of permanence of continents and oceans is right or wrong. Oldham clearly thought it might be wrong.

Mr. F. Debenham commented that Mr. Lake's exposition of Wegener's theory amounted to an explosion, and that Wegener's proposed fitting of geological formations as well as his geodetic evidence of the motion of Greenland "would not hold water." He remarked, however, that geographers and geologists generally have too short a vision:

> Now, not for the first time perhaps, but for the first time boldly, Wegener has come forward with a theory which deals with the distribution of the continents in a bold way and offers himself for sacrifice; and he is certainly getting it. So that in addition to thanking Mr. Lake for his very clear undermining of the theory, I think we certainly ought to thank Professor Wegener for offering himself for the explosion.

The next speaker was Dr. Harold Jeffreys, who said that he came to the meeting intending to answer the physical arguments for the theory but that none had been offered. His main complaint against the theory was that the causes Wegener proposed for continental migration were ridiculously inadequate—something like a millionth or less of what would be required. He suggested that a change in rate of the earth's rotation would be sufficient to cause migrations of land and sea, but he did not know whether any geological evidence shows a split and motion of the right kind. At present most of the land is in one hemisphere and as this distribution is unstable the landmass must be tending to break up and the pieces to separate as widely as possible. The real problem, however, is why and how the land ever became concentrated in one hemisphere. For this, the best suggestion extant, he thought, was Osmond Fisher's theory of the Pacific being the scar left when the moon was torn from the earth.* But, concluded Dr. Jeffreys, "it would be a very long business to go into that theory in detail. There is a great deal to be said for it and something to be said against it, and I must not take up any more of your time."

Two other participants commented briefly. Mr. Evans quoted Lake as saying that the geological map of South America on which Wegener relied was based on very imperfect data, and Evans agreed. It was largely his own map, Evans said, and he knew better than anyone else how imperfect it was. Despite several objections to Wegener's details, however, Evans believed there is solid geological evidence that Africa and South America have drifted apart. Mr. C. S. Wright also felt that Wegener's hypothesis should be considered on its general merits rather than on a few points of detail. He said that the hypothesis was meant to explain facts and, although these might include some fictions, one must still compare his hypothesis with any others designed to explain these same facts.

The meeting ended with the president of the Society, the Earl of Ronaldshay, saying: "The impression left on my mind by the discussion is that geologists, as a whole, regret profoundly that Professor Wegener's hypothesis cannot be proved to be correct." If so many participants had not implied this, he would have thought from the vigor with which they destroyed the theory that they were finding keen satisfaction in doing so. He summarized the general feeling as he saw it: "Some theory of this kind is required to explain facts which have long been known to geologists and while they feel bound to condemn this particular hypothesis as being one which is not capable of meeting this long-felt want they still hope some other hypothesis of a kindred nature will be discovered which will satisfy their requirements."

In short, the sense of the meeting was clear: a theory of continental drift was needed—but not Wegener's.

*Dr. Jeffreys rejected earth fission a few years later when he calculated that the distortions of the earth would amount to only about a thousandth of those needed to rip away the moon, and that any detached mass would immediately fall back to earth.

The Solid Solid-Earth

The men who discussed Wegener's theory that January afternoon in London were clearly dissatisfied with the current hypotheses of contraction and of permanence. They were interested in new ideas. Wegener's hypothesis was a fresh departure that offered seemingly endless possibilities for research. One might have expected an explosion of geological investigations. The tone of the meeting, however, was set by the speaker, Philip Lake, and it was essentially hostile. The favorable comments made during the discussion period may be seen in this context: men who are in agreement that they are against a hypothesis can afford to be generous.

A few stout defenders emerged on the continent, particularly in Switzerland, France, and Holland, where geologists were impressed with the magnitude of crustal motion signaled in the structures of the Alps, the Moroccan Atlas Mountains, and the Malay Archipelago. In 1922, Emile Argand, founder of the Geological Institute of Neuchatel, Switzerland, outlined his own views to the 13th International Geological Congress at Brussels in a paper with the deceptively simple title *La Tectonique de l'Asie*. Far from limiting his discussion to Asia, Argand described the evolution of the earth's crust in terms of "mobilisme" as opposed to "fixisme." He favored Wegener's concepts of floating continents and of the plasticity of rocky materials under long-term stress. Fixism, according to Argand, was not a theory at all but a negative view common to many theories. Wegener's mobilism, on the other hand, was a comprehensive theory, supported by a large and diverse body of evidence to which Argand added many structural details. Wegener, declared Argand, had raised important issues which his opponents had never answered.

The fixists, however, failed to see their view as the simple lack of a theory. Seismological data of ever increasing discrimination were indicating that the earth's interior possesses significant strength to depths of tens if not hundreds of kilometers. Powerful opposition to Wegener's hypothesis was voiced in England by Dr. Harold Jeffreys who, in 1924, established geophysics on a firm mathematical basis with the publication of his book *The Earth, Its Origin, History, and Physical Constitution*. In that book Jeffreys presented a critical review of half a century of observations on the earth's interior and developed his own theory of the evolution not only of the earth but of other bodies in the solar system.

Dr. Jeffreys was not wholly against the idea of the crust moving over the interior of the earth under certain circumstances, but he was very dubious indeed about the principles as well as the mechanism proposed by Wegener. He pointed out that the continental sial is more radioactive than the oceanic sima and should, if anything, be the weaker of the two layers. Yet Wegener began by floating the Americas through the yielding sima which, nevertheless, exerted enough resistance to crumple their margins into cordilleras against the combined forces of their westward impetus and of gravity.

It is an impossible hypothesis, said Jeffreys, that a "small force can not only produce indefinitely great movement, given a long enough time, but that it can overcome a force many times greater acting in the opposite direction for the same time." He made the point that if the sima is the weaker layer and will allow continents to plough through it like ships sailing before the wind, then it will not crumple their prows—even less will it crumple their prows and let them keep sailing. And he added that even Wegener's ice floes are never deformed by the resistance of open water but only by collision with other ice fields or by running aground in bottom sediment, a truly resistant medium.

This aspect of the problem was strictly academic, however. Dr. Jeffreys saw no evidence that the sima is ever the weaker medium. He said that the continental crust is strong enough to maintain Mount Everest, and the oceanic crust is strong enough to hold down the Tuscarora Deep without slowly smoothing out under the pull of gravity, which is by all odds the strongest force operative on the surface of the earth. The Eötvös and tidal forces postulated by Wegener are, according to Dr. Jeffreys, about one-millionth as powerful as required to move continents and build mountains. In all of geophysics Jeffreys knew of no force acting with a component tangential to the earth's crust that approached the required magnitude.

Dr. Jeffreys' own explanation of earth evolution embodied a new, highly quantitative version of the thermal contraction theory. He postulated that the earth began as a molten globe and that, up to the present time, cooling has progressed from the outside inward to a depth of about 700 kilometers (one-tenth of the radius). The nucleus below 700 kilometers has not cooled significantly and therefore has not changed in volume. It possesses little or no strength to resist long-term stress but is rigid with respect to short-term stress, as shown by the transmission of earthquake waves. The outermost "crust," about 100 kilometers thick, has cooled as much as it ever will and is strong as well as rigid. The intermediate zone is actively cooling and contracting. This layer is under tension as it is shrinking in volume, yet its inner circumference is constrained to fit over the uncooled nucleus. The thinning of this layer removes support from the cooled "crust," which is thereby thrown under compression and frequently gives way by folding or thrusting. Between the zones of tension and compression there is a level of no strain, or zone of weakness, where contraction keeps pace with cooling and stress cannot accumulate.

In reply to Osmond Fisher, Clarence Dutton, F. B. Taylor, and the other critics who had doubted that thermal contraction has been sufficient to account for mountain-building, Jeffreys stated that it has in fact been about double what is required. According to his calculations, the total reduction of the earth's surface area to date has been about 4×10^{16} cm^2, whereas the estimated amount of shortening indicated in the mountain ranges of the world is only about 2×10^{16} cm^2.

After publication of *The Earth*, advocates of continental drift were under permanent challenge from the solid earth school of geophysics, which flourished as seismology became an ever more exact science and as experimentalists such as Percy Bridgeman and Francis Birch of Harvard University began to probe the behavior of rocky materials under high pressures. Strictly defined, the term solid earth geophysics refers to the geophysical study of the earth as distinct from that of the ocean waters and the atmosphere. Very soon, however, it acquired a secondary connotation—the solid, solid earth, strong and reverberant as steel. This view of the earth developed in the 1920s and persisted into the late 1960s. "Far from responding to shocks after the fashion of a mudball, sometimes suggested as a mechanical model, the globe rings like a fairly good bell," wrote Francis Birch in 1965. To many scientists a globe of this character would hardly seem the kind of medium in which to postulate the easy slipping about of continental blocks.

If a mechanism does not exist for moving continents, continents have not moved; and a force is not likely to exist and remain wholly unknown to geophysics in the 20th century. Such logic appeared unanswerable, and for the majority of geophysicists that argument alone was sufficient to kill Wegener's hypothesis. Geologists and biologists would have to look for other solutions to their problems.

Dr. Jeffreys argued that since we know that the earth is solid, rigid, and elastic, and we also know that the atmosphere is turbulent and unpredictable, we should take the distribution of Permocarboniferous tillites as a lesson in meteorology rather than take the vagaries of climate as evidence for continental drift. Many geologists agreed, particularly those who lived and worked in the northern countries and had to contend with no problems so spectacular as the Gondwana Series.

The American Symposium

The first international symposium on continental drift was held on November 15, 1926, in New York. It was sponsored by the American Association of Petroleum Geologists under the chairmanship of W. A. J. M. van Waterschoot van der Gracht, a Dutch geologist and vice president of the Morland Oil Company. Alfred Wegener himself attended the meeting and so did Frank B. Taylor. The eleven other participants included John Joly from Ireland, J. W. Gregory from Scotland, H. B. Molengraaf from Holland and eight Americans: Professors Chester Longwell and Charles Schuchert from Yale University, Edward Berry and Joseph Singewald from Johns Hopkins University, Rollin T. Chamberlin from the University of Chicago, Bailey Willis from Stanford University, and two scientists from outside the academic world, David White of the National Research Council and William Bowie, chief of the division of geodesy, U.S. Coast and Geodetic Survey.

The chairman, Dr. van der Gracht, favored Wegener's hypothesis. He opened the sessions with a detailed review of the evidence in its favor and closed them with a reply to each of the main arguments that had been presented against it. In the published proceedings Dr. van der Gracht takes up 105 of the 226 pages, and Wegener, who spoke very briefly, uses seven pages. Thus, the "pro" side of the argument is well represented. A few participants either were favorable toward certain aspects of the concept or were neutral. Those who spoke most persuasively, however, were decidedly against Wegener's hypothesis, and they used geological evidence together with logic, wit, and sarcasm with telling effect. One of the copies of the proceedings in a library at Harvard University once belonged to a young man who was to become a professor of geology at that institution. He inscribed the volume with his name along with the place and date: "Hugh E. McKinstry, Timmins, Ontario, Canada, January, 1930." He also penciled numerous comments in the margins; so, in effect, he becomes an active participant from whom we shall hear from time to time.

The completeness of this iconoclasm

"The mere fact that a group of American geologists has undertaken a serious discussion of possible continental drifting indicates a change in viewpoint within comparatively few years," said Professor Chester Longwell of Yale University. Longwell was skeptical of the evidence but, he stated, not hostile to the hypothesis:

> Perhaps the very completeness of this iconoclasm, this rebellion against the established order, has served to gain for the new hypothesis a place in the sun. Its daring and spectacular character appeals to the imagination both of the layman and of the scientist. But an idea that concerns so closely the most fundamental principles of our science must have a sounder basis than imaginative appeal.

Longwell pointed out that continental drift would help to solve numerous troublesome geological enigmas; yet, in analyzing the arguments for it, point by point, he found them wanting in substance. The proposed causal mechanisms were too weak, the geological and paleontological controls poorly established, and the matching of the continental margins accidental. To illustrate this point, Longwell presented a map (Figure 36) on which he had followed Philip Lake's suggestion of fitting Australia and New Guinea with "a certain freedom" (Wegener's famous phrase) into the Arabian Sea. The match is as good as most of those in Pangaea, and Longwell suggested that it is about as meaningful. In conclusion, however, he proposed that the possibility of continental drift should be kept in mind as geological work proceeds in all continents. In subsequent years Longwell remained skeptical but eminently fair-minded on the subject. In 1958, when one of the first of the more recent symposia on continental drift was held, at the University of Tasmania, Chester Longwell was the guest of honor. In a later section we will review the ideas he expressed at that time.

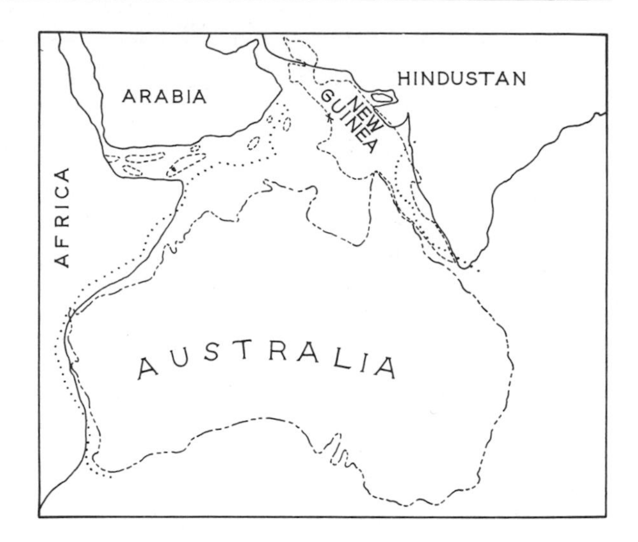

Figure 36. Longwell's map showing Australia and New Guinea fitted into the Arabian Sea. The dotted lines mark the 100-fathom line bounding the Australian platform. New Guinea has been rotated 30° clockwise about point X. (Used with permission of The American Association of Petroleum Geologists.)

A hypothesis of the footloose type

Far less tolerant than Longwell was Professor Rollin T. Chamberlin, of the University of Chicago, who said: "Wegener's theory, which is easily grasped by the layman because of its simple conceptions, has spread in a surprising fashion among certain groups of the geological profession." Chamberlin then proceeded to attack the theory on eighteen separate counts which may be summarized as follows: What was happening throughout most of geologic time? Wegener has not given us a general theory of the earth but only a description of one episode that began very recently. The geological framework of the present continents was formed during the Precambrian, and subsequent earth history has been marked by rhythmic cycles of mountain-building and quiescence which Wegener entirely ignores. Folded mountain systems of Precambrian and early Paleozoic age lie parallel to the present platform margins of all continents. [!] The rimming shelf seas of the Paleozoic are inconsistent with an arrangement in one great mass like Pangaea. The fit of the continental margins is very bad indeed; there is no genuine transatlantic matching of co-magmatic provinces, and the attempt to match Pleistocene terminal moraines is ludicrous. If the sima were the weaker layer, why did not the ocean floor compress into folds ahead of the blocks of sial? Given a landmass like Pangaea, why is there not an even greater similarity in flora and fauna among the fragments?

"Wegener's hypothesis in general," Chamberlin said, "is of the foot-loose type in that it takes considerable liberty with our globe and is less bound by restrictions or tied down by awkward, ugly facts than most of its rival theories." Chamberlin ended with a plea for the planetesimal hypothesis, originated in part by his father, T. C. Chamberlin: "That hypothesis, instead of being detached and free-floating, is an integral part of a comprehensive geological philosophy which extends from the birth of the earth through the various stages to the present. Wegener seems to be entirely oblivious of its existence." [McKinstry, in marginal note: "In other words, 'We the Chamberlins have been snooted.' "]

Chamberlin obviously felt that a good theory had been slighted whereas a symposium was being held to discuss a nonsensical one. At that time, however, the planetesimal hypothesis was probably enjoying its period of highest prestige.

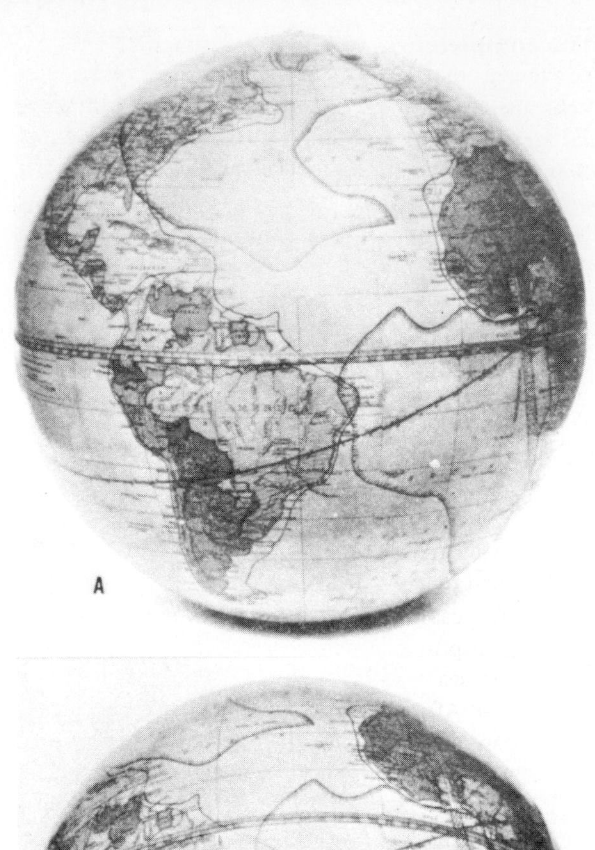

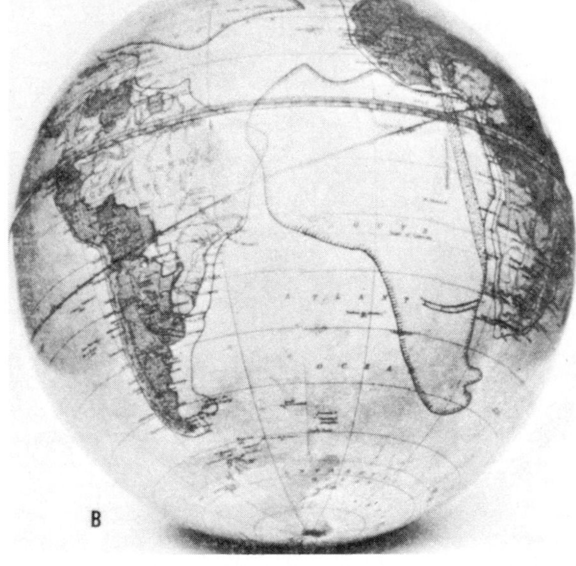

Figure 37. Two of Schuchert's eight-inch globes. In A, North America has been moved eastward until Newfoundland touches Ireland; a wide gap is created between Siberia and Alaska; and there is no semblance of a fit south of Ireland. In B, South America is fitted to Africa but the San Franciscan geosyncline (stippled band) ends abruptly at the Brazilian coastline. The impression that it extends into Nigeria is a photographic illusion. Schuchert called special attention to the misfit between the Americas. (Used with permission of The American Association of Petroleum Geologists.)

". . . the dream of a great poet"

Charles Schuchert, a paleontologist and professor emeritus at Yale, reviewed the hypothesis in detail. We are to believe, he pointed out, that this immense rifting and drifting went on during one of the earth's most marked times of crustal quietness, the early and middle Cretaceous, when almost no mountains were made in the world—a time almost devoid of volcanic activity, when the continents were about as pene-planed and low as they ever have been and when they were flooded with one of the greatest oceanic transgressions of all times. On the other hand, he said, Pangaea must have endured unbroken all through the late Precambrian and Paleozoic; yet two of the earth's greatest times of mountain-making occurred at the close of the Precambrian and Paleozoic. Why was Pangaea not broken in periods of crustal unrest? Why did it break in a time of great crustal stability? Of course, Schuchert commented, all of this is as determined by orthodox geology. [But so was Wegener's geological evidence.]

Schuchert illustrated his talk with pictures of an eight-inch globe on which he had covered the continents with plasticene, cut them out, and experimented with their fit. The results (shown in Figure 37) were dismal in the extreme. Schuchert was convinced that no significant matching of margins was possible without taking unconscionable liberties. Furthermore, he said that by any scheme of fitting the Americas to Euro-Africa a very large gap was left at the Bering Strait, thus destroying the only Tertiary land bridge that every biologist agrees is absolutely essential for the observed distribution of fauna. Wegener had once answered this criticism by claiming that the Americas rotated in such a way that Alaska always lay near Asia at the Bering Strait. Schuchert's most generous rotation left a fatal gap of at least 950 kilometers. [But, Wegener replied later, Schuchert rotated North America around an axis at the north pole instead of one in Alaska.]

[McKinstry, in the margin, asks: "Why not move Asia northward at the same time? We know it moved *south* when the Himalayas were formed."]

In any case, Schuchert asked, how could one expect a ruptured coastline to retain its shape very long? The sea waves have been continuously pounding against Africa and Brazil, rivers have brought to the ocean great amounts of eroded material, yet everywhere the geographic shorelines are said to have remained unchanged. Schuchert added that it apparently made no difference to Wegener how hard or soft the rocks are, how often the strand lines have been elevated or depressed, or how great were the fluctuations of sea level during the Pleistocene when the lands were covered by millions of square miles of ice made from water subtracted from the oceans. Wegener, he said, wants us to believe that the original fracture lines have retained their original shape during 120 million years. Schuchert: "Is there a geologist anywhere who will subscribe to this startling assumption?" [McKinstry: "In other words;—the continents could not be expected to fit. Since they actually do not fit the hypothesis is disproved!"]

Schuchert reviewed the evidence of geologic features across the Atlantic and, while conceding a tolerable match for some, he concluded that none demanded an origin in close contact. In fact, given Wegener's model, he found the geological and faunal similarities surprisingly few. Why, for example, Schuchert asked, is there no mountain range in Newfoundland known to have the same date of orogeny and intensity of folding as the Caledonides of Ireland? [McKinstry: "Maybe different structural geologists did it."]

Schuchert described a previously unknown feature, the Paleozoic Franciscan geosyncline of Brazil, which strikes toward the Atlantic and, by Wegener's match, should continue northeastward into Nigeria. Schuchert found, however, that the geosyncline would abut the ancient Precambrian shield of Africa, a fact which, he said, "deals a crushing blow to the displacement hypothesis."

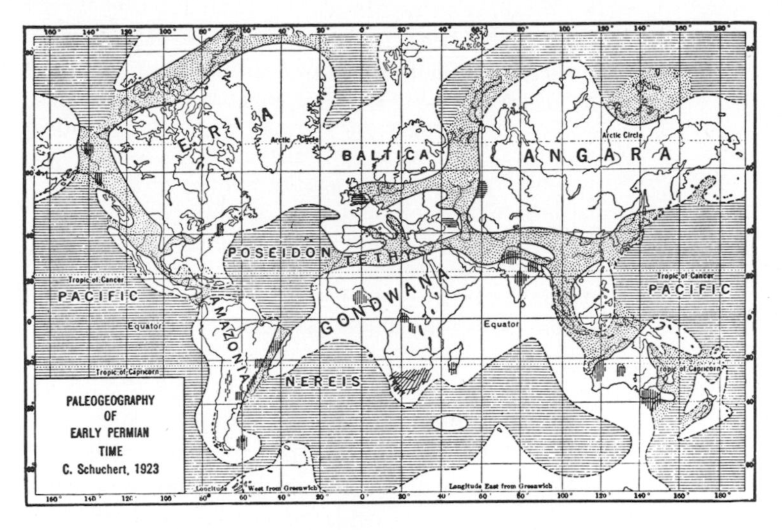

Figure 38. Continents and land bridges of the early Permian according to Schuchert. Glacial deposits are indicated by hatching, geosynclines by stippling. (Used with permission of The American Association of Petroleum Geologists.)

To account for the distribution of late Paleozoic and more recent flora and fauna, Schuchert, who was a paleontologist, favored the system of Permian landmasses shown in Figure 38. He criticized Wegener's scheme of polar wandering and Permian climatic zones, saying that Wegener had lumped together on a single diagram events that took place during a lapse of 50 million years and had made all facts fit his hypothesis by overgeneralizing. It is not, he said, as claimed by Wegener, that the detailed worker cannot see the forest for the many different trees, or that the paleontologists need a geophysicist to show them the road on which they should travel. "Facts are facts," Schuchert said, "and it is from facts that we make our generalizations, from the little to the great, and it is wrong for a stranger to the facts he handles to generalize from them to other generalizations."

Schuchert was a convinced uniformitarian, and he believed fundamentally in the permanence of continents and oceans:

We are on safe ground only so long as we follow the teachings of the law of uniformity in the operation of nature's laws. The battle over the theory of the permanency of the earth's greater features introduced by James D. Dana has been fought and won by Americans long ago. In Europe, however, this battle is not yet fought to a conclusion, since there are leading geologists who still follow Lyell and believe in the impermanence of the continents and oceans, and others who do not hesitate to push the earth's poles anywhere in order to explain single floral or faunal peculiarities.

With respect to his own land bridges, Schuchert admitted that the foundering of such structures presents a difficulty for which no solution is in sight, but he expressed confidence that "the geophysicists will in time find the way in which this was accomplished." Manifestly, sunken land bridges were no worse for geophysics than were continents that drift over vast horizontal distances.

Having said all this, Schuchert reminded his audience that movement of landmasses does, nevertheless, take place on the earth, and he cited the estimates of crustal shortening for the Alps, 560 to over 1,600 kilometers; the mountains of central Asia, up to 2,900 kilometers; the American cordilleras—very large, but, he said, still wholly unknown. Faced with such figures he quoted the statement of Galileo in regard to the earth: "And yet it does move."

Schuchert closed with an outline of earth history as he saw it, beginning with a molten earth that differentiated into a sialic crust that either was continuous (but of variable thickness) or was discontinuous, in very large slabs overlying a mobile, largely glassy, basaltic substratum. Crustal movements, beginning in the early Precambrian (then believed to be about 1,500 million years ago), resulted in folding, thrusting, and the welding of the sial into continental islands between which lay seas that made possible the cycle of evaporation, rain, erosion, and sedimentation. In the late Precambrian, more crustal shifting resulted in the development of three large protocontinents—Holarctis, Antarctis, and Equatoris—which included large areas of the present continents and several important land bridges. The Pacific was already in its present position, where it evolved into ever greater proportions by the crustal shortening involved in the folding of its continental borders. This, Schuchert said, is progressive geology which includes both large protocontinents and the permanency of the earth's greater features. In contrast, Wegener's concept is (as described by Termier, the director of the Geological Survey of France) "a beautiful dream, the dream of a great poet. One tries to embrace it, and finds that he has in his arms but a little vapor or smoke; it is at the same time both alluring and intangible."

The unyielding sima

Several participants questioned Wegener's geophysics. Bailey Willis stated that a close matching of ruptured coastlines on Wegener's model should not be expected. How, he asked, can the sima be the more yielding layer and still resist the sialic rafts until they are forced into folded cordilleras? If indeed the sialic continents do yield to compression on their forward margins, they must also, like any body in motion, be subject to tension in the rear, and large amounts of normal faulting should have utterly destroyed the configuration of the original rift line along the east coast of the Americas. Willis added that the forces causing the westward drift are, according to Wegener, the pull of the sun and moon on the viscous earth. Such forces would, however, act more effectively on the denser sima than on the lighter sial.

Willis: "How is it, then, that the lighter continental mass is drawn forward by a lesser force through the denser sub-oceanic mass, which remains stationary under a stronger pull?" [McKinstry: "I'll bite."]

Dr. William Bowie of the U.S. Coast and Geodetic Survey also argued the geophysical aspects of the problem. The earth is rigid with respect to short-term stresses, as shown by earthquake waves, but it has so little overall strength that it is everywhere in a state very close to isostatic equilibrium under the constant pull of gravity. If the sima is a weak layer, how can the ocean floors maintain ridges and trenches instead of smoothing out like a mud flat? How can the sima possibly crumple the frontal rims of the sialic rafts? Why do violent earthquakes occur under the ocean floors? If the sial floats under an impulse sending the blocks equatorward, why are continents concentrated in the northern hemisphere and on only one side of the earth?

With respect to polar wandering, Bowie pointed out that major changes in climate result from variations of only one degree to two degrees in the angle between the equator and the ecliptic. This angle oscillates by as much as three degrees above and below its average position every 22,000 years. A slight increase results in hotter summers and colder winters in both hemispheres; and a decrease produces colder summers and warmer winters in both. The broad pattern of latitudinal zoning, however, remains unchanged.

Bowie did not believe that the mass of the earth is likely to shift significantly with respect to the rotation axis. He cited the calculations of his colleague, Dr. Lambert, who showed that a migration of the vast landmass of Eurasia through 45° of latitude would not shift the mass of the earth relative to the rotation axis by more than about one-half a degree. Clearly, then, the earth's rotational stability is not threatened by minor events such as the rise of a mountain range or the growth of an ice cap.

A Pacific continent?

"The verdict on Professor Wegener's theory will depend on whether it explains more difficulties than it creates," wrote Dr. J. W. Gregory in *Nature* in 1925. At the New York symposium, in 1926, Gregory said that he was by no means hostile to the idea of some horizontal migration of continental masses. Indeed, he himself had postulated such motion as early as 1915. In general, however, Gregory—who had made an exhaustive study of the East African rift valleys—believed that vertical earth movements predominate. Some of Gregory's arguments at the symposium echoed ideas he had discussed in more detail in his *Nature* article where he pointed out that if two pieces of wood with parallel grain are seen in a sheet of water, it does not follow that they were in contact and have floated apart—they may be the ends of a warped plank. The Pyrenees and the Caucasus are regarded as parts of one mountain system even though they are separated by the full width of Europe; and so it may be with the Armorican folds of Europe and the Appalachians of North America. Have the "matching" structures, Gregory asked, been separated by 3.2 kilometers of subsidence of the ocean floor, or by 4,000 kilometers of horizontal drift opening a new ocean floor? He discussed faunal resemblances at some length, pointing out that in some instances the correspondences between western North America and Asia are as striking as those across the Atlantic. Wegener had cited the manatee as a creature inhabiting the tropical estuaries of South America and western Africa. But, said Gregory, the alligator lives only in tropical America and in the Yangtze River of China! The main flora of China matches that of western North America, and the Pritchardia palms of the southwest Pacific are found elsewhere only in Hawaii and in Cuba!*

Although it is true, as pointed out by Wegener, that one species of earthworm gardens the soils of Eurasia and eastern North America, another one, said Gregory, performs the same function in India, Australasia, and western North America. Gregory concluded that all of these anomalies demonstrate the need for a Paleozoic Pacific continent or, at least, for a substantial trans-Pacific land bridge. However, such a continent would contradict Wegener's hypothesis because it would demonstrate the actuality of profound vertical subsidence and would imply that the Pacific has grown larger, not smaller, since the Carboniferous.

"Continental movement . . . not improbable"

Several other speakers opposed Wegener's hypothesis at the symposium. We have outlined the main types of objections that were voiced then and for many years thereafter. One of the neutral participants was John Joly, who had published the second edition of his book *The Surface History of the Earth* a year earlier. In it Joly explained the periodic nature of mountain-building as resulting from the generation of radioactive heat in the mantle. He believed that in oceanic areas this heat moves upward by convection and is readily dissipated. The continents, in contrast, act as thick insulating blankets which prevent the escape of heat and so promote liquefaction of

*Earlier investigators thought the palm seeds were catapulted to Cuba by volcanic eruptions.

subcontinental sima. The substratum has a lower density when molten than when it is solid; therefore the sialic blocks subside more deeply into the liquid and, as a consequence, the seas encroach upon the lands. In time, the heat is dissipated by volcanism and igneous intrusion, accompanied by the tectonic deformation and metamorphism of geosynclinal sediments. Once rid of the excess heat, the substratum becomes solid again, the continents rise, the seas retreat, and the geosynclinal materials are uplifted to mountain ranges. A period of tectonic quiescence follows until the renewed buildup of subcontinental radiogenic heat initiates another cycle.

Joly had not included continental drift as one of the manifestations explained by his hypothesis. At the symposium he presented a 14-sentence statement that began: "I think continental movement is not improbable during periods of fluid substratum." The pro-drifters were overjoyed.

Frank B. Taylor, always an original thinker and lucid writer, reviewed his theory of the creep of continental masses from the poles toward the equator. It was at this symposium that Taylor finally ventured to suggest capture of the moon in the Cretaceous period as the event which, by causing an increase in the earth's rate of rotation and hence of the centrifugal force, initiated the equatorward sliding of the landmasses. Still, Taylor remained relatively immune from public attack—possibly because Wegener, an involuntary lightning rod, was drawing the fury of the blast to himself.

"Two notes on my theory . . ."

Alfred Wegener himself said very little. In his talk entitled "Two Notes on My Theory of Continental Drift" he discussed the tillite problem and the geodetic measurements which, he was fully confident, would soon persuade all scientists of the actuality of drift. With respect to the tillites, Wegener doubted the glacial origin of every reported deposit of Permocarboniferous age in Siberia, the United States, and other sites north of India. In that period, he said, North America and Europe should have occupied a warm climatic belt, as indicated by lush coal seams and coral reefs, which gradually gave way to evaporites and desert sands. Wegener predicted that the "tillites" would all prove on reexamination to be pseudoglacial.*

Otherwise, he rested his case on geodetic measurements of continental movement: "The hypothesis of continental drift has one advantage over all other geological theories, that it involves the possibility of checking its truth by repeated astronomical observations of latitude and longitude." Wegener's faith in these measurements never faltered.

Perhaps Wegener had no need to defend continental drift in more detail because he had a very skillful advocate at the meeting. Chairman van der Gracht undertook to answer the objections in a final summary of the proceedings.

Summation

A great deal was happening throughout most of geologic time, the chairman replied to Professor Chamberlin and all of the others who had protested against the idea that there had been only one late episode of crustal movement. Old Paleozoic or earlier continental drift is fully possible, he said; it simply was not discussed in detail by Wegener because the relevant facts are too little known. Actually, Wegener had discussed this problem in a general way. Wegener thought it "not inconceivable" that the primeval earth was covered by a thin sialic layer enveloped by a universal sea (panthalassa) about 2.4 kilometers deep. In this case he saw most of geologic time taken up with the folding and thrusting of the sial into an ever smaller landmass. Wegener suspected, however, that the sial might always have been patchy and that its original extent was both unknowable and unimportant as long as it took the form of Pangaea by the late Carboniferous. It is true that he brushed lightly over the greater part of earth history and began his story with the recognizable continental drift that started only 40 million years ago.† It is equally true that he could never have built a convincing pattern of earlier drift on the basis of the geological record—he faced trouble enough with the most recent episode.

*He was right. By 1961 even the famous Squantum tillite near Boston was pronounced pseudoglacial by R. H. Dott, Jr., then at the Massachusetts Institute of Technology.

Dr. van der Gracht believed that an original crust of universal sial rolled up into continental nucleii in remote Precambrian time and later coalesced into Pangaea. He thought that a major geosyncline—some kind of old Paleozoic Atlantic—formed between Europe and America and was closed by the Caledonian orogeny, leaving traces on both sides of the present ocean. This old line of weakness might explain the pre-Cretaceous mountain chains of Euro-Africa and the Americas and their close relationship to the present Atlantic coastlines, a feature pointed out by Dr. Chamberlin and others.

In reply to Dr. Bowie, Chairman van der Gracht suggested that the sima beneath the continents has little residual strength, but that beneath the cold oceans (with bottom temperature of about 0° C.) it has an upper rind of considerable strength—enough to maintain relief, propagate earthquake waves, and crumple the advancing prows of sial. The concentration of continents in the northern hemisphere is a fact today but it does not hold for the periods before mid-Tertiary. At that time the equator divided Pangaea about equally, and the present asymmetry is due to subsequent polar wandering together with continental drift.

Dr. van der Gracht freely admitted the lack of a plausible explanation of the mechanics of continental drift. The objection still holds and is serious, he said, unless the hypothesis of periodic melting of the sima by radioactive heat, proposed by Dr. Joly, stands the test of time. However, he added, a similar objection stands for the great thrust sheets of the Alps: "Yet they exist!" Van der Gracht suggested that we might refer the defense to Galileo.

Despite Dr. van der Gracht's point-by-point defense, the sense of the American symposium was negative with respect to Wegener's hypothesis. The globes displayed by Dr. Schuchert, with their spectacularly misfitting continents, the geological discrepancies cited by Dr. Chamberlin and others, the geophysical objections of Drs. Bowie and Willis, and the multitude of floral and faunal anomalies that seemed to contradict as often as they confirmed Wegener's scheme were, in the aggregate, all too impressive. Sarcasm, a deadly weapon, had been used freely. It is a rare scientist of however well-established reputation who is willing to be caught-out favoring a hypothesis that others laugh at. He knows the derision will not stop there; presently he himself will be laughed at. Partly as a result of that symposium, more than 35 years were to pass before American geologists would meet again for the purpose of seriously discussing continental drift.

†The 40 million years (before the present) estimated for the beginning of the Jurassic when Wegener introduced his hypothesis in 1912 had lengthened to 65 million years by 1926. The estimate is now about 180 million years. (When Schuchert implied that the breakup began 120 million years ago he was erroneously extending the time of the breakup backward to what was then estimated as the beginning of the Carboniferous.)

Alternative Hypotheses

The critics of Wegener's hypothesis enjoyed a majority in numbers and the leadership of eminent individuals. Nevertheless, a few scientists of unchallenged prestige saw Wegener's hypothesis as worthy of serious consideration. One of these was Professor Reginald A. Daly of Harvard University. In 1926 Daly published a book, *Our Mobile Earth*, and on its title page appeared the words *E pur si Muove*, the apochryphal phrase from Galileo. Geologists have been mightily impressed with the story of Galileo and have transferred from astronomy to geology his lonely plea for a mobile rather than a static planet.

"The continents appear to have slid downhill"

Speculating on the causes of mountain-building, Daly stated that the older theories, all readily accessible in textbooks, are all more or less unsatisfactory, and he said:

> Less widely published is a new, startling explanation, first announced by an American geologist, Frank B. Taylor, and independently worked out by Alfred Wegener, a German meteorologist. Many geologists have found their idea bizarre, shocking; yet an increasing number of specialists in the problem are already convinced that it must be seriously entertained as the true basis for a sound theory of mountain-building. The subject has been under discussion only a few years and is laden with difficulties, so that a full and objective treatment is still impossible. Nevertheless, every educated person cannot fail to be interested in this revolutionary conception.

The curved chains of mountains bordering the Pacific basin and Eurasia had led both Taylor and Wegener to suggest that mountain ranges are crumpled along the forward margins of moving continents. Daly agreed, but he proposed a causal mechanism different from theirs. He called his own version the "down-sliding" or "landsliding" hypothesis:

> The continents appear to have slid downhill, to have been pulled down over the earth's body by mere gravity; mountain structures appear to be the product of enormous slow *landslides*. Each chain has been folded at the foot of a crust-block of continental dimensions which was not quite level, but slightly tilted.

Why should the surface of the earth be tilted? According to Daly the earth is distorted because it is contracting, because its speed of rotation is changing, and because its lands are being eroded—resulting in a dimunition of load over the highlands and an increase in the basins. Distortions alone, however, would never

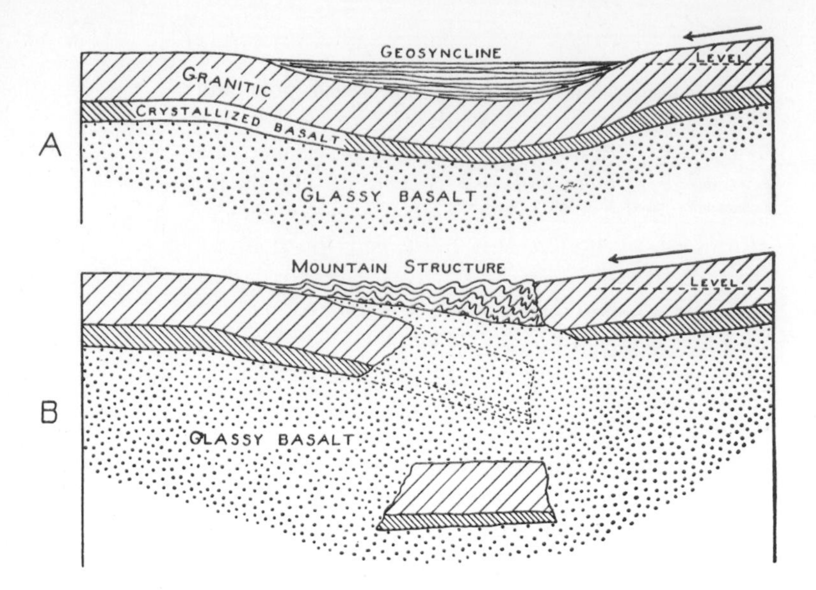

Figure 39. Continental sliding according to Daly, who estimated that the layers would have the following densities, in grams per cubic centimeter: granite, 2.65; crystallized basalt, 3.00; glassy basalt, 2.80-2.85. (From *Our Mobile Earth* by R. A. Daly, 1926; used with permission of Charles Scribner's Sons.)

lead to continental sliding were it not for the presence of a zone of weakness underlying the earth's strong outer crust. Daly believed that the granitic rafts of the continents are underlain by a world-encircling layer of crystalline basalt, and below that there is a substratum of hot, eruptible, basaltic glass that acts as a lubricant beneath the sliding crust. Daly pointed out that this glassy layer not only is slippery but that it must be of lower density than the overlying crystalline basalt, a situation which is gravitationally unstable; therefore, whenever the crust fractures along lines of weakness, such as those occupied by geosynclines, great "blocks or plates" of it plunge into the substratum. Daly illustrated his idea in diagrams (Figure 39).

Daly concluded that continental sliding has occurred throughout geologic history and has built the world's mountain ranges from early Precambrian times to the present. Examining the earth as a whole brought him "face to face with a principal mystery of nature": the existence of one great land hemisphere overlooking a deep ocean. The molten earth should have formed a smooth spheroid of revolution covered by a layer of water. How and when, asked Daly, did the planet acquire its asymmetry? How, indeed, did it form dry land at all? Daly speculated that the asymmetry may be a primeval feature resulting from the manner in which the gaseous earth condensed to a liquid, or that it may have been imposed at an early date by the ripping away of the moon. The true cause, he concluded, is unknown, but the fact of asymmetry has contributed to the restlessness of the earth's figure and to continental sliding.

Daly pictured the young planet with a thin crust of granite which was broken into slabs that were crumpled and thrust together into thick buoyant masses which finally broke the surface of the ocean to form the first continents. He thought that by the late Paleozoic, after a long history of motion and growth, all of the landmasses coalesced into one supercontinent. In the Mesozoic this immense, high dome of sial broke under its own weight and our familiar continental fragments began sliding toward the Pacific basin.

Later in his career Daly came to feel that both he and Wegener had overestimated the magnitude of continental migration, but he retained his faith in gravitational sliding as the prime cause of mountain-building. By the 1940s he focused much of his attention on the nature of the substratum, or asthenosphere (a name he did not use in *Our Mobile Earth* or in *Architecture of the Earth*, published in 1938). He finally concluded that the asthenosphere is neither strictly basaltic nor glass but is a hydrous, two-phase mixture of crystals and glass with the bulk composition of peridotite.

Daly's later researches led his colleagues all over the world to associate his name with the problem of the asthenosphere. That term, however, had been coined in 1914 by Joseph Barrell, a professor of geology at Yale, who wrote a series of papers on the strength of the earth that appeared in the *Journal of Geology* of that year. Barrell pictured the earth's outermost layer, or lithosphere, as a very strong shell, tens of kilometers thick, that incorporated both granitic and basaltic materials. Underlying the lithosphere was what he described as a "thick, hot, basic, rigid yet weak shell, the asthenosphere, or sphere of weakness" extending to depths of hundreds of kilometers. Between the asthenosphere and the core lay a strong shell which he called the centrosphere but which Daly called the mesosphere.

By introducing the name and the concept of the asthenosphere into geologic literature, Barrell, in effect, founded modern tectonic geology. His own interpretation of crustal tectonics was, however, never acceptable to many of his American colleagues. Like Suess and other Europeans, Barrell believed in the process he called continental fragmentation—the progressive enlargement of the ocean basins by the foundering of slabs of continental lithosphere that were made heavy by the intrusion and crystallization of dense materials from the asthenosphere. (Barrell pointed out that J. D. Dana himself had finally reinterpreted his own doctrine of permanence loosely enough to allow for certain former connections between the southern continents.)

Interestingly enough, Joseph Barrell and Reginald Daly, beginning from similar views of the lithosphere and the asthenosphere, developed totally divergent theories of the evolution of continents and oceans. Barrell sustained the ideas of Suess; Daly began an approach to the current model of plate tectonics.

Die Fliesstheorie

In 1927 Beno Gutenberg, in Germany, concluded, because of seismic evidence and the requirements of isostasy, that the Atlantic, Arctic, and Indian oceans have sialic floors. The antagonists of continental drift, many of whom had never doubted this, were happy to see the point confirmed by a prominent geophysicist. Gutenberg, however, was proposing a new theory, his "Fliesstheorie" of continental spreading.

Gutenberg's reading of seismic evidence indicated a sharp density contrast between continental and oceanic crust at the margins of the Pacific but none at the borders of other oceans. Furthermore, new soundings revealed in ever more impressive detail the rugged ridge and valley topography of the Atlantic floor, which, it seemed, could only be reconciled with Airy's model of isostasy if the bedrock were sialic. Gutenberg was persuaded that all of the lesser oceans have floors of sial but that the Pacific floor is of sima.

According to the earliest version of his Fliesstheorie the earth was enveloped by a layer of sial until the moon was ripped away at a very early period and carried most of the crust along with it.* The remaining sial lay concentrated in one large, somewhat lopsided mass covering the south pole and extending northward towards the tropics in the hemisphere opposite the Pacific scar. Under the combined impulses of gravity and of "Polfluchtkraft," the sial began to thin and to spread radially northward, most effectively in the hemisphere opposite the Pacific where it already predominated (Figure 40). As the moving sial arrived in the equatorial regions, mountain-building was initiated and east-west ranges were formed in response to periodic global contraction. As spreading continued, these ranges moved farther north and new ones formed along the equator. As a result, Europe is now crossed by three ranges—the Caledonides, Hercynides, and Alps—with ages that are successively younger toward the south. Crustal spreading was accompanied by some clockwise or counterclockwise rotations and by the development of weaknesses that caused large areas to subside below sea level. Eventually the sial acquired its present distribution—thicker in the continents, thinner under the Atlantic, Indian, and Arctic oceans, but never covering the Pacific floor.

Gutenberg's theory, set forth here in all too brief an outline, attempted to account for the distribution of mountain ranges, for the shapes of the continents, which taper toward the south, and for the difference between Pacific and Atlantic types of coastlines. In addition, by postulating a generally northward motion of all continental masses (except Antarctica), Gutenberg explained successive climatic changes without an appeal to polar wandering.

Six years after it appeared in Germany the Fliesstheorie was reviewed for English readers by Philip Lake in the March 1933 issue of *Geological Magazine*. Contrary, perhaps, to our expectations, Lake was very favorably impressed. One of his main criticisms, however, was that Gutenberg had relied too much on Wegener and Köppen for his geology and paleoclimatology. Also, he pointed out that Gutenberg's sial sheet had started in the southern hemisphere but now lies mainly in the northern, where neither gravity nor "Polfluchtkraft" could have carried more than half of it. To accomplish this, Lake

*In 1936, in a discussion of the Fliesstheorie published in the *Bulletin of the Geological Society of America*, Gutenberg rejected the idea of the moon having been ripped from the earth and left the cause of the asymmetrical distribution of sial an open question.

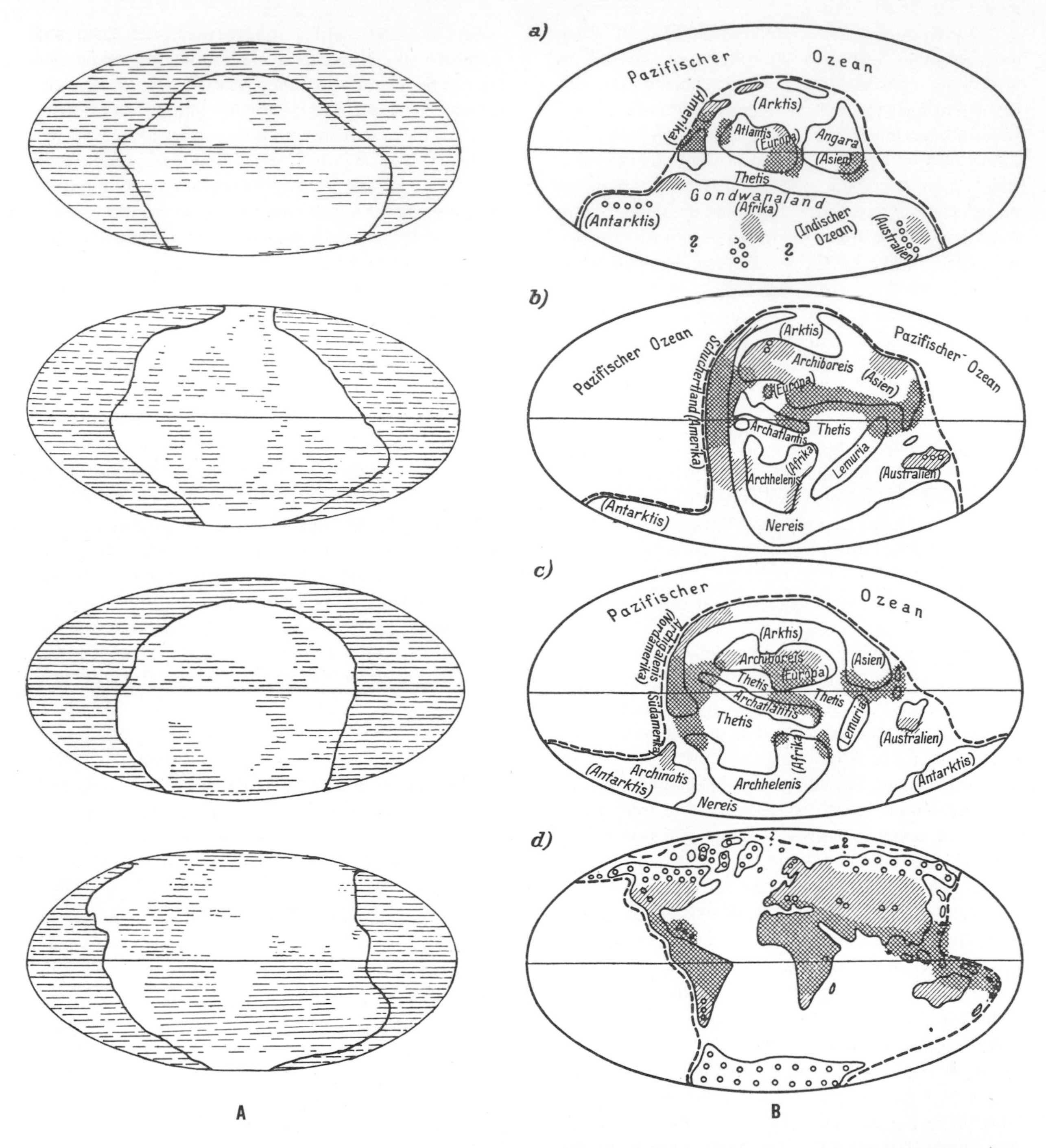

Figure 40. Beno Gutenberg's illustrations of his Fliesstheorie. **A:** Sketches, published in 1927, showing the present continents and oceans slowly taking shape as a large mass of sial spreads northward. **B:** Detailed drawings that were published a few months later. The dashed line represents the sial-sima boundary; circles, regions of cold climate; oblique hatching, temperate climate; cross hatching, tropical climate. In both series, the periods represented are: a) Carboniferous, b) Cretaceous, c) Eocene, d) Recent. (From Gutenberg, 1927, in *Gerlands Beitrage zur Geophysik:* **A,** from volume 16, page 244; **B,** from volume 18, page 283. Used with permission of *Gerlands Beitrage zur Geophysik.*)

commented, some other force is needed that Gutenberg did not "seem to explain." On the whole, nevertheless, Philip Lake liked the Fliesstheorie. He was particularly taken by the idea of England starting out at the antarctic circle in the Precambrian, moving to the south temperate zone for deposition of early Paleozoic strata, passing through the horse latitudes during formation of the Devonian Old Red Sandstone, and then arriving at the equator for accumulation of the coal swamps in the Carboniferous. After spending the Mesozoic and early Tertiary in the tropics and subtropics, England arrived at its present latitude in the Pliocene. Lake wrote:

> We need not conclude that it went to the Arctic regions in the Glacial period and came back again, for other causes may have temporarily lowered the temperature. . . . It is an impressive history, more consistent than the lawless wanderings imagined by Köppen and Wegener.

Continental genesis and isthmian links

In 1929, in an article entitled "Continental Genesis," Bailey Willis proposed a model of earth evolution that supported the idea of permanence of continents and ocean basins. Willis felt no need to refute the concept of continental drift. He reserved his scorn for collapsed ocean basins, a concept he traced to Buffon. Buffon himself could be forgiven, said Willis, because he was groping in the dim light before the dawn of earth science: "Yet Suess, Haug, and others of the modern school who regard the collapse of hypothetical lands and continents as a condition precedent to the formation of mediterranea and ocean basins also see strange things, even in the morning light. . . . Thus, in the twentieth century, the views of Buffon appear."

Willis believed that the earth's asymmetry, with one land and one water hemisphere, is a primeval feature derived from the inhomogeneous nature of the planetesimals from which the globe accreted. Following an early suggestion of T. C. Chamberlin, Willis argued that the main mass of the earth consists of one large bolt shot from the sun along with a host of smaller bodies. The stem end of the bolt, from deep within the sun, was of denser material than the outer end and this asymmetry was reflected in the newborn planet. Willis thought that the earth was originally cold and began heating at the center. He pictured immense dikes—produced by strain melting—carrying molten materials outward along pathways that were diagonal (not radial) to the core region. The rising materials differed in composition depending on the nature of their source rock and on the amount of assimilation that occurred en route toward the surface. Willis described great blisters (asthenoliths) forming where the hot dikes encountered the base of the crust, and he attributed uplift, depression, and mountain-building to the energy from the magmatic processes involved in the growth and collapse of these asthenoliths. Willis explained the distribution of sial and sima by his theory which, with all its originality, demonstrated beyond question that his adherence to the conservative doctrine of permanence derived from no lack of imagination on his part.

Although Willis first presented this theory in his presidential address to the Geological Society of America and supplemented it with numerous publications, it received very little attention, judging from the paucity of references to it by other writers. Three years later, however, Willis published a paper entitled "Isthmian Links" which received favorable notice around the world. In it Willis reiterated his belief, first expressed in 1893, that orogenic forces originate beneath the ocean basins. He proposed that these forces cause deformation not only along continental borders but also within the ocean floors themselves. Ridges between two ocean basins are upthrust thousands of meters and form great cordilleras which rise above sea level as a chain of islands or a narrow isthmus such as Panama. Willis pointed out that these cordilleras, because they are composed of oceanic sima and have no buoyancy, will subside to deep levels when pressure is relieved. Thus, these truly oceanic mountains form land links that are intermittent by nature. Willis sketched several late Paleozoic isthmian links, two of which are shown in Figure 41. He traced in detail how the presence of these links could have altered the circulation of ocean waters, changed the world's climates without changing the overall distribution of landmasses, and brought on the Gondwana glaciations.

To many scientists Willis' isthmian links were the perfect compromise between the doctrine of permanence and the Suessian idea of submerged continents.

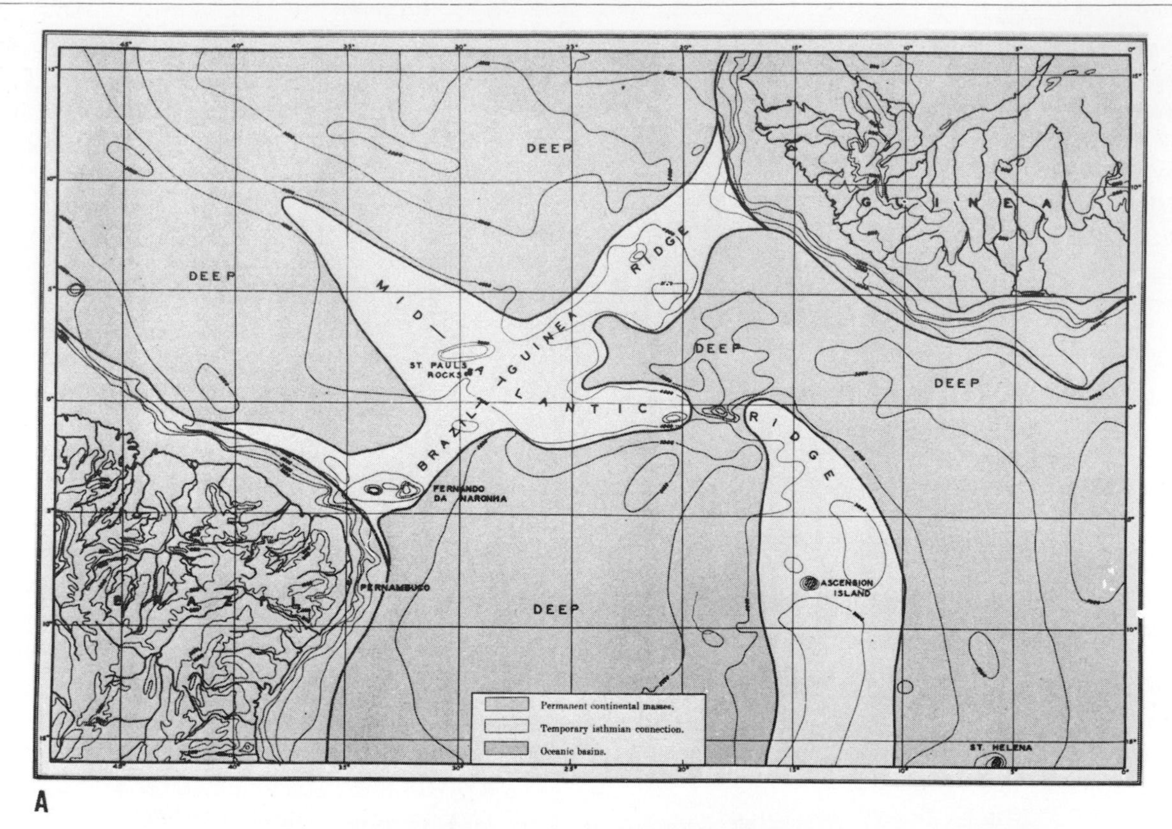
DEEP
DEEP
MID-ATLANTIC RIDGE
GUINEA
BRAZIL-GUINEA RIDGE
ST PAULS ROCKS
DEEP
DEEP
FERNANDO DA NARONHA
PERNAMBUCO
DEEP
ASCENSION ISLAND
ST. HELENA
Permanent continental masses.
Temporary isthmian connection.
Oceanic basins.

A

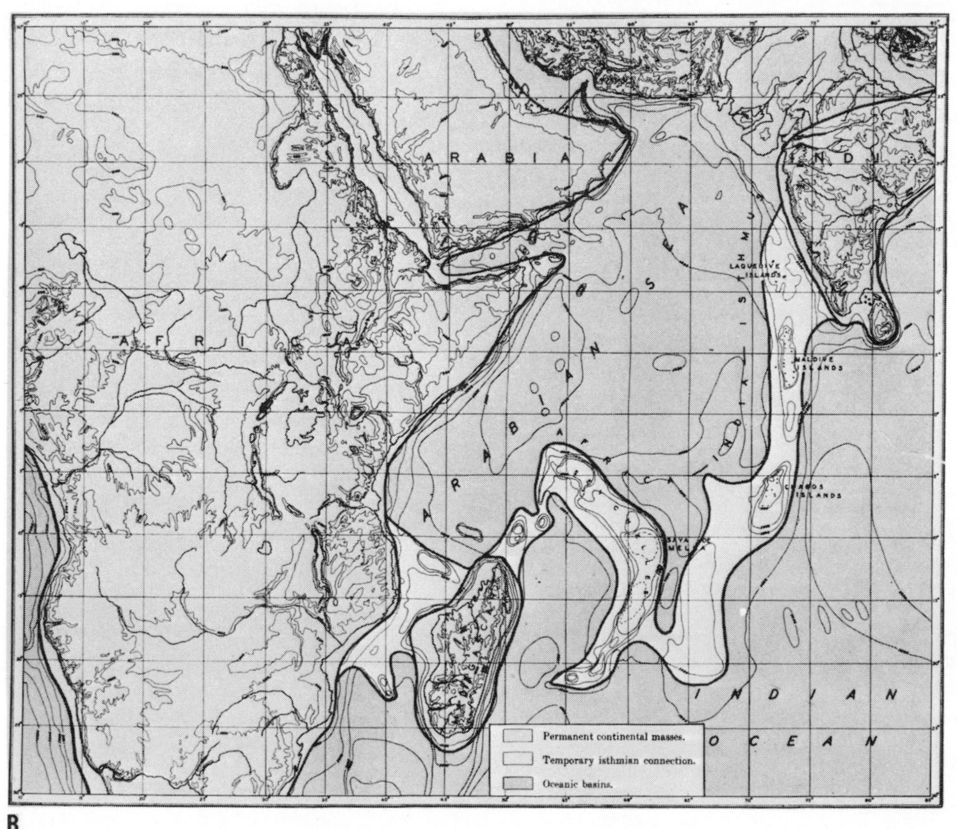
ARABIA
AFRICA
INDIA
LAQUEDIVE ISLANDS
MALDIVE ISLANDS
CHAGOS ISLANDS
INDIAN OCEAN
Permanent continental masses.
Temporary isthmian connection.
Oceanic basins.

B

While conforming to the requirements of isostasy, the isthmian links allowed for migrations of land and shallow marine flora and fauna and altered the world's climatic zones, but they were too small to upset the water balance in the oceans. Above all, their existence was reversible—they were easily built and destroyed. Finally, they seemed to remove all need for continental drift.

". . . there remains a far stronger case for continental drift"

Arthur Holmes of Edinburgh and Cambridge universities became a lifelong supporter of the hypothesis of continental drift. He was among the earliest and greatest pioneers in developing the radiometric methods for determining the ages of rocks and minerals and the age of the earth itself. He was also a total scientist, whose grasp of essential principles and store of factual knowledge were all-embracing. The final edition of his book *Principles of Physical Geology*, published in 1965, is an encyclopedic compendium of the earth sciences written with grace and humor. "Holmes was an anachronism in our time," said a recent postdoctoral fellow from Princeton, "because he knew so much."

After some early qualms about the subject, Holmes decided that on balance the geological evidence around the globe favored continental drift. In a review of the proceedings of the American symposium on continental drift that appeared in the September 1928 issue of *Nature* he wrote:

> The impression that remains with me after considering all the adverse criticism is that the latter is mainly directed against Wegener, and that when all has been said, there remains a far stronger case for continental drift than either Taylor or Wegener has yet put forward. . . . As van der Gracht insists again and again, the details of the picture, and particularly the mechanical and physical explanation, will require generations of further research.

As a specialist in radioactivity Holmes was unimpressed by the arguments for a functionally solid earth. He believed in convection currents in the mantle, and in 1931 he outlined such a theory in an article entitled "Radioactivity and Earth Movements" that was published in the *Transactions of the Geological Society of Glasgow*. His main ideas were summed up in the two diagrams reproduced in Figure 42. One diagram illustrates a planetary pattern dominated by two convection cells occupying the entire mantle; the other shows a local pattern—superimposed on the planetary one—that is governed by the distribution of continents and oceans. The local pattern shows a current rising under a slab of sial, spreading in opposite directions, rifting the sial, and opening new oceans by pulling the continental fragments away from a stranded ridge. The convergence of two currents at the continental borders would, according to Holmes, exert sufficient pressure to transform basalt to eclogite and the resulting increase in density would promote a localized sinking of the ocean floor and the formation of deep trenches. The combined effects of outward and downward motion cause mountain-building along the leading edges of the continental fragments.

Convection of the earth's interior had been proposed to explain mountain-building and crustal motions by Osmond Fisher in 1881, by Otto Ampferer in 1906, and by Rudolph Staub in 1928. Although Wegener cited the work of Fisher in his original publication of 1912 and in the first three editions of his book, he did not include convection currents among the possible causes of continental

Figure 41 (facing page). Two of the temporary isthmian links, composed of oceanic sima, described by Bailey Willis. **A**, the Brazil-Guinea ridge; **B**, the Africa-India isthmus. These narrow, intermittent connections met with widespread approval from scientists who opposed sunken continents and continental drift. (Used with permission of the Geological Society of America.)

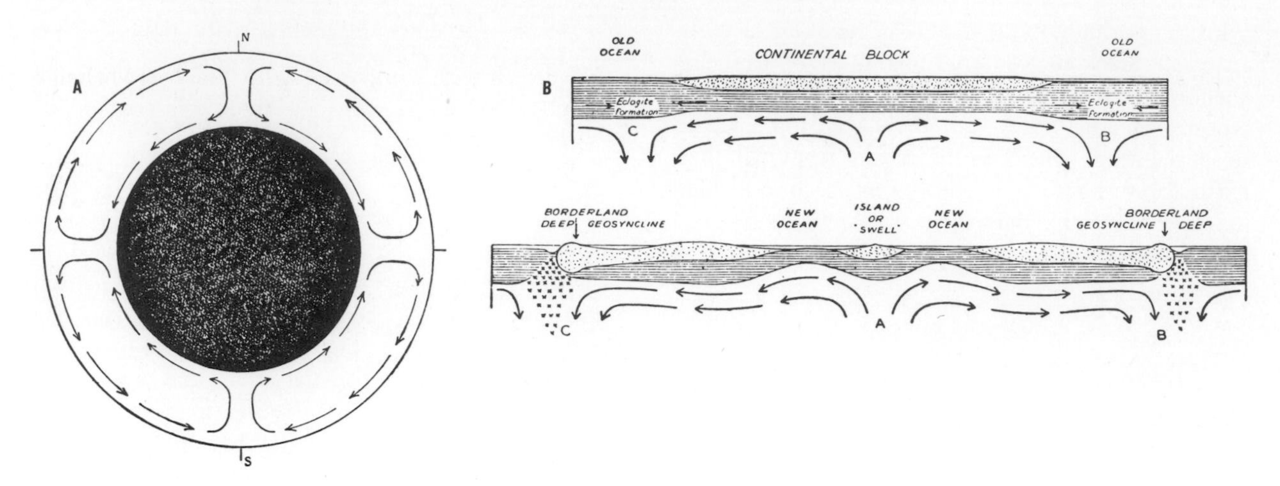

Figure 42. Two of Arthur Holmes' diagrams illustrating subcrustal convection. **A**: Pattern of broad planetary circulation with currents rising at the equator and sinking at the poles. Such a pattern would carry landmasses away from the equator and must have predominated, according to Holmes, throughout the lifetime of the Tethys. **B**: A subcontinental convection pattern that is superimposed on the planetary circulation. The slightly higher temperatures at the base of the continental crust cause currents to rise and spread outward and this leads to the rifting and separation of sialic fragments. An island is stranded in the "dead" space between diverging limbs, a mode of origin Holmes postulated for Iceland. Where two currents converge, Holmes speculated that basalt is transformed to eclogite and the increase in density results in subsidence and the formation of deep trenches. (Used with permission of The Geological Society of Glasgow.)

drift per se to the likelihood of mantle convection as a significant factor in crustal tectonics.

Convection: the porridge pot problem

When Alfred Wegener added convection currents to the forces causing continental drift, some scientists felt that he raised his hypothesis to the status of a fully fledged theory. Others were less enthusiastic. To Harold Jeffreys, for example, convection currents in the solid mantle, moving in the directions and at the rates required by Wegener, were simply one more deus ex machina that made no physical sense. It was as though having argued for years that toys are not brought down the chimney on Christmas Eve someone had "solved" the problem by inventing Santa Claus as the mechanism for bringing them down the chimney. One absurdity does not confirm another.

Probably the first scientist to discuss convection as a process operative in the earth's interior was William Hopkins of Cambridge University. In 1839 Hopkins wrote that there are two distinct processes of cooling. One of these is by conduction, which occurs in solids and materials that are imperfectly fluid. Of the other he said: "[in] masses in that state of more perfect fluidity which admits of a free motion of the component particles among themselves . . . cooling is said to take place by *circulation* or *convection* . . . of the exact laws of cooling by the latter process we are comparatively ignorant." Hopkins spoke a home truth more permanent than he knew. Of the exact laws of cooling by convection we are, to this day, comparatively ignorant.

Hopkins could not have stated more explicitly the rule—which has been generally accepted from his day to ours—that convection occurs in fluids but not in solids. It also has been generally assumed that convecting fluids must be, or will soon become, well mixed and homogeneous. Convection involves the dynamic flow of materials under the combined influences of a temperature gradient and gravity. It commonly is visualized in terms of regularly circulating currents with warm materials rising over a heat source and being displaced by the sinking of cooler, and hence heavier, materials. A symmetrical pair of currents rotating to either side of a rising limb is called a convection cell. The development of such a cell, or an array of such cells, requires very special conditions which, as pointed out by Dr. Jeffreys, is familiar to every cook who has heated a pan of porridge on a stove. Whether the contents convect, boil, or burn depends on a sensitive balance between the temperature gradient and the viscosity. Porridge can easily char at the bottom while it is still cool a few centimeters away at the top. Any tendency for convection to even-out the temperature gradient is prevented by the high viscosity.

How do conditions in a pot of porridge apply to the interior of the earth? By the mid-1920s numerous geophysicists reasoned that convection requires fluidity; hence, since the earth's mantle is solid, convection in the mantle is impossible. This proposition gained increasing support during the next three decades as new seismic and gravity data indicated that the mantle is not only solid but also markedly inhomogeneous in density, which means it must be inhomogeneous in physical state or chemical composition, or both. Such inhomogenieties may signal instability and motion in the mantle, but not necessarily in an orderly pattern of convection cells. Thermally driven currents were not expected to persist across phase boundaries or chemical discontinuities.

The clues to the nature of the interior were, however, read differently by different scientists. All could agree that the mantle is rigid with respect to short-term stress as shown by its transmission of seismic shear waves. But some materials which are rigid and elastic in response to short-term stress possess so little strength that they deform plastically when subjected to long-term stress. That the mantle is strong as well as rigid was suggested to some scientists by the departure from isostasy of numerous topographic features of great age and by the apparent disequilibrium figure of the earth. Others saw evidence for a weak but rigid mantle in the isostatic rebound of Scandinavia and in the isostatic balance between the main crustal features (continents and oceans). While advocates of a strong, solid mantle argued against convection on physical grounds, advocates of a weak mantle argued in favor of it as the most rational explanation for crustal deformation. Aside from Arthur Holmes and Alexander du Toit, however, few scientists associated mantle convection with continental drift. The overwhelming majority viewed it primarily as an explanation for mountain-building.

Proponents of convection disagreed as to whether the number and distribution of convection cells were controlled by the thickness of the convecting layer or by the distribution of continents and oceans. Nevertheless, there was general agreement—contrary to the pattern proposed by Osmond Fisher in 1889—that the main currents tend to rise under the continents and sink under the oceans. For example, in the 1930s the Dutch geophysicist Felix A. Vening Meinesz proposed that mantle convection occurs in an octahedral pattern in which the four rising limbs maintain the gross tetrahedral arrangement of the continents. At about that time, numerous advocates calculated theoretical models or experimented with laboratory models of convection. David L. Griggs of Harvard University simulated folded mountain ranges by means of rotating drums in plastic materials. He pointed out that if cooling at the surface is inefficient, the heat carried upward by the rising limbs will destroy the temperature gradient and convection will cease. Griggs suggested that such an event has occurred several times in earth history and that the resumption of convection after a quiet period accounts for the periodicity of geologic revolutions.

On the whole, the concept of mantle convection has had a curious history. Geophysicists have remained in total disagreement as to whether convection in the mantle is possible, while many nongeophysicists have accepted it as an established fact without realizing there is any problem involved.

Greenland, 1930

On April 1, 1930, Alfred Wegener sailed from Copenhagen as leader of a party of 21 scientists and technicians who were to spend a year and a half making geophysical and meteorological observations in Greenland. He set up three year-round observing stations at about latitude 71° north, in a section of Greenland not previously explored. One station, named Eismitte, was located in the center of the ice cap at an elevation of 3,000 meters. Here instruments were deployed and underground chambers hollowed into the ice where men could live and—for the first time—take measurements throughout the winter in central Greenland. The other two stations were located at the eastern and western edges of the ice cap. A broad series of investigations was carried out at numerous points along the line between stations. Seismographs, excited by dynamite explosions, measured ice thickness and yielded a maximum value of about 2 kilometers; deep shafts were sunk for measurements of glacial stratigraphy and temperatures; and barometric and astronomical measurements of altitudes, combined with simultaneous gravity determinations, were taken in the hope of settling the question of whether the Greenland massif is perceptibly rising. Meteorological observations were aimed in part at obtaining a year-round cross section of the high-pressure area over the ice cap.

Late in October Wegener visited the camp at Eismitte. There, on November 1st, the party celebrated Wegener's 50th birthday—with a precious apple for each man—before Wegener left for the west coast camp in the company of Rasmus Villumsen, a Greenlander. The two men departed with two sledges and 17 dogs. The morning was half overcast, the temperature had warmed up to –39° Centigrade and there was a slight breeze blowing from the south-southeast, an ambience described by Ernst Sorge, one of the three men who wintered at Eismitte, as "splendid sledging weather." Wegener and Villumsen never arrived at the coast.

On May 7 of the following spring a party from the western station visited Eismitte and each group of men learned with dismay that Wegener and Villumsen had not wintered with the other. A search was begun and signs were soon found of a difficult trip with severe loss of dogs. One hundred kilometers west of Eismitte was an abandoned box of pemmican; 30 kilometers farther was an abandoned sledge. At a site 195 kilometers from Eismitte and 189 kilometers from the west coast station, Wegener's skiis were found standing upright in the snow, about 3 meters apart with a broken ski stick midway between them. Digging into the snow at this point the search party found the body of Alfred Wegener, carefully sewn between two sleeping bag covers. He was fully dressed, lying upon his sleeping bag and a reindeer skin, and he was covered by his fur clothing and another reindeer skin. His clothing was in perfect order and he showed no ill effects from hunger or cold. He clearly had not died on the march, but in camp and, in all likelihood, from sudden heart failure due to overexertion. Wegener had always believed that men on skiis should keep pace with dog sledges, and the race over an uneven surface in the dim November light evidently had exhausted him. The search party placed his body back in the snow just as they had found it, and they erected over it a large mound of ice blocks topped with a Nansen sledge. A small cross was made from Wegener's broken ski stick and black cloth was tied to each of his skiis.

Rasmus Villumsen, who had so carefully laid Wegener's body to rest and marked the grave, had picked up Wegener's diary and small bag of personal effects and started for the west coast station with the remaining sledge and dogs. Villumsen, 22 years old, robust, and independent, should have had about a four-day journey. The search party found signs that he and his dogs had spent several days near the marker at 168 kilometers from the station. Signs of dogs were also found at the 153-kilometer cairn. Beyond that, all traces vanished.

"No drifting . . . of these islands"

The death of Alfred Wegener probably had no immediate effect on the fate of his hypothesis. The concept had been proposed; the critics had replied. Details were easy to demolish. The broad picture violated geophysics. By about 1930 those favoring Wegener's hypothesis and those opposed to it had hardened their positions and no longer listened to each other. Thus, a long, sterile twilight set in, with the hypothesis held in disrepute by the majority of scientists.

Continental drift soon began to supply comic relief in geology and geophysics classrooms. Ridiculous "matches" and species distributions proved an endless source of merriment. Percy E. Raymond, professor of paleontology at Harvard University, told of finding half of a Devonian pelecypod in Newfoundland and half of one in Ireland; voilá, the two matched perfectly! They were the two halves of the same pelecypod, which had been wrenched apart by Wegener's hypothesis in the late Pleistocene!

Walter Bucher, professor of geology at the University of Cincinnati, illustrated the parallelism of

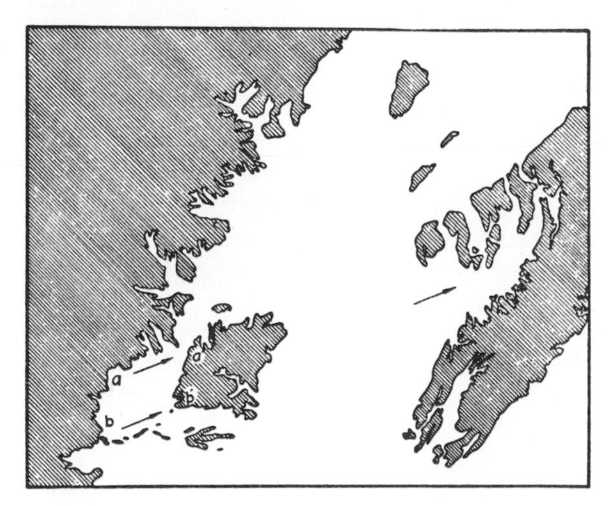

Figure 43. Bucher's hypothetical map of the central United States with all of the area from the Allegheny Plateau to the Great Plains flooded up to the 1,000-foot contour. Higher elevations are shaded. Arrows indicate direction of apparent "continental drift." Bucher wrote: "This little map contains food for useful thought. The correspondence of curvatures and dimensions on opposite shore lines is quite as striking as in the case of Greenland, with which this region corresponds somewhat in scale. The outlines are notably straight and angular. Wave-cutting prolonged through geological periods would make them more so, strongly suggesting faulting." (From *Deformation of the Earth's Crust*, by Walter H. Bucher, copyright 1933 by Princeton University Press. Used with permission.)

coastlines that would result from flooding the central United States up to the 1,000-foot contour. In his sketch map (Figure 43) the eastern margin of the great plains appears to have drifted away from the Alleghenies, leaving behind an island, the Ozark Plateau. Wave erosion, Bucher said, would make the shorelines smoother and even more strikingly matched. Yet what, he asked, does this parallelism mean? "One thing is certain: There was no 'drifting' in the case of these 'islands.' "

Bucher's own theory of the earth was described in a detailed, tightly reasoned, and thoroughly documented book, *The Deformation of the Earth's Crust*, published in 1933. Bucher believed that the crust had been subject alternately to periods of tension in which rifting has occurred along linear belts and periods of compression characterized by the folding and thrusting of the same belts. He ascribed these cycles to fluctuations in the heat content and the consequent expansion and contraction of the asthenosphere. Each pulse, he believed, is felt everywhere simultaneously and this accounts for the periodicity of geologic revolutions. Bucher argued that all parts of the crust possess approximately the same strength and therefore the continents and ocean basins display a similar tectonic fabric of stable areas laced with mobile belts. Although he outlined in detail how the distribution of mobile belts has changed during geologic time, he believed that the gross arrangement of continents and ocean basins has been permanent. He concluded that, despite the alternating expansions and contractions of the asthenosphere, the planet has experienced a net loss of heat since the early Precambrian; hence, it has contracted and assumed a tetrahedral shape which governs the distribution of the continents and oceans.

Gondwana reassembled

Unperturbed by the laughter in the north, Alexander du Toit, of Johannesburg, South Africa, became a forceful advocate of continental drift. Du Toit lived in the heart of ancient Gondwana, where the evidence of continental drift is most compelling. In 1937 he published *Our Wandering Continents*, one of the classic books on the subject. It is dedicated "To the memory of ALFRED WEGENER for his distinguished services in connection with the geological interpretation of OUR EARTH."

Why, with all its advantages, has the "new Geology" received so few wholehearted supporters, asked du Toit. The opposition, he concluded, arises from (1) the deeply entrenched conservatism in geological thinking which renders new departures unwelcome and (2) the failure of scientists to conceive of a mechanism competent to displace continents. Du Toit believed the mechanism derives ultimately from the periodic buildup of radiogenic heat, which makes possible both the sliding of sial over the sima and the displacement of sial by subcrustal convection currents.

Du Toit's theory of the earth differed from Wegener's in several important respects. Du Toit pictured the oceans, including most if not all of the Pacific, as floored by sial which may rupture in the brittle zone near the surface but undergoes much stretching, creeping, and flowing at deeper levels. And he replaced Pangaea by two protocontinents, Laurasia in the northern hemisphere and Gondwana in the southern, separated by the Tethys. The rest of the earth, before the close of the Paleozoic, was occupied by "panthalassa," the primeval ocean.

The idea of one northern and one southern continent was not, of course, new. It had been proposed by F. B. Taylor, who left his landmasses nameless, and also by the German geologist Rudolph Staub, who, in 1924, outlined a theory in which two siliceous masses underwent alternate periods of flight from the poles (due to centrifugal force) and drift toward the poles (due to subcrustal streaming, or convection). Borrowing from Suess, Staub named his continents Laurasia (Laurentia + Asia) and Gondwana. Du Toit adopted these names for his own continents and also pursued Staub's idea of the reversible motion of landmasses.

In reconstructing Laurasia and Gondwana (see Figure 44) du Toit left gaps up to 350 kilometers wide between the present coastlines. Most of these gaps were occupied, he believed, by dry land until the Jurassic, but since then much territory has been removed by erosion. Du Toit's Gondwana included the southern continents plus Arabia and India, and it was largely through his efforts that the name Gondwana came to be associated with continental drift. In subsequent years, however, a confusion has arisen in the minds of many scientists who, forgetting du Toit, associate continental drift (Wegener's hypothesis) with Gondwanaland (Suess' name for a vast, nondrifting southern landmass).

Partly at the urging of Reginald A. Daly, who had visited South Africa to investigate the Gondwanaland problem for himself, du Toit spent five months in Brazil, Uruguay, and Argentina on a grant from the Carnegie Institution. In 1927 his observations were published in a paper entitled "A Geological Comparison of South America with South Africa." Although he did not contemplate as close a union of these two continents as Wegener did, he nevertheless found so many matching geological features that Wegener, in the 1929 edition of his book, commented: "du Toit's book made an extraordinary impression on me, since up till then I had hardly dared to expect so close a geological correspondence between the two continents."

In addition to those between Africa and South America, du Toit found geological continuities between other southern continents and reconstructed a Paleozoic feature he called the Samfrau (S-AMerica-aFRica-AUstralia) geosyncline, stretching from Argentina to Australia across the Cape Province of Africa.

Du Toit never pictured geological processes as occurring simultaneously from end to end of such a structure. With respect to the Permian glaciations, for example, he believed that the earliest ice cap was formed in Argentina and that it was followed by a succession of ice caps in South Africa, India, and Australia. He conceded a point raised in 1925 by A. P. Coleman that ice caps must be fed by nearby sources of moisture, and so one huge ice cap could never at any one time have covered all the glaciated areas of Gondwana.

Du Toit outlined separate histories of orogeny and of polar wandering for each of his two continents, but he concluded that they tended to behave as parts of one double landmass moving discontinuously

> with many tremblings along their respective and interfering paths, their erratic progress being punctuated by volcanic eruptions and earthquake shocks. Sheets of sial of colossal extent, mass, and momentum were involved and the mere slowing-up of any large portion of one might, as in an ice-field, have been sufficient to develop pressure-ridges, and the reverse to have produced depressions or even rifts.

Du Toit pictured both landmasses as undergoing counterclockwise rotation and a southerly drift from the Devonian to the late Carboniferous, when ice caps formed on Gondwana while Laurasia lay in the north temperate and tropical zones. However, he believed that from the end of the Paleozoic onward the double landmass was impelled northward and dispersed into fragments that crowded toward the north pole as well as toward the rim of the Pacific basin. He spoke of Laurasia as undergoing internal compression as it moved toward the north pole, where the meridians converge, and of Gondwana as undergoing tension as it moved toward the equator, where the meridians diverge. Unfriendly critics seized upon this language to taunt du Toit with believing that tectonic forces are controlled by meridians, a geographical fiction. He never fully succeeded in dispelling this impression although he explained repeatedly that he had been

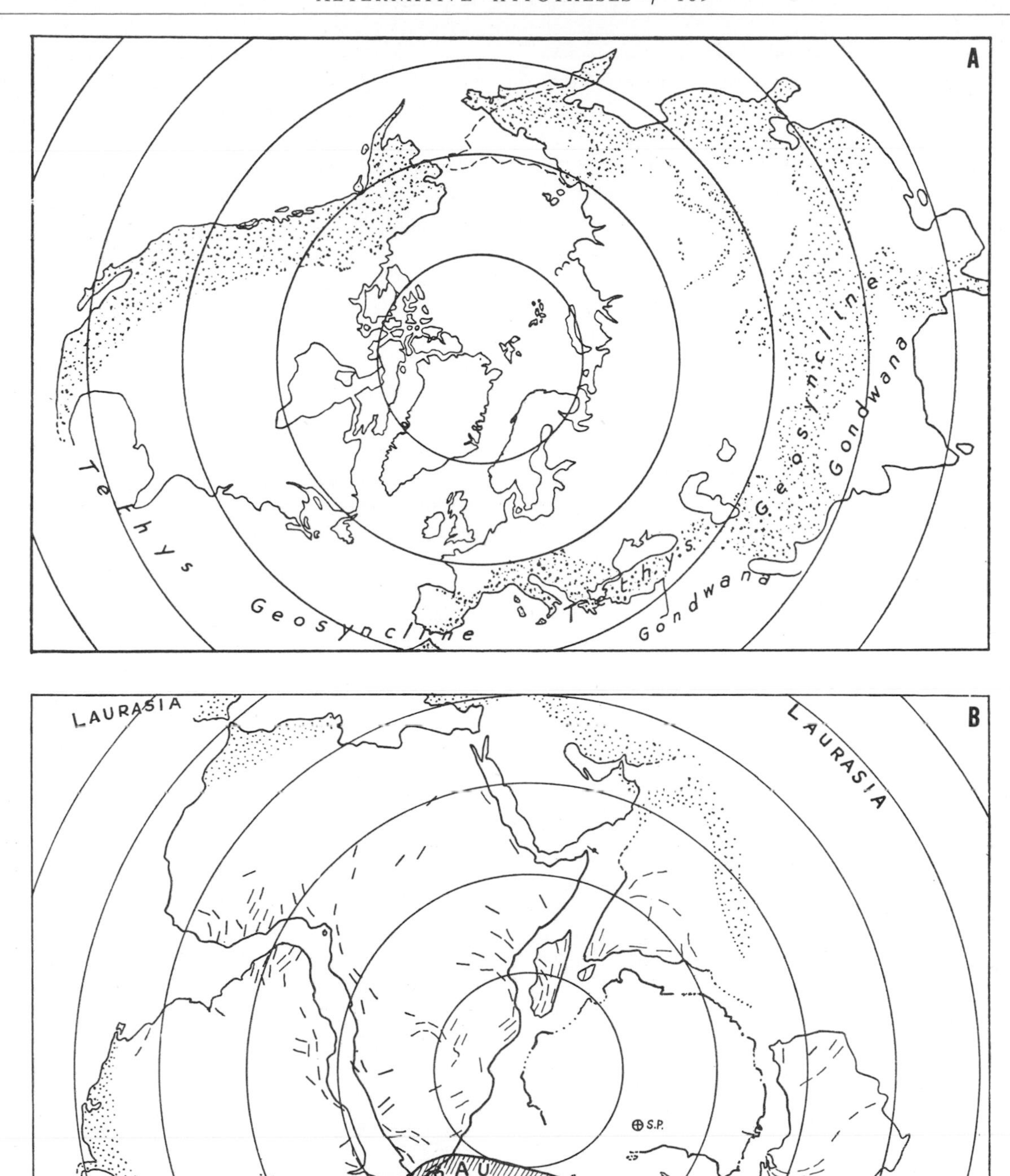

Figure 44. Du Toit's (1937) continental reconstructions. The stippling marks areas subject to compression in the late Cretaceous and Tertiary. **A**: Laurasia in the early Mesozoic. The space between present coastlines is partly land and partly sea. **B**: Gondwana in the Paleozoic. The space between coastlines is mostly land, making Gondwana a more cohesive unit than Laurasia. (Used with permission of Oliver and Boyd, Ltd., Edinburgh.)

referring to the differential stresses on landmasses moving through successive latitudes on a rotating spheroid.

After considering each mechanism previously proposed, du Toit concluded that continental drift and orogeny must be viewed as functions of radioactivity. He supported John Joly's theory that the buildup of radiogenic heat beneath the continental crust periodically weakens the substratum and induces mountain-building. In addition, he believed that as the high-standing continents have always been bordered by "fossae" (foredeeps and geosynclines), the continents slide toward the deeps during periods of subcrustal weakness. This motion of the sial over the tranquil sima promotes folding of the continental margin and the outward migration of the foredeeps. Under other circumstances, portions of the sial are carried, pulled, or stretched by subcrustal convection currents in the sima. Du Toit drew a diagram (Figure 45) of continental stretching and published it alongside Holmes' diagram (Figure 42). Unlike Holmes, du Toit closed his convection cell, thus creating a broad, horizontal tongue of circulating sima, and he placed his geosynclines beyond the continental borders. His diagram accounts for continental accretion as well as folded mountain ranges and interior rift valleys.

In reply to those critics who cited the lack of a mechanism to cause horizontal motion, du Toit pointed to the lack of one to explain vertical motion:

> In pointing out the difficulties introduced by isostasy, Schuchert voices the pathetic appeal, "Geologists must find a way to sink land bridges." Why? Save to extricate orthodoxy from an impasse! For the words "sink land bridges" we might with at least equal fairness substitute "move the continents."

Much as he admired Alfred Wegener, du Toit made two errors in reporting Wegener's ideas. He stated that Wegener had matched the continents along their present coastlines and that he had proposed the tilting of the earth's axis in space as the mechanism for polar wandering. In *Our Wandering Continents*, du Toit wrote:

> Most persons view the continental shelf as an integral part of the continental block, and criticise Wegener for endeavoring to fit together the masses by their present coastlines instead of by the submerged margins of the shelves. It should be noted that in his fourth edition he has recognized the desirability of doing so.

Faced with such doubly compounded nonsense, what may we conclude? That du Toit never read anything of Wegener's except the fourth edition (in which Wegener simply repeated what he had said from the first about the continental shelves)? That "most persons" never read anything at all of Wegener's? That du Toit listened to "most persons"?

We remember that Philip Lake took issue with Wegener as early as 1922 for matching continents at the top instead of at the bottom of the continental shelves. In Wegener's third edition we find Wegener himself criticizing his compatriot H. Wettstein for a matching of continents "the submarine shelves of which . . . he did not consider."

With respect to polar wandering, du Toit stated: "The apparent changes in the positions of the poles at various epochs are interpreted by Wegener as due to actual shifting of the polar axis, the feasibility of which has been stoutly denied by most physicists." Elsewhere he stated: "Throughout this discussion the term 'Polewandering' or 'polar shift' will explicitly refer to the creeping of the thin and distorting crust over its rigid core that has been rotating upon a *fixed axis* and not to a change in the direction of that axis in stellar space."

We have already outlined in detail Wegener's views of polar wandering. Wegener rejected the idea that crustal creep of the type described by du Toit constitutes polar wandering because he doubted that the crust ever rotates about any axis different from the rotation axis. It is true that he did examine the possibility that the earth's axis may tilt beyond its mean angle with the plane of the ecliptic. But the polar wandering mechanism that Wegener favored most strongly is the internal shift of the earth's mass with respect to the rotation axis.

Du Toit was writing his book less than seven years after the death of Wegener, whose reputation by then was clearly in need of defense from his friends as well as from his enemies.

Paleozoic Pangaeas

From Peking, China, another voice was raised in support of polar wandering and continental drift.

Amadeus W. Grabau, an American paleontologist and author of several textbooks on stratigraphy and index fossils, moved to China in the mid-1920s and devoted 26 years to teaching and to field work there and in Mongolia. Earth history, he concluded, has been dominated by two phenomena: (1) pulsation, a rhythmic rise and fall of sea level in each geological period as shown by the record of simultaneous marine transgressions and regressions in the strata of all continents; and (2) polar migration, which he defined as the shifting of the sialic crust with respect to the poles of the rotation axis.

Grabau's book entitled *The Rhythm of the Ages, Earth History in the Light of the Pulsation and Polar Control Theories* was published at Peking in 1940. In it Grabau described the planet as originally having a continuous crust of sial covered by an ocean, "the great primitive shoreless sea of the earth." What caused the lands to emerge from this sea? What is the origin of ocean basins?

Grabau rejected the earlier ideas that density differences in the primitive crust had caused humps and hollows in the surface, or that sialic slabs have foundered into the sima. "Finally," he said, "the theory that the Pacific basin is the scar left when the moon was torn from the earth can now be discarded, though it has recently been revived in Germany." The only viable alternative, he said, was the "Theory of Continental Drift, first proposed by F. B. Taylor in America, and by Alfred Wegener in Austria."*

To Grabau, Pangaea's formation was a more difficult problem than its breakup. How was the primitive sial to be ruptured and compressed from a spherical shell of granite to a hemisphere of crushed and contorted granitic gneisses and schists? He concluded that the crust could not break up and push to one side, exposing the raw sima on the other half of the globe, without the aid of an external force. For this force Grabau tentatively invoked the agency of a stellar body that approached the earth and remained centered over the south pole long enough to draw the light sial into one large southern landmass before the beginning of the Paleozoic. The stellar body then moved away, and throughout the Paleozoic the great

*Wegener's last academic appointment was at the University at Graz, Austria, and this has led to a widespread impression that he was an Austrian.

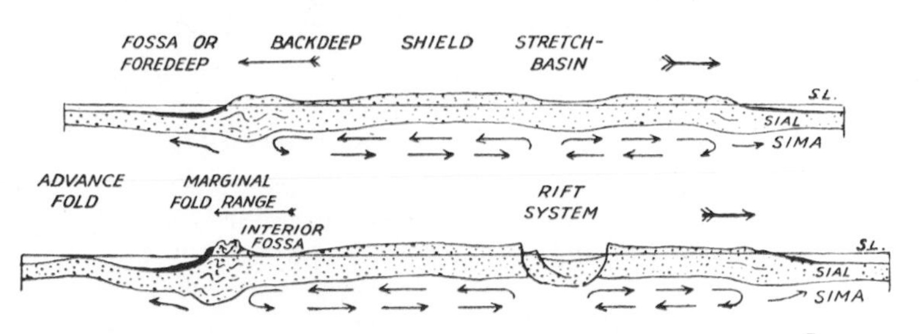

Figure 45. Du Toit's (1937) conception of the stretching and deformation of the sial due to the subcrustal drag of a shallow but fully closed convection cell. Du Toit shows the ocean underlain by a thin layer of sial; black wedges represent orogenic sediments. This diagram illustrates continents growing outward partly by stretching and partly by the assimilation of marginal sediments. (Used with permission of Oliver and Boyd, Ltd., Edinburgh.)

landmass (Pangaea) slowly rotated over the pole and crept northward, meanwhile undergoing alternate episodes of geosynclinal sedimentation and orogenesis.

The oscillations of sea level (which caused simultaneous flooding of all the geosynclines of Pangaea) followed by withdrawal, Grabau ascribed to the cyclic buildup and dissipation of radioactive heat through the ocean floor. John Joly had estimated that 60 million years or more are required for a heating and cooling cycle, and this rate of global pulsation or "heart-beat" seemed to fit well with Grabau's reading of the worldwide stratigraphic record.

Grabau's outline of earth history from the Cambrian to the Recent is splendidly illustrated in a series of twelve fold-out maps and eleven single-page maps with the land areas printed in yellow ochre, the geosynclines in light blue-gray, and the ocean in vivid cerulean. Two of these maps are reproduced here (in wholly inadequate halftones) as Figure 46. His maps are unusual in that they trace the history of Pangaea many millions of years further than was attempted by Wegener or du Toit. According to Grabau, Tertiary and Recent history was marked by the breakup of Pangaea followed by continental drift and mountain-building in the familiar pattern.

The evidence for oscillations of sea level had puzzled many geologists and led to various theories of rhythmic changes in the figure of the earth and of radioactive heating and cooling in the asthenosphere

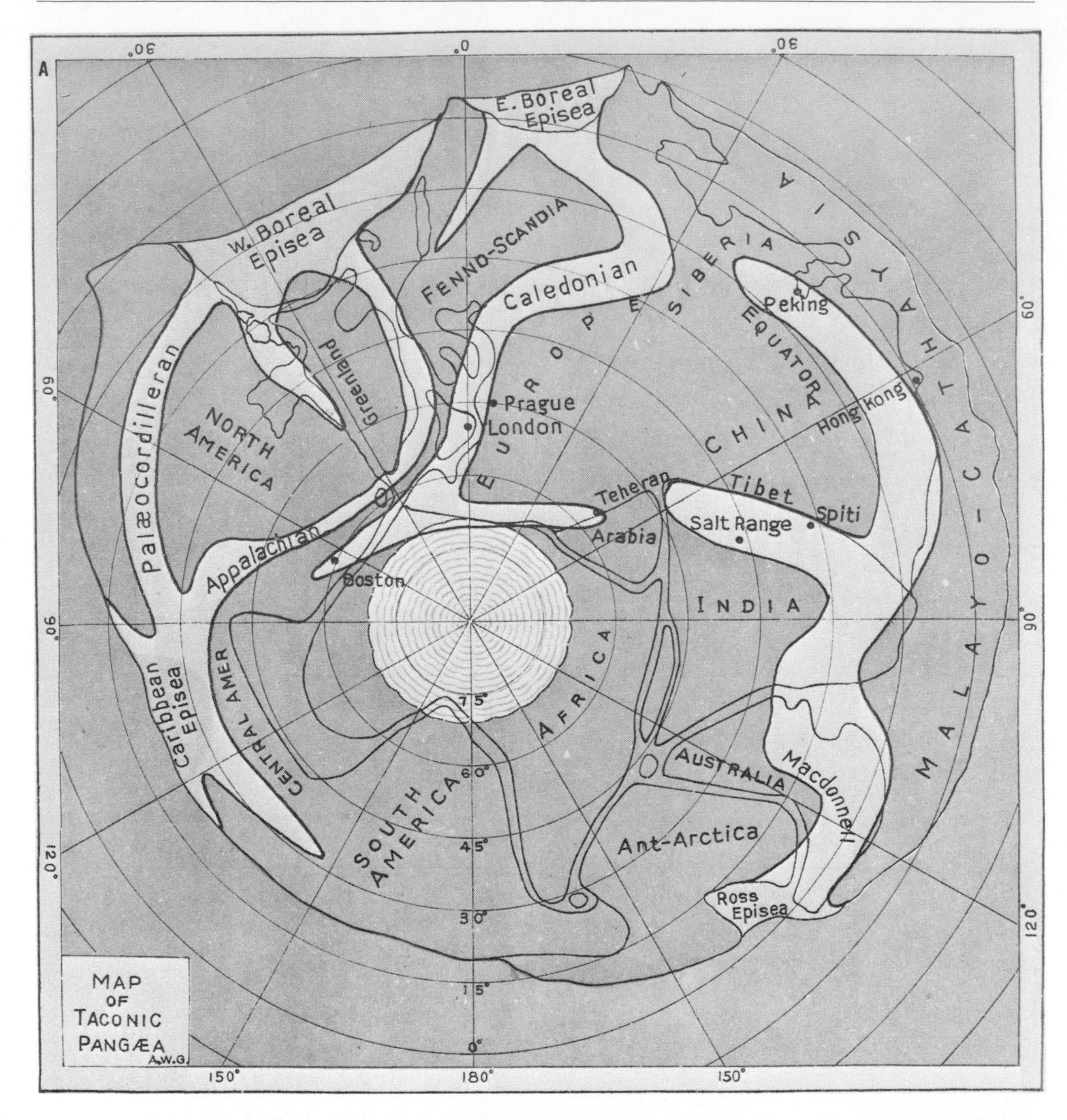

Figure 46. Two of Grabau's Paleozoic Pangaeas. **A**: In Taconic time, which Grabau equated with earliest Cambrian, when all of the sial was massed toward the region of the south pole and an ice cap was centered on the Sahara region. The subparallel Appalachian and Caledonian geosynclines extend across New England, to the British Isles and Fennoscandia. **B**: In Permian time, after Pangaea has rotated and drifted northward. Du Toit's Samfrau geosyncline has been extended to a system of troughs stretching from California across South America and Antarctica to Hong Kong and into Eurasia. (From Grabau, 1940, published at Peking by Henri Vetch.)

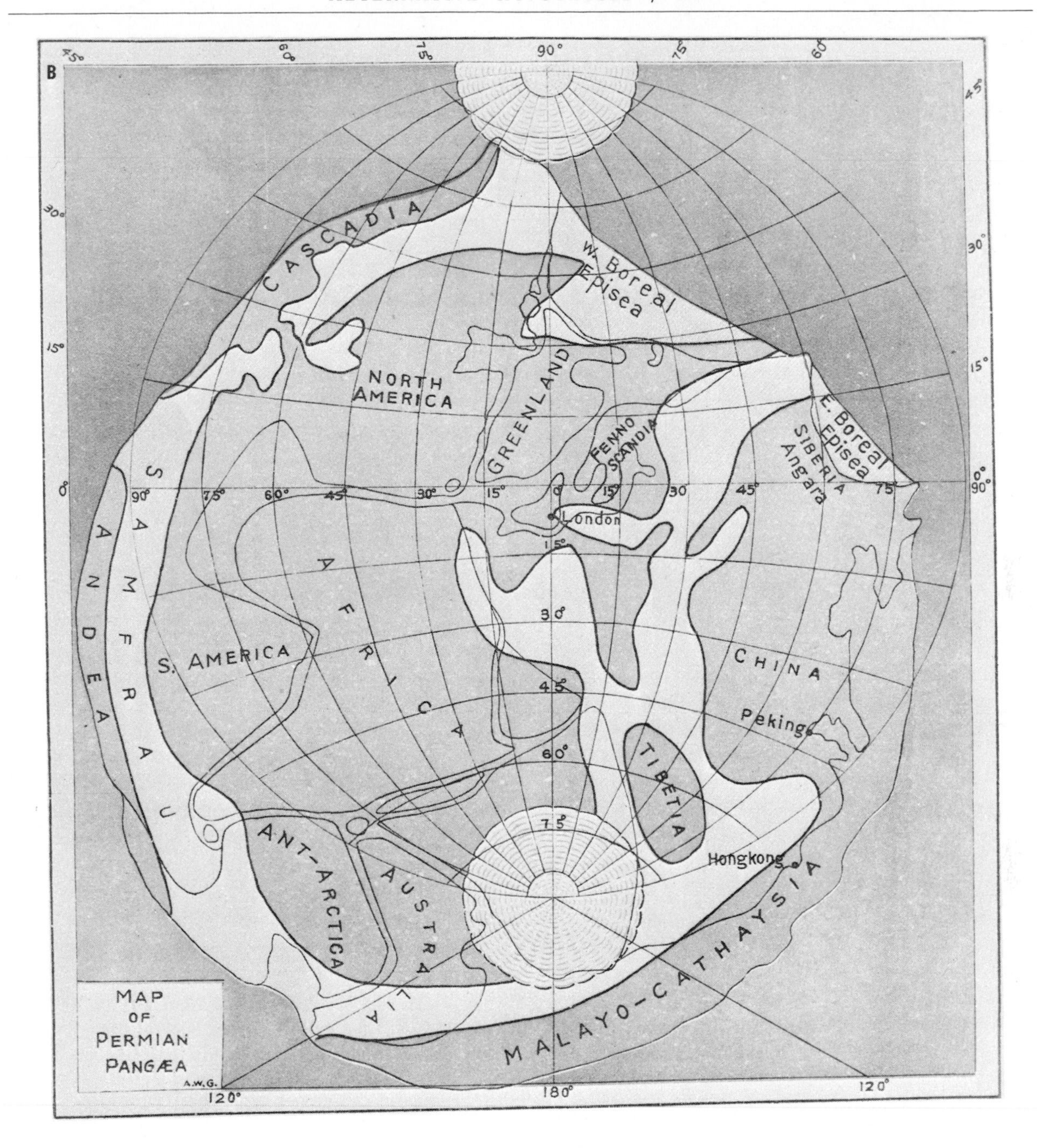
B
CASCADIA
W. Boreal Episea
NORTH AMERICA
GREENLAND
FENNO SCANDIA
E. Boreal Episea
SIBERIA
Angara
London
SAMFRAU
ANDEA
AFRICA
S. AMERICA
CHINA
Peking
TIBETIA
Hongkong
ANT-ARCTICA
AUSTRALIA
MALAYO-CATHAYSIA
MAP OF PERMIAN PANGÆA
A.W.G.

or in the oceanic crust. Some investigators agreed that the oscillations were synchronous throughout the world; others thought that the sea waters have alternately risen in high latitudes at the same time they were falling in the tropics, and vice versa. Still others (a small minority) denied the apparent periodicity of geological revolutions and sea-level oscillations. They maintained that mountain-building activity and marine invasions have been going on somewhere in the world throughout geologic time.

Grabau's story of Paleozoic Pangaea was poorly received. Elaborations on the theory of continental drift were not welcome in 1940, and he seriously weakened his argument by his appeal to a stellar force in moving the sial. He was by no means alone in contemplating cosmic forces as causes of tectonic deformations. Ten years earlier Erich Haarmann, in Germany, had described extraterrestrial forces as the basis of his so-called "Oszillationstheorie" of earth history. To most geologists, however, cosmic forces appeared too fortuitous and catastrophic to be acceptable as a fundamental cause of continental evolution.

During a long career, Grabau published numerous volumes of geology and stratigraphy that proved to be of lasting value. In studying the earliest Paleozoic geosynclines of North America and Europe, for example, Grabau had described identical trilobite faunas in slates of the Hudson River Valley, western Newfoundland, western Greenland, and the northern tip of Scotland. A separate and distinct trilobite fauna occurs within a short distance to the south and east of these localities–in Weymouth, Massachusetts, eastern Newfoundland, Shropshire, England, and southern Norway. This striking evidence for two separate, more or less parallel Paleozoic geosynclines extending from Europe to North America (as shown in Figure 46) proved invaluable to later advocates of continental drift.

The verdict of paleontologists

The paleontological and biological evidence respecting continental drift was subjected to a searching analysis in 1943 by George Gaylord Simpson, a vertebrate paleontologist. In an article entitled "Mammals and the Nature of Continents" published in the *American Journal of Science*, Simpson recalled a statement by W. A. J. M. van Waterschoot van der Gracht: "There are few subjects where there exists greater diversity of opinions regarding practically everything than in paleontology." Simpson pointed out that

> The remark was made in the course of a symposium on continental drift that exemplified greater diversity of opinions than paleontology can offer. Doctor van der Gracht's dictum becomes amusing when it is noticed that on his particular subject the verdict of paleontologists is practically unanimous: almost all agree in opposing his views, which were essentially those of Wegener.

Du Toit, Simpson recalled further, had canvassed for opinions at length and had found no paleontologists who were active protagonists of continental drift and only one paleontologist who was sympathetic to it. Given this overwhelming agreement among specialists, Simpson said, it would seem reasonable to cease discussion of the paleontological arguments were it not for the fact that adherents of continental drift all claim that the paleontological data support them. And he added: "It must be almost unique in scientific history for a group of students admittedly without special competence in a given field thus to reject the all but unanimous verdict of those who do have such competence."*

Simpson reviewed the problems posed by the distribution of mammals, both fossil and living, and concluded that they could be solved most logically and completely by assuming that continents have remained essentially stable throughout the time involved in mammalian history–which includes the entire Cenozoic and part of the Mesozoic. He said that mammal migrations have taken place in intermittent waves, with some tendency toward cycles, and this indicates a repeated opening and closing of the migration routes. The evidence is compatible with the formation of isthmian links of the type described by Willis but incompatible with continental drift, which assumes one definitive break followed by ever widening gaps between some blocks while other blocks approach each other and collide.

With respect to existing faunal resemblances on separate continents, Simpson found that none is close

*Simpson wrote his article before Grabau's book had reached Europe.

enough to demand former contiguity. If anything, he said, the opposite is true; the resemblances are far too few to admit of former continental union: two parts of one rifted continental block might be expected, for example, to share faunas that are as similar as those found today in Ohio and in Nebraska, areas that are 800 kilometers apart in the same faunal zone and have complete continental union with only slight climatic and geographic barriers. Simpson reported that among the Recent mammals occurring in Ohio 100 percent of the families, 82 percent of the genera, and 65 percent of the species also occur in Nebraska. "*But no fossil land faunas resembling each other to a degree at all comparable with this have ever been found on continents now separated.*" [Italics Simpson's.]

Failing this degree of correspondence, Simpson compared the Recent mammals of Florida and of New Mexico (1,600 kilometers apart) that occupy different zones of the same faunal region and have complete continental union with distinct climatic differences and minor geographic barriers. Of Florida mammals, 92 percent of the families, 67 percent of the genera, and 18 percent of the species also occur in New Mexico. Interestingly, an almost equal similarity is found between the mammals of France and of northern China—countries which are 8,000 kilometers apart, are in different zones of the same faunal region, and have complete continental union with minor climatic differences but marked geographic barriers. Of French mammals, 89 percent of the families, 64 percent of the genera, and 26 percent of the species occur in China. A similarity of this magnitude should, Simpson believed, be about the least one could expect between continents formerly united according to the drift theories. However: "*Resemblances of this degree are altogether exceptional among fossil vertebrate faunas of continents now distinct.*" [Italics Simpson's.] The one exception he knew of involved the Eocene mammals of Europe and North America where 95 percent of the families, 54 percent of the genera, and zero percent of the species occur in common. This distribution could best be explained, he said, by a migration route across the North Atlantic through Greenland.

Simpson's own explanation for mammalian distributions was straightforward. He required intermittent periods of flooding and of dry land between several pairs of continents: Europe-Asia, where the connection is now at a maximum; Eurasia-Africa, where the connection formerly was greater; North America-South America, where the connection, formerly broken, is now below a maximum; and Asia-North America, where the connection is now broken at the Bering Strait. In addition to these connections which are universally admitted, he tentatively favored a fluctuating connection between North America, Greenland, and Europe. For all of the other hypothetical land bridges across the Atlantic, Indian, and Pacific oceans he could see no compelling evidence. They were as unnecessary to the mammals as were drifting continents.

Simpson found the literature full of mistaken identities, unsubstantiated claims, internal inconsistencies, and worthless conclusions. Wegener, for example, had cited percentage figures, taken from Theodore Arldt, on "identische" mammals and reptiles on the two sides of the Atlantic. Arldt himself (whose figures, said Simpson, were unreliable to begin with) had been talking not about species but about families or subfamilies. "Such looseness of thought or method amounts to egregious misrepresentation and it abounds in the literature of this perplexing topic," Simpson said.

As an object lesson in the workings of the scientific method gone awry, Simpson traced the history of a transatlantic feature called the hipparion bridge running from Florida and the Antilles across to North Africa and Spain. According to Simpson, the hipparion bridge was invented in January 1919 by the French geologist Leoncé Joleaud for the convenience of a fossil genus related to the horses. Joleaud's evidence (not collected in the field but cited from the literature) was that the teeth of a European species had been found in South Carolina and Florida. The identification was tentative in the first place, was incorrectly cited, and eventually was proved to be erroneous. Meanwhile, having erected the bridge for the passage of *Hipparion* in the late Miocene, Joleaud broadened it to include all the territory from Maryland to Brazil and across to northern France and Morocco, lengthened its effective lifetime throughout most of the Tertiary, and added eight other migrants as passengers.

By 1924 Joleaud had come to favor the idea of continental drift except for its inability to account for intermittent faunal exchanges. Joleaud therefore proposed that drifting continents have moved back and forth with an accordion motion ("un mouvement en accordéon"). This, said Simpson, "seems to me the climax of all drift theories."

Joleaud may have been the first to propose a repeated opening and closing of the Atlantic by means of reversible continental drift. He was not to be the last, however, despite Simpson's obvious exasperation. A more recent review of the fossil evidence–some of which was first described by Grabau–has led to the supposition that the North Atlantic was open at the beginning of the Paleozoic, that it closed before the end of the Devonian, and that it opened again in the Mesozoic. Today, however, no one supports Joleaud's idea that a series of reverses took place as late as the Tertiary.

Joleaud's theory of the hipparion bridge, if not that of his accordion-like continents, achieved high respectability when it was accepted by John W. Gregory in a presidential address on "The Geological History of the Atlantic Ocean" that he presented to the Geological Society of London in 1929. Gregory, we remember, had favored sunken land bridges over drifting continents at the American symposium in 1926. On Gregory's authority the hipparion bridge became one of the few land bridges to be counted by du Toit as absolutely necessary to explain disjunctive faunal distributions. Therefore, by the time Simpson wrote his article the hipparion bridge was firmly established in the literature although it had never existed in the Atlantic. "The plain fact is that this bridge was a fantastic blunder from the moment of its birth in Joleaud's mind," wrote Simpson.

With respect to the "Gondwana" continents, Simpson commented that the claimed resemblances of floras and faunas are "sometimes erroneous, some of them do not necessitate land connections, and many have been grossly exaggerated." He found, however, that genuine cases do exist of terrestrial life forms with disjunctive distributions. To explain most such distributions Simpson developed a theory based in part on the 19th-century insights of Alfred Russel Wallace and in part on the ideas of W. D. Matthew published in 1915 in a paper entitled "Climate and Evolution." The theory held that common ancestral forms developed in the Holarctic faunal region of Eurasia and North America where a broad uniformity existed by the early Tertiary. The temporary rise of land links led to migrations of these ancestors to the southern continents, and the subsequent closing of the links encouraged the development in each isolated landmass of peculiar faunal assemblages sharing certain haunting similarities.

In the absence of isthmian links, land animals may accidentally cross wide marine barriers by extraordinary feats of floating, swimming, or rafting in tangled masses of debris. The low probability that a species will survive the crossing and reproduce in the new habitat led Simpson to describe such paths of travel as sweepstakes routes where success is rare and governed mainly by chance. He believed that the unique, unbalanced faunas of Madagascar, the Galápagos Islands, the Hawaiian Islands, and other such isolated landmasses developed by sweepstakes routes (see Figure 47).

Madagascar, according to both Wegener and du Toit, was wedged between India and Africa until the onset of continental drift. Wegener depicted Madagascar as separating from Africa as early as the Jurassic, but he kept it within 48 kilometers of the mainland until the late Tertiary in order to allow a bush pig and a hippopotamus to swim the channel. Du Toit, on the other hand, split Madagascar from Africa in the early Cretaceous but, mainly on dinosaurian evidence, left it attached to India until late Cretaceous. He was forced, however, to briefly reunite Africa with Madagascar in the Oligocene to explain the alleged affinities among lemurs, monkeys, and hippopotamuses. The mechanics of this reverse drift were not explained and the faunal evidence for it is wanting. Simpson pointed out that there are, in fact, no monkeys on Madagascar and the gross imbalance in the island's fauna reflects the chance introduction of species.

A close examination of the Australian fauna convinced Simpson that Australia has maintained its present relationships with other continents at least since the Cretaceous, that the marsupials entered by island-hopping in the early Tertiary, and that the rodents followed the same route at a later period. The evidence for similar marsupials in South America is so slight as to rule out any direct connections between that continent and Australia and to suggest the introduction of ancestral lines from northern continents.

Simpson was particularly critical of du Toit, who had found it necessary to add land bridges and reverse drift to his system in order to explain intermittent migrations. In fact, du Toit postulated two bridges

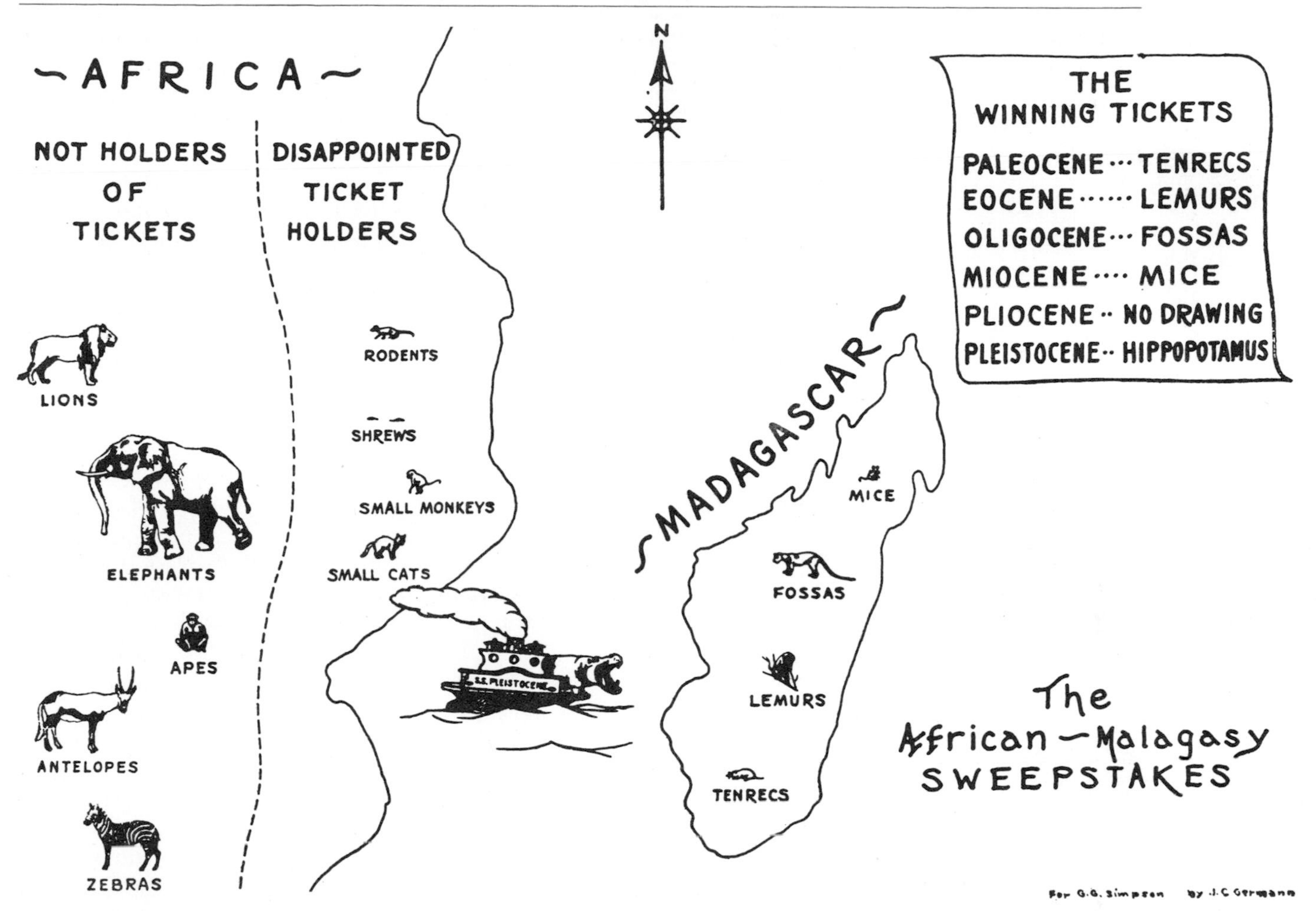

Figure 47. George Gaylord Simpson's illustration of a sweepstakes route. (From *Journal of the Washington Academy of Sciences*, volume 30, 1940; used with permission of the author and the Washington Academy of Sciences.)

across the mid-Atlantic during the Tertiary. One, the manatee bridge, connected Brazil with North Africa; the other, the hipparion bridge, stretched between the West Indies and Spain. Both bridges postdated the fragmentation and drifting apart of Gondwana and Laurasia. Simpson posed the very penetrating question: if so ardent a continental drifter as du Toit requires hypothetical land bridges to explain the distribution of life forms, what is the biological evidence that uniquely favors continental drift? Upon rechecking the arguments of du Toit on this subject, we find in *Our Wandering Continents* the following statement opening a section headed "Life in Relation to Past Land Connections":

> The undoubted likenesses between the fossil or living forms of certain ocean-parted lands has, and not unreasonably, been ascribed by some workers to *parallel development* or *convergent evolution*, but, the moment that other lines of evidence point definitely towards Drift, such objections fall away and the observed biological resemblances constitute instead an important argument in favor of that hypothesis.

The italics are du Toit's, but we would italicize "*the moment that other lines of evidence point definitely towards Drift*," for here du Toit is admitting that biological organisms do not of themselves constitute sufficient evidence for drift. Simpson concluded that no paleontological evidence demanded—and the mammalian evidence opposed—an explanation in terms of continental drift.

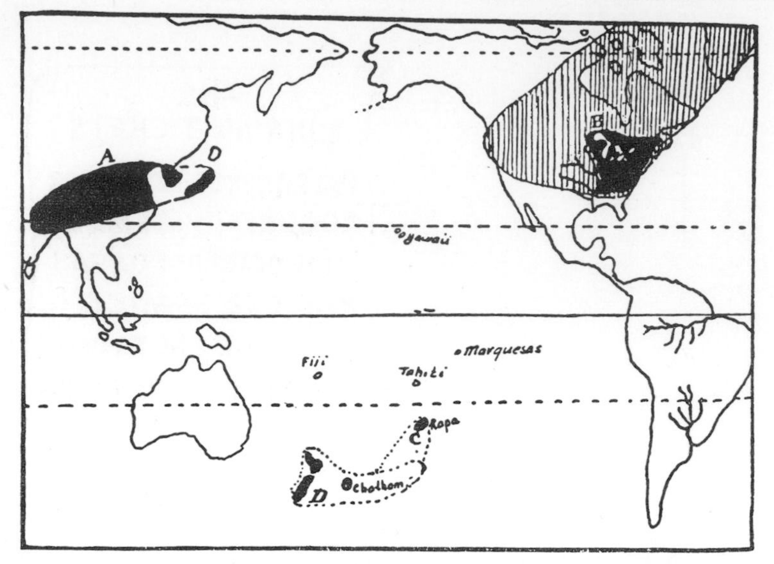

Figure 48. The distribution of primitive dogwoods, genus *Cornus*. Black areas indicate the present distribution; vertical stripes mark the known distribution in the Cretaceous. (Map by F. B. H. Brown; used with permission of The Bernice P. Bishop Museum, Honolulu.)

The building of a body of dependable, significant data

Du Toit was not slow in replying to Simpson. His first rejoinder, "Tertiary Mammals and Continental Drift," was published in the March 1944 issue of the *American Journal of Science*. In it he charged that Simpson was unwarranted in limiting his discussion to mammals because they arrived on the scene too late in geological history to be of critical importance in the argument. Meanwhile, he said, Simpson, throughout his paper, had entirely ignored the vital stratigraphic and tectonic evidence. Simpson's percentage figures for faunal resemblances might not apply to fossil vertebrates because of the chance nature of discovery. And du Toit regarded the hypothesis of continental drift "as essentially established by the Paleozoic and Mesozoic evidence." Although he claimed to have been misquoted in some instances, du Toit readily admitted that he had erred in accepting the hipparion bridge, now that the history of that feature had been sorted out by Simpson.

In the April issue of the same journal, Chester Longwell replied to du Toit on behalf of George Gaylord Simpson—who was then in the armed services and would, Longwell supposed, have no opportunity to give further attention to continental drift until the end of World War II.

Simpson was fully justified in limiting his discussion to mammals, wrote Longwell. In so doing he had two clear objectives: to correct a number of gross errors that had appeared in published statements regarding mammalian distribution and to use the established facts of mammalian distribution as a test in comparing the merits of three hypotheses: sunken continents, permanent continents and ocean basins, and continental drift. Longwell added that he hoped that specialists in other fields would emulate the example set by Simpson—the result would be the elimination of errors and other weaknesses from the literature on drift and the building of a body of dependable, significant data.

Longwell himself discussed several arguments relating to continental drift. He outlined the distribution of primitive dogwoods (genus *Cornus*) which occur in eastern North America, Japan, China, Bengal, and some islands of the South Pacific but not in Australia or South America. A glance at the illustration he used (Figure 48) reveals an enigma more perplexing by far than that of the *Glossopteris* flora. The dogwoods are dispersed over wider oceans and more formidable climatic barriers, yet within any one region they seem strangely confined. Their distribution cannot easily be explained by any appeal to continental drift—least of all to Wegener's pattern with its shrinking Pacific Ocean. The dogwoods had previously been "explained" by seeds floating in currents of the Mesozoic oceans or by the reconstruction of a sunken Pacific continent. Longwell offered no explanation. He simply made the point that disjunctive floral distributions are not reliable as indicators of continental drift.

Longwell also reviewed the geological evidence as presented by both Wegener and du Toit. He found that the dating and matching of lithologic units were far too imprecise. Furthermore, the geodetic evidence for the westward drift of Greenland and North America was now contradictory. While one report showed a widening of the distance between Paris and Washington of 0.32 meters per year from 1913 to 1927, another showed a decrease of 0.7 meters per year for the same period between the Greenwich Observatory and Washington. [Nothing was reported about any change between Paris and Greenwich.] Longwell proposed to withhold judgment of geodetic evidence until astronomers and geodesists could agree on the results.

Finally, Longwell commented on Wegener's celebrated matching of the Pleistocene glacial moraines from New England to Old England and across to Belgium. According to Wegener, this part of the old and new worlds was supposed to have remained in contact until after retreat of the ice caps which began about 10,000 years ago. Yet, on the far side of North America the Rocky Mountains were folded 70 million years ago, presumably as the continent in its westward course of drifting met the cold Pacific sima. Since Wegener was a meteorologist, said Longwell, surely he must have realized that the isothermal lines of today indicate that cold temperatures penetrate farther south in North America than they do in Europe. The glacial moraines simply indicate that this was also true in the Pleistocene.

Longwell, always the skeptic but always calling himself "not unfriendly" to the idea of drifting continents, wanted to be shown, and the proof had to be positive.

It is interesting that almost no critic from whom we have heard so far regarded Alfred Wegener as a colleague of his own discipline. Köppen, the meteorologist, and Schuchert, the paleontologist, called him a geophysicist; Jeffreys, the geophysicist, and Longwell and Daly, the geologists, called him a meteorologist; and Termier, director of the Geological Survey of France, implied that he was a poet. W. B. Harland, in 1969, reviewing four books on continental drift, called Wegener a total scientist. Harland was close to the mark.

We have not, of course, reviewed the opinions of all proponents of continental drift, nor of the much larger roster of its opponents. During the interchange between Simpson, du Toit, and Longwell, Bailey Willis entered the fray with his article "Continental Drift: Ein Marchen," from which we quoted in our introduction. Willis felt that the time had come to call a halt to the use of scientific time, talent, and periodical space for further consideration of this fairy tale.

Willis was certainly wrong in his judgment, yet little or no new evidence for continental drift was presented between about 1935 and 1955. The debate languished, with all but a tiny minority of workers believing the issue was closed for good and all. Continental drift as an idea declined into a state of limbo, not a little disreputable, and perilously close to the status of science fiction.

Before the subject was raised again and the wall of silence breached, many new discoveries were made that altered scientific conceptions of the earth. In order to view the reopening of the continental drift debate in its proper perspective, we should outline the advances made between 1945 and 1965 in radiometric geochronology, in exploration of the ocean floor, and, above all, in the new science of paleomagnetism.

Renaissance in Earth Science

Geochronology

Experiments in radiometric dating began soon after the turn of the 20th century with Bertram Boltwood's attempts to date uranium-bearing minerals. Decades of work were required before the properties of the various radioactive isotopes and their daughter products were adequately understood, their decay rates accurately determined, and the problems of contamination solved. The dating methods were brought to a high degree of sophistication after World War II when numbers of brilliant young students began applying the newly won knowledge of nuclear physics and chemistry to geologic problems. Within a few years the young scientists had transformed radiometric geochronology into an essentially new branch of earth science. "There are no old men practicing the art," commented Gerald J. Wasserburg, one of the preeminent practitioners, in 1966.

The radiometric method

Stated briefly, the radiometric method depends upon the spontaneous decay, in a closed system, of a parent isotope to one or more stable daughter isotopes. (Two isotopes of the same element contain equal numbers of protons but different numbers of neutrons; therefore, they differ in mass.) Every radioactive isotope has a unique decay scheme that operates at a rate independent of temperature, pressure, or chemical bonding to produce radiation, heat, and one or more daughter products. The decay rates of different isotopes vary tremendously—from a fraction of one second to tens of billions of years. Radioactive decay is expressed in terms of half-lives, or the time required for the atoms of a given radioactive species to decay to half of their original number. For example, it takes 1,310,000,000 years for one-half of the potassium-40 (^{40}K) in existence to decay spontaneously to its daughter products; another 1,310,000,000 years for half of the remaining ^{40}K to decay; another 1,310,000,000 years for half of *that* remainder, and so on. We seem to be involved with Zeno's paradox, which says we can never come to the end because we are always halving what remains. And mathematically, the process of radioactive decay is described as infinite. To all intents and purposes, however, radioactive isotopes do decay completely and so vanish from existence. This has been observed with respect to short-lived isotopes created in cyclotrons, and it clearly has happened to some of the isotopes that were created in the original nucleosynthetic event that produced the elements of the solar system. We know this because, for one example, iodine-129, an isotope with a half-life of only 17 million years, long ago decayed completely. In so doing, however, it left behind traces of a stable daughter product, xenon-129, which is found in meteorites. The presence of xenon-129 indicates that these meteorites must have been cold, crystalline bodies within a span of about 200 million years (approximately ten half-lives of iodine-129) after the time of nucleosynthesis.

Of all radioactive isotopes, those of potassium, rubidium, uranium, and thorium have proved the most useful in dating geological events. All four are sufficiently widespread in crustal rocks and have half-lives that are long enough to help unravel the igneous and metamorphic record of the Precambrian.

In inorganic systems, radiometric measurements date two types of events: the cooling of rocks or minerals to temperatures low enough to trap and retain radiogenic gas isotopes produced within their lattices; and the crystallization or "differentiation" of a molten magma or aqueous solution into an array of rocks or minerals, each of which inherits its own unique portion of the total supply of a given radioactive isotope.

The first type of age measurement, that of the so-called gas-retention age, records the last time when a rock or mineral containing uranium or potassium cooled to a low temperature. ^{238}U and ^{235}U produce a long series of unstable daughters leading finally to ^{206}Pb and ^{207}Pb respectively. As they decay they also produce helium-4 (^{4}He),* a gas isotope that easily diffuses from the rock as long as the temperature remains above about 100° C. At lower temperatures the helium is trapped in the crystal lattices, where it accumulates as the uranium isotopes continue to decay. As the decay rate is known, the elapsed time since cooling can be calculated from measurements of the uranium and helium contents. The same principle applies to the potassium-argon system, as ^{40}K decays to two stable products, ^{40}Ca and ^{40}Ar.

Gas retention ages are very useful in dating the terminal phases of orogenic cycles or the final cooling of lavas. They bear no necessary relationship to the time when a rock originally crystallized, as some rocks are reheated again and again over geologic time, and each time this happens the gases leak out and the atomic clock is reset upon cooling. The clock is not quite infallible, however. As it has long been known that gases may in fact leak from cold minerals, gas retention ages have generally been interpreted as minimum ages. More recently the discovery was made that argon can be retained or even concentrated in some minerals, or fluid inclusions in minerals, during heating events. This results in age measurements that are spuriously old. Whenever possible, therefore, gas retention ages are determined on numerous samples from the same geological environment.

The second type of age measurement, the dating of an epoch of crystallization, is more complicated in concept than that of the gas-retention age. This type of dating is exemplified by the rubidium-strontium (Rb-Sr) method, first investigated about 1948 by Louis H. Ahrens, then at Massachusetts Institute of Technology. This method depends upon the fact that Rb, a minor element in potassium-bearing minerals, has a radioactive isotope, ^{87}Rb, which, with a half-life of about 47 billion years, decays to ^{87}Sr. Admixed with the radiogenic ^{87}Sr is ^{86}Sr, a stable isotope that is neither created nor destroyed by radioactivity. Unlike the gases helium and argon, ^{87}Sr is in little danger of diffusing out of a mineral lattice. All Sr, however, contains some ^{87}Sr in the first place, so the absolute abundance of this isotope gives no immediate indication of the time-lapse since crystallization. In order to determine this time-span, measurements are made of the ^{87}Rb, ^{86}Sr, and ^{87}Sr not in one sample at a time but on groups of samples having a common origin: several minerals from the same rock, or several rocks from the same parent magma.

For these groups of samples the assumption is made that "in the beginning," when cooling and crystallization of a homogeneous parent magma were initiated, each mineral received a different portion of the available rubidium and strontium but every portion of strontium had within it the same identical ratio of ^{87}Sr to ^{86}Sr. Over the millions of years since crystallization, the amount of ^{86}Sr has never changed; the ^{87}Sr, however, has gradually increased—more in some minerals, less in others, according to their varying ratios of $^{87}Rb/^{86}Sr$. Today, therefore, the ratio of radiogenic to stable strontium-87 differs in each mineral. As the rate of decay of ^{87}Rb is known and the amounts of ^{87}Rb and ^{87}Sr can be measured and compared with the unchanging ^{86}Sr, the time that has elapsed since all strontium ratios were identical can be calculated. A diagram of the results (such as the one in Figure 49)

*By recent international agreement the superscripts precede the element in isotopic symbols. However, nobody recites the terms that way orally.

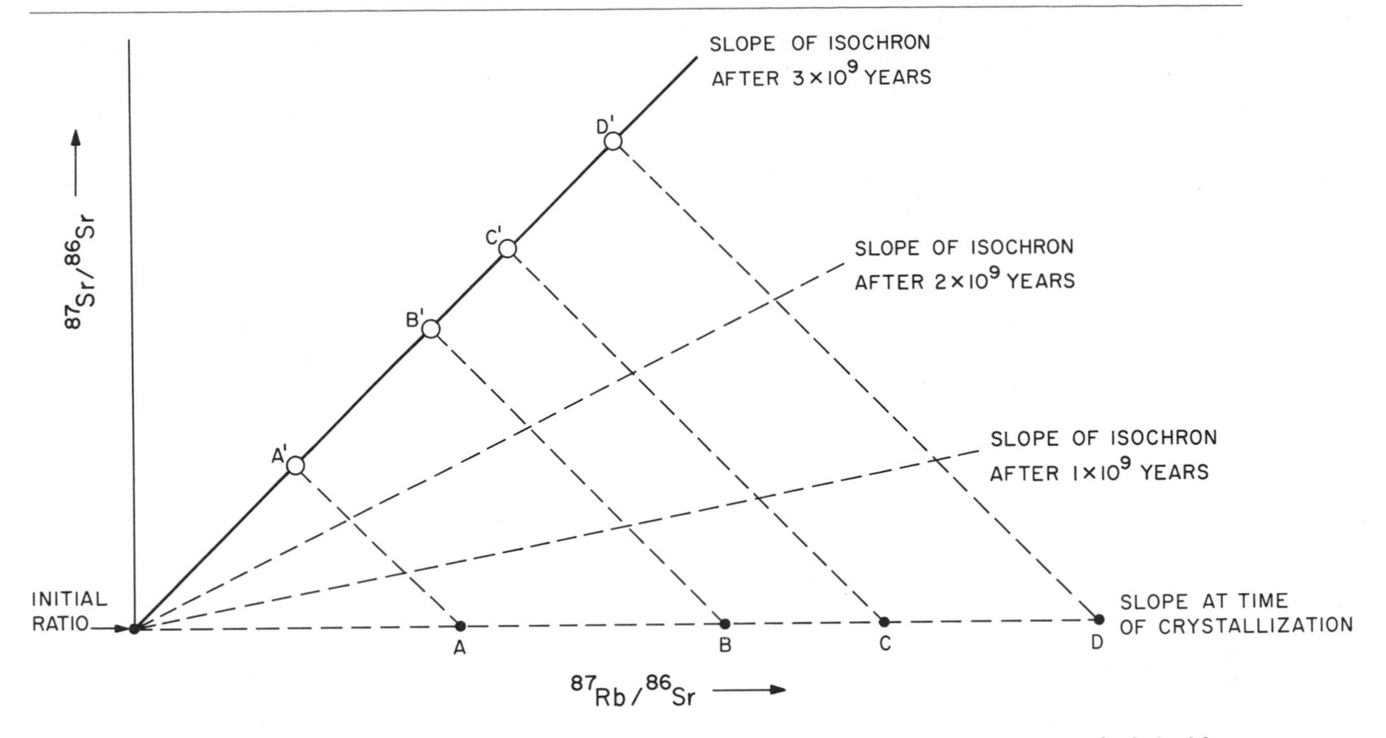

Figure 49. Schematic drawing of a rubidium-strontium (Rb-Sr) isochron. A-D represent four different igneous rocks derived from the same parent magma. At the time of crystallization each rock contains the same initial ratio of ^{87}Sr to ^{86}Sr but different amounts of ^{87}Rb. As time goes by, the ^{87}Rb decays to ^{87}Sr and the slope of the isochron increases as shown. The same principle would apply if A-D were four minerals in the same rock.

illustrates these principles. The $^{87}Sr/^{86}Sr$ ratios of all minerals that began with identical ratios—and so belong to the same system—will plot along a straight line connecting them with their initial strontium isotopic ratio. This line, called an "isochron" or line of "equal time," can be used not only to measure the time since crystallization of the rock but also to distinguish any minerals belonging to another period of crystallization. The ratios of the extraneous minerals will fall off the isochron.

This method of dating applies not only to Rb and Sr but also to the uranium-lead system, where the case is complicated by the presence of two radioactive isotopes of uranium having different half-lives and decay schemes that produce two different isotopes of radiogenic lead. Fortunately, as with ^{86}Sr, there is one stable lead isotope, ^{204}Pb, that is neither created nor destroyed by radioactivity. It therefore acts as the constant reference value for this system. The decay of two uranium isotopes should produce concordant results that would plot on separate isochrons and indicate the same elapsed time since crystallization. Instead of using two isochrons, geochronologists generally simplify the procedure by constructing a single diagram in which the ratios of the radiogenic lead isotopes with respect to ^{204}Pb are plotted against each other, giving a so-called lead-lead age determination.

The earliest experiments in radiometric dating made use of the uranium-lead-helium system. Problems with contamination by atmospheric helium were not fully solved until after 1950. The other two most powerful methods, K/Ar and Rb/Sr, were developed after 1947.

The age of the earth, dated with confidence

Once these principles of measurement were well understood and the experimental techniques were developed to a high degree of precision and reproducibility, it was possible for the first time to confront with confidence the hoary problem of the age of the earth and to date the oldest rocks of the crust. We have seen that the calculations of earlier investigators led to estimates of the earth's age ranging from the 6,000 years of Archbishop Ussher, through the 75,000 years of Buffon and the 20 million years of Lord Kelvin, to the unlimited spans of time favored by T. H. Huxley and Andrew Ramsay. With the development of radiometric techniques some rocks were found to be over a billion years old.* By 1945 the age of the earth was calculated at about two billion years. In 1956, however, Claire Patterson, of the California Institute of Technology, announced an age for the earth of $4.55 \pm 0.07 \times 10^9$ years. In a triumphant mood, Patterson began a classic paper with the statement: "It seems we now should admit that the age of the earth is known as accurately and with about as much confidence as the concentration of aluminum is known in the Westerly, Rhode Island, granite." [The Westerly, Rhode Island, granite had been chosen as a standard rock, and samples of it were shipped to laboratories all over the world for comparing the accuracies of analytical methods. Its aluminum content is known, with great confidence, as $14.12 \pm 0.18\%$ Al_2O_3.]

Patterson and several colleagues had been working for nearly five years on this problem. They had measured the ratios of radiogenic to primeval lead in samples from two sources in the solar system—the earth and meteorites—on the assumption that both were originally derived from the same fundamental solar material in the same event. Even the oldest rocks of the earth's crust are much younger than meteorites. Nevertheless, the lead isotope ratios of samples from both sources plotted along a single isochron, the slope of which indicated that it had taken about 4.55×10^9 years for the radiogenic lead ratios in samples of stone meteorites and in terrestrial rocks (basalts, lead ores, and sea sediments) to build up to measured values relative to the primeval lead measured in a uranium-free iron meteorite. The final result tied the earth and meteorites to the same chronology and helped to demonstrate the unity of the solar system as well as to measure its age.

The figure of 4.5 billion years was soon accepted by the majority of investigators as the answer to the intriguing problem of the age of the earth. Later measurements have raised the best estimate to about 4.7 billion years, a figure just above the upper limit of error listed by Patterson. The finality of this calculated age has, quite understandably, been questioned by both scientists and laymen who feel more than justified in asking if new information may not arise that will once again double or triple the age of the earth. Within the past 15 years some very old ages, ranging from 6 to 10 billion years, have been reported for certain samples of shield rocks and iron meteorites. These dates have not been substantiated, however, and most geochronologists regard them as measurements erroneously interpreted.

The reliability of the age of 4.7 billion years was recently affirmed, in response to close questioning, by George W. Wetherill, of the University of California at Los Angeles, who helped develop the systematics of the K/Ar, Rb/Sr, and U/Pb methods of dating. "Some problems in geology do get solved," he said. "The age of the earth is one of them."

Wetherill pointed out that minor adjustments will be made—there is still some uncertainty, for example, about the precise half-life of ^{87}Rb—but that neither this nor any other correction is likely to change the age by more than a few hundred million years. We may safely assume, therefore, that only with the passage of time (another 4.7 billion years) will the age of the earth double again.

The oldest rocks—the oldest traces of life

What, exactly, does this figure date? It indicates that about 4.7 billion years have elapsed since the earth attained its present mass and began acting as a closed system with respect to uranium and lead, and that the same conditions apply to each of the parent bodies from which meteoritic fragments have been derived. It says nothing about the time when the chemical elements were created in some great nucleosynthetic event, such as the explosion of a supernova in this part of our galaxy, and nothing at all about the formation of the galaxy itself—or of all galaxies, which now appear to be fleeing from each other in an expanding universe.

If we try to reconstruct the past by reversing the estimated rate of this expansion, we find all matter contracting toward a point source it would have

*In American usage, followed in this book, a billion is 1×10^9; in Europe, a billion is 1×10^{12}. To avoid confusion, Harold C. Urey has proposed the term "one aeon" for 1×10^9 years.

occupied about 10^{10} years ago. Is this the date of the absolute "beginning" when all matter was created out of "nothing" in one "big bang"? Or was it only the start of the latest expansion in an eternal series of universal expansions alternating with contractions?

Whatever the time and manner of the ultimate origin of the universe, our own solar system—including the planets, meteorites, and comets, all of which follow orbits that are gravitationally tied to the sun—is about 4.7 billion years old. The process of burning hydrogen to form helium in the sun will probably lead to a phase of greatly expanded luminosity which will render the earth too hot to sustain life, about 4.5 billion years from now. Therefore, barring cosmic accidents, the earth is nearly half way through its period as a possible abode for life.

When did the earth begin to be a suitable abode for life? The earliest traces of living organisms known at present are fossilized primitive algae and bacteria found in the Fig Tree chert of South Africa, a formation that is about 2,400 million years old. Until very recently a diligent search of the continents had uncovered no rocks older than 3,500 million years. Thus it appeared that the oldest surviving patches of crust were a full billion years younger than the planet itself and that all tangible record was lost of that first long interval of earth history. But in November 1972 an age of about 3,700 million years was established for granitic rocks from the Godthaab district of West Greenland. These rocks, collected by V. R. McGregor of the Geological Survey of Greenland, were dated by S. Moorbath and his colleagues at the isotopic laboratory at Oxford University. In recent years many petrologists had come to expect the oldest rocks to occur among the mafic and ultramafic complexes of the shields. The thickness and extent of ancient granites in Greenland and elsewhere remain unknown. However, the fact is now established that granites were among the earliest rocks to form in the earth's crust.

Continental accretion

Continental accretion is a concept dating back to James D. Dana, who also fathered the principle of permanence of continents and ocean basins. There is no gross inconsistency between the two ideas as long as ancient continental nuclei are seen as fixed in position and growing marginally at the expense of the oceans. According to Dana's version of accretion, processes operate, continuously or periodically, to transform coastal sediments into mountain ranges which are eventually eroded down to provinces of shields.

The evidence for such growth was first deciphered from fossiliferous strata, all of which are Paleozoic or younger. When radiometric dating methods began to open up some of the mysteries of Precambrian geology, the results seemed to substantiate the idea of accretion. The most ancient rocks, more than 2.5 billion years old, were found in the heartlands of the continental shields. Surrounding and intruding these rocks are Precambrian formations that are 0.6 to 2.5 billion years old, and these are bordered, in places, by Paleozoic to Recent rocks less than 0.6 billion years old. A broadly concentric age zoning in North America is shown on a map (Figure 50) published in 1963 by A. E. J. Engel. Here the bedrock is six times older in two patches deep in the Precambrian shield than it is at the present margins.

The type of pattern shown in Engel's map implies that the continents have grown from small isolated patches of sial surrounded by sima. If the earth never had a continuous crust of sial, disposal problems would not arise and there would be no need to postulate the ripping away of the moon, the explosive impact of meteorites, or the foundering of enormous slabs to get rid of half of the primeval crust. The problem, indeed, is how to account for the appearance of isolated sialic continents at an early date in earth history.

A catastrophic solution to this problem envisions the continents as fallen asteroids or moons that bombarded the earth in its final stage of accretion. In 1914 the German geologist H. Simroth described Africa with its high plateaus and lobate form as resulting from the fall of the earth's second moon. Simroth was a uniformitarian in that he believed the earth should have had more than a single moon. In 1965, W. L. Donn, B. D. Donn, and Wilbur Valentine, of Lamont Geological Observatory, ascribed all continents to the infall of sialic bodies. These scientists argued that the oldest rocks then known in the Precambrian shields included sediments that were metamorphosed as early as 3.6 billion years ago. The

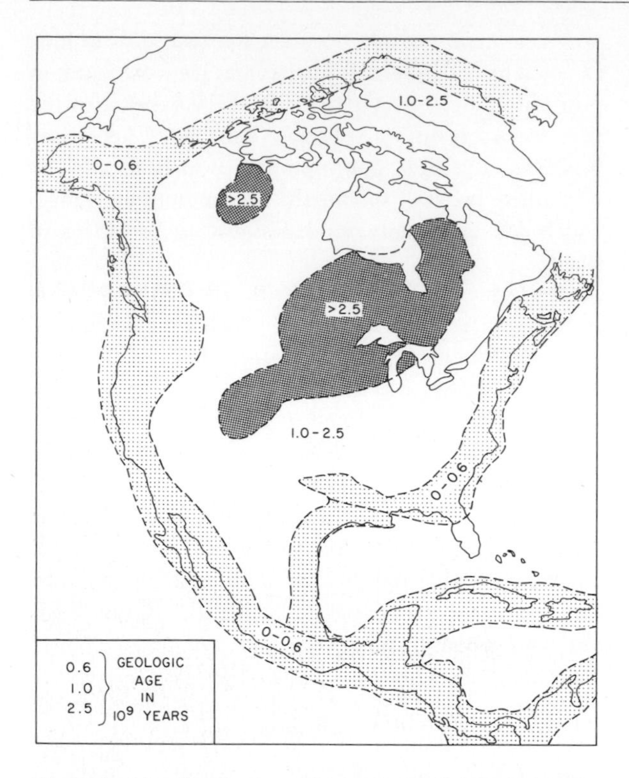

Figure 50. The distribution of bedrock ages in North America according to A. E. J. Engel, 1963. (From *Science*, volume 140, number 3563, page 145. Copyright 1963 by the American Association for the Advancement of Science. Used with permission of author and publisher.)

metamorphism marked the closing stages of a geosynclinal-orogenic cycle which must have begun with the erosion of preexisting highlands several hundred million years earlier. Tracing this cycle backward brings one ever closer to the birth date of the planet itself 4.7 billion years ago and poses the problem of whether enough time elapsed between that event and the appearance of the first landmasses for the sial to have been generated by internal processes. Pointing to the moon, Donn, Donn, and Valentine reminded their readers that a large, light-colored, low-density body occurs nearby, and the presence of one such body establishes, within reason, the possibility that there once may have been others.

The argument for "extraterrestrial" continents had a certain logic as well as strong imaginative appeal; yet, it gained few serious supporters. Today we know from the samples returned by the Apollo missions that the light-colored crust of the moon does not resemble continental sial either in chemical composition or in mineralogy. Both the elemental analysis performed by the Surveyor 7 spacecraft at a site near the crater Tycho and the small, light-colored samples of rock collected by the astronauts and by the unmanned Luna missions of the Soviet Union indicate that the lunar highlands are a feldspar-rich rock unlike any of the widespread varieties of terrestrial rock. Our closest equivalent, mineralogically, is a rock called anorthosite, or gabbroic anorthosite, which occurs in large, coarse-grained bodies within the Precambrian shields. All terrestrial anorthosites are old (1.5 to 3.5 billion years) and enigmatic; their origin has never been adequately understood. No direct connection is suggested between the lunar crust and the anorthosites of the shields, but the newly discovered, and totally unsuspected, presence of rocks of this general category on the moon is one of the most valuable new clues we possess toward unraveling the early history of the earth-moon system.

Now that we know for certain that the lunar rocks do not resemble the bulk composition of our continents, we must turn back to the tortuous problem of deriving the sial from the earth itself. Sial has traditionally been equated with granites (or granodiorites), rocks that contain quartz (SiO_2) and are coarse-grained enough to indicate crystallization at depth rather than volcanic extrusion on the surface (see Figure 51). Granites are among the most common rocks of the Precambrian shields and they also form the cores of folded mountain ranges, where they occur in large bodies called batholiths. As shields and mountain ranges are eroded, the deeper the exposure the more granitic is the bedrock, a general rule which seems to indicate that the origin of continents is simply a larger aspect of the origin of granites.

In the time of James Hutton geologists began to view granites as igneous rocks that rise from chambers of molten magma (or, as some believed, from a worldwide reservoir of granitic melt) and intrude the preexisting country rock. The ultimate source of the magma and the mechanics of intrusion were in dispute, but to followers of Hutton the plutonic nature of the rock was not.

Early in the 19th century, however, geologists began to notice that while some granites have abrupt,

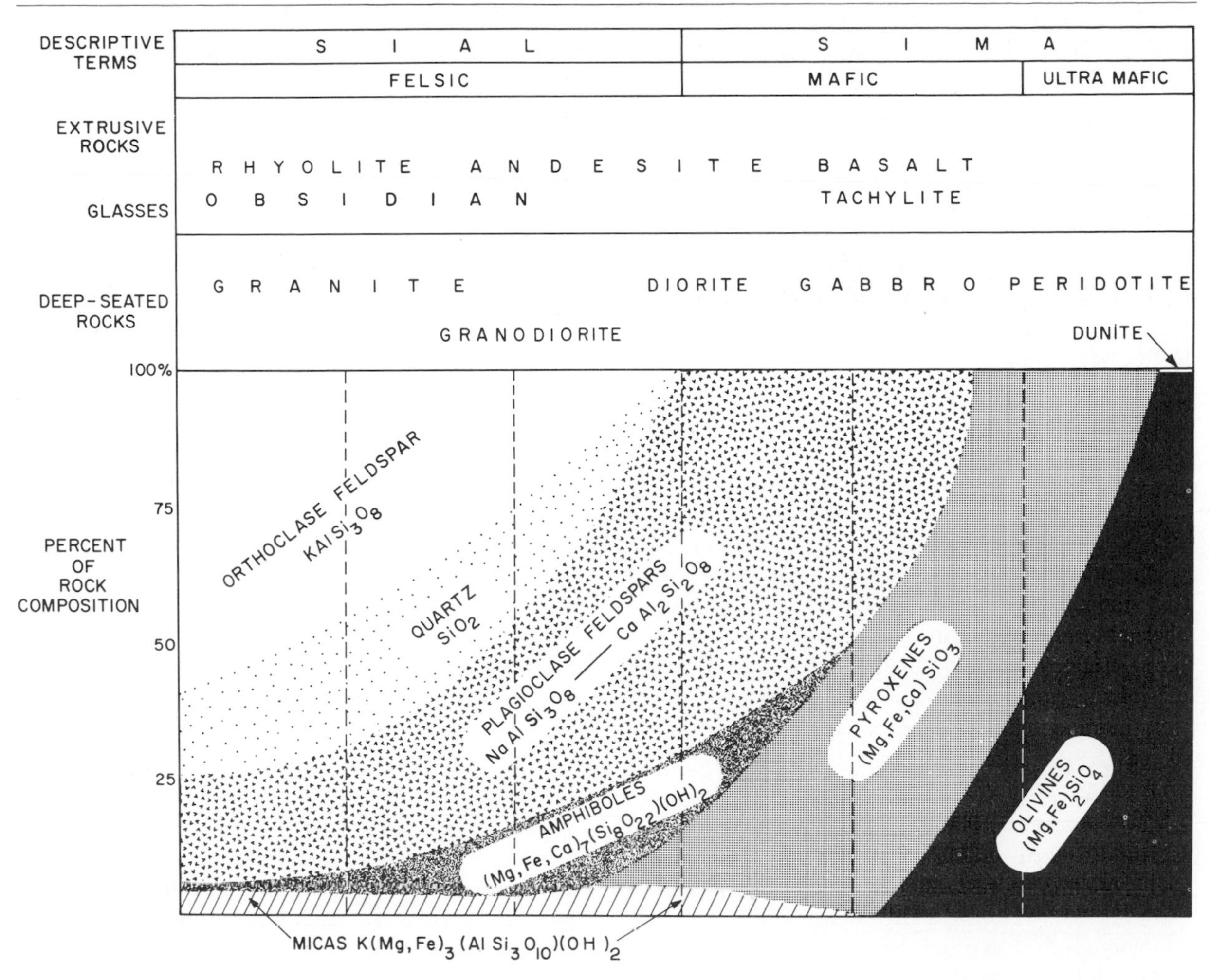

Figure 51. Generalized diagram indicating the mineralogical composition of the major igneous rock types discussed in the text.

cross-cutting contacts with their country rocks, others have no clearly defined contacts at all but grade imperceptibly into metamorphosed sediments. This led the Norwegian geologist B. Keilhau to suggest, in 1836, that siliceous sediments are transformed to igneous-looking bodies by "granitification." Thus began the granite controversy which preoccupied petrologists for the next 120 years. Over this period proponents of the igneous origin, who predominated in Germany and America, became persuaded that granitic magmas are residual solutions left over from the fractional crystallization of basaltic (or gabbroic) parent magmas. In their view, granites are new rocks intruded into the crust and therefore are an important factor in continental accretion. On the other hand, the champions of granitization, who dominated petrological thinking in Scandinavia, Finland, and France, viewed granites as secondary rocks formed from sediments (which had, in turn, been eroded from preexisting highlands) by one or more of several possible processes: melting and mobilization in the depths of geosynclines, the solid state recrystallization of sediments either by permeation with hot, chemically active solutions or by a massive degree of cation diffusion without the aid of aqueous solutions.

The split along national lines was never absolute. Among the geologists who visualized the growth of sialic continents in terms of granitization was Andrew Lawson, of the University of California. In 1932

Lawson proposed that the earliest "crust" was basaltic sima covered by an ocean. Domes of basalt rose above the water and formed the first landmasses. As these were eroded the mafic minerals decomposed while the more resistant, siliceous materials concentrated in adjacent basins and eventually were reworked into mountain ranges with granitic cores. Repetition of this cycle built the continents and made them progressively younger toward their margins. Plausible as this process sounds, it is readily seen to be self-limiting: without the addition of new material, how much can a continent grow by reworking its own mass?

Opposed to the view that granite is a recrystallized sediment was Norman L. Bowen, a skilled field investigator and one of the foremost American experimental petrologists. By 1928 Bowen had proposed the theory that the granites of the crust were formed by the fractional crystallization and upward migration of light elements from the mantle, which he supposed has the chemical composition of olivine basalt. Bowen showed experimentally that, when basaltic melts cool, the earliest minerals to crystallize are olivines rich in magnesium and feldspars rich in calcium, both minerals that are water-free. As cooling progresses, the pyroxenes, amphiboles, and feldspars of typical mafic rocks crystallize out, leaving the remaining magma ever more enriched in silica, soda, potash, and water. The final liquid to crystallize has the composition of a granite. Interestingly, the small traces of uranium and thorium in the original magma tend to remain in the liquid until the end, and these elements, together with potassium, make the continental granites at least six times more radioactive than the oceanic basalts.

According to Bowen, the volume of the residual granitic magma equals about 7 percent of the volume of the original basalt. Some investigators challenged Bowen to show them the whereabouts of the remaining 93 percent of mafic rocks that should surround every granite batholith. In any given mountain range this problem seemed acute. Bowen pointed out, however, that the total volume of continental sial equals less than 0.01 percent of the volume of the mantle. Hence, all of the world's granites form patches of the thinnest scum on the ultramafic layer at depth.

In the 1940s the granite controversy rose to a crescendo. In 1947 at a meeting of the Geological Society of America three or four schools of thought joined a battle which ended in a draw. There are granites of at least two kinds, was the prosaic conclusion, and they must be distinguished mainly by their field occurrences. This consensus ruled out the extreme views on both sides. Few if any petrologists retained the belief that granites are all strictly magmatic rocks; few were willing to argue that granitic magmas are impossible and that every granite is formed in situ by wet or dry chemical diffusion.

The middle ground was prepared by H. H. Read in England, where petrologists were always in profound disagreement on the origin of granite. When Read began his researches in the 1930s, magmatists predominated: "It is true [he wrote] that the transformation of a pre-existing rock into granite as it stood—that is, granitization—was advocated in some measure at certain unfashionable academies, but respectable opinion held that granites were the dregs of basaltic crystallization."

After more than a decade of exhaustive study Read concluded that a granite series does exist and that it includes igneous bodies as well as products of granitization. But he denied that the igneous granites are the dregs of basaltic crystallization and proposed instead that granites and basalts are separate products, both derived from the underlying sima but unrelated to each other by any process that occurs within the crust. In Read's view, the discontinuous sialic rafts of the continents may have developed by the granitization of sediments, augmented and transformed by emanations from the sima.

That the granite problem, important as it is, might not be either the sole or the proper key to the evolution of continents was perceived by J. Tuzo Wilson of the University of Toronto. In the 1950s Wilson developed a theory that all of the earth's crust overlying the Mohorovičić discontinuity has been built during geologic time by the eruption of basalts and of andesites from the mantle. Andesite, strictly defined, is the extrusive equivalent of diorite and is intermediate in composition between basalt (sima) and granite ("true" sial). (Refer to Figure 51). Over the protests of numerous petrologists it has recently become common practice to define andesite more loosely so as to include a range of volcanic rocks more siliceous than basalt. Regarded in this way, andesite is closer than is granite to the bulk chemical composition of the continents.

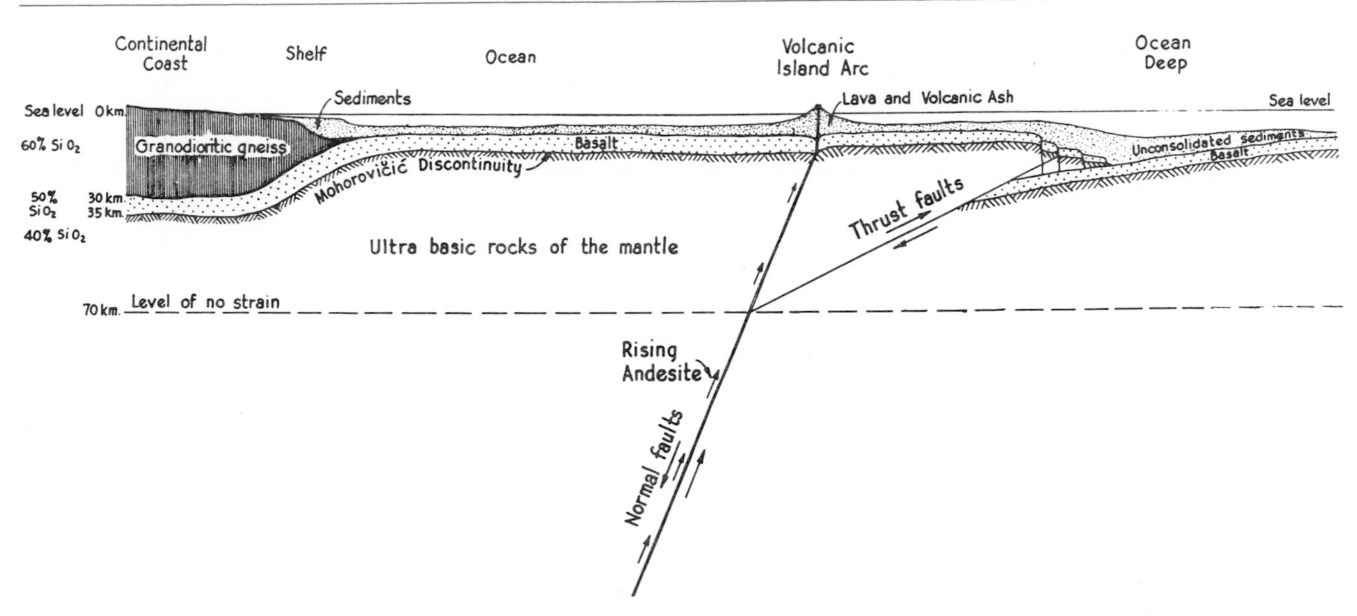

Figure 52. Crustal layers beneath a continent, ocean, island arc, and trench, according to J. Tuzo Wilson in 1959. (Used with permission of the author and the *American Scientist*.)

Andesites are the principal volcanic rock of the Andes Mountains—from which they derive their name—and in the island arcs of the oceans. Wilson noted that andesites, metamorphosed to greenstones, also are common among the oldest rocks of the Precambrian shields. From tectonic and petrologic considerations he concluded that the generation of andesitic magmas, deep in the mantle, and their extrusion on the surface have resulted in the growth of the continents from small volcanic islands that broke the surface of the world ocean more than 3.5 billion years ago. He suggested that these islands, the first continental nuclei, originated at widely separated sites along two fracture zones, several hundred kilometers deep, that cut the primeval planet along what are very nearly great circle courses. One of these deep fractures is now marked by the circum-Pacific ring of fire; the other is the east-west Tethyan mountain belt. The present continents have grown by the addition of island arcs and associated geosynclines—first on one side, then on the other—of the original island nuclei.

Wilson postulated that a shallower fracture zone, penetrating to less than 100 kilometers and marked by the oceanic ridges, has served as a permanent channel through which the entire earth has been covered with a layer of basalt a few kilometers thick.

Wilson described both basalt and andesite as low-temperature fractions derived by partial melting of mantle rock at different depths (Figure 52). He calculated that the melting and extrusion of as little as 5 percent of the volume of the mantle would be sufficient to produce the entire 30-kilometer thickness of the continental blocks. The resulting contraction of the upper mantle would be of the right order to cause the thrusting and folding observed in mountain ranges. Wilson's theory went very far in interpreting earth history in terms of global contraction, the permanence of ocean basins, and the accretionary growth of continents. In 1959 he wrote:

> . . . no other theory can explain so much. Continental drift is without a cause or a physical theory. It has never been applied to any but the last part of geological time.

The date of that statement is of special interest because early in the next decade, calling himself "a reformed anti-drifter," Wilson developed some of the most fundamental evidence in support of continental drift. He thereby provided an all too rare example of a scientist who, stimulated by new insights, turned away from his own elegantly developed theory in favor of a totally new approach.

Other persuasive advocates of continental accre-

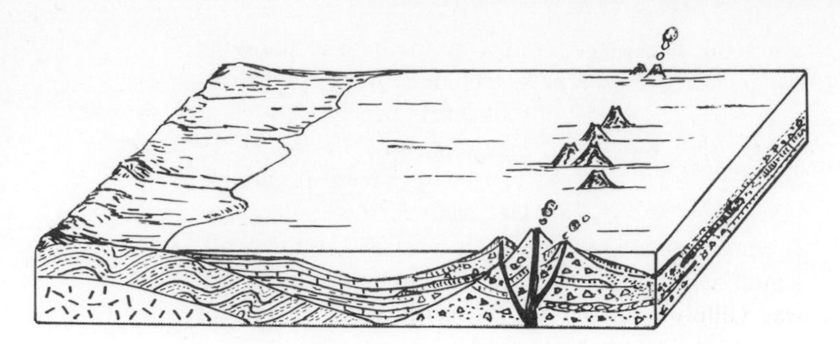

Figure 53. A miogeosyncline (basin filled with stratified sediments derived in large part from an eroded continent) and a eugeosyncline (volcanic arc, where deposits are a disorderly accumulation of sediments, lavas, and pyroclastic detritus).

tion included Hans Stille, in Germany, whose publications date from about 1910 through the 1950s, and two Americans, Marshall Kay of Columbia University and Patrick Hurley of the Massachusetts Institute of Technology, who are pursuing active research today. Stille pointed out, from the standpoint of tectonics, that the earth's surface is subdivided into stable areas bordered by mobile belts. The stable areas, which he called cratons, occur on continents and in ocean basins. Stille believed that the cratons have grown in area since the early Precambrian. The continental cratons, for example, include the Precambrian shields and large areas of younger formations. The mobile belts also occur on continents and in ocean basins and include geosynclines and mountain ranges in all stages of formation. Stille worked out a detailed history based on the idea that the continent of Europe has grown by the accretion of mountain belts; also, he correlated mountain-building episodes around the world and demonstrated more persuasively than had anyone before him that orogenesis has occurred simultaneously in many parts of the globe in short periods (geological revolutions) separated by intervals of comparative quiet.

By midcentury the subject of geosynclines as factors in continental accretion had waxed very complex. Numerous workers perceived that geosynclines are by no means simple catchment basins for sediments but are often compound structures consisting of two parallel belts, one characterized by more magmatic activity than the other. Technical terms proliferated as geologists proposed separate names for each portion of a geosyncline in each stage of its development. Most of these terms may be disregarded by the nonspecialist except for two—miogeosyncline and eugeosyncline—which were proposed by Hans Stille. Stille applied the name miogeosyncline to troughs containing stratified sediments more or less on the classical model and the name eugeosyncline to belts characterized by chaotic sedimentation accompanied by magmatic intrusion and active volcanism. The continental shelves and some interior basins may be identified as miogeosynclines, whereas the offshore island arcs, which supply large volumes of volcanic ash and lava as well as rapidly eroded detritus to the adjacent sea floors, may be identified as eugeosynclines. Thus, island arcs became the modern equivalent of the geanticlines of Dana and the borderlands of Schuchert. A schematic view of a miogeosyncline and a eugeosyncline is shown in Figure 53.

P. M. Hurley's support for the idea of continental accretion derived from the determination of Rb-Sr isotopic ratios in suites of igneous and metamorphic rocks from the shields and surrounding areas. His results indicated that new sial has been differentiating from the mantle and contributing to the growth of continents over geologic time.

Continental accretion, like any other geological theory, always had opponents. Some pointed out that few continents display a convincing pattern of age zoning. In Africa, for example, Precambrian rocks extend almost to the strandlines of both the Atlantic and Indian oceans. Across the South Atlantic ancient crystalline rocks, deformed at the end of the Precambrian, are sculptured into the spectacular scenery of Rio de Janeiro. Even in North America the zoning is not truly regular. Exposures of Precambrian rocks (not shown in Figure 50) occur along the coast of California and in other "young" areas such as Arizona, Nevada, and Idaho. Furthermore, geochronological studies outlined in detail in 1966 by Gerald Wasserburg showed that substantial portions of certain North American orogenic belts consist of reworked older continental materials. Many geochronologists were convinced that at least half and perhaps much more of the North American continent was already in existence as early as 2.5 billion years ago. Indeed, one school of petrologists maintains that a thin skin of sial formed very quickly and covered the entire planet at the beginning of earth history.

As a result of many complications, some loss of faith in marginal accretion was beginning to be felt before the renewal of interest in continental drift. About 1964, Hurley, like Wilson, reoriented his own thinking in favor of continental drift. Unlike Wilson, Hurley concluded that the two processes are not mutually exclusive.

The continental "freeboard"

Whether or not the continents have grown larger by marginal accretion, radiometric geochronology shows that they are vastly old and record a restless geological history dating back 3,700 million years. Sedimentary strata occurring throughout this record demonstrate that shallow seas often have flooded the continents and received detritus eroded from highlands. After every such invasion the seas have retreated again, and the marine beds have themselves been exposed to subaerial erosion. How have the continents maintained their "freeboard" throughout 3.7 billion years of base leveling? How have the oceans remained "oceanic" despite the colossal load of sial deposited in them?

If the relief were not renewed, the continents would be reduced to sea level, once and for all, in less than 10 million years, wrote James Gilluly, of the U.S. Geological Survey, in 1955. Gilluly estimated that the present volume of land above sea level is about 130 million km^3 and that the volume of continental rock transferred to the ocean every year is about 13.6 km^3. At present rates of erosion, all of the sial to a depth of 12 kilometers in the continental crust could have been transferred to the oceans in the 225 million years since the Late Paleozoic. Eight or nine times as much would have been deposited there in just two billion years. Yet the continents appear to stand as high as they ever did, and the ocean is floored by sima all the way to the continental slopes. What causes the mysterious disappearance of the vast tonnages of eroded sial? Neither continental accretion nor varied rates of erosion, within reasonable limits, begins to account for the great bulk involved.

Gilluly concluded that the sial deposited in the oceans is returned to the continents, not along their margins but on their undersurfaces. He proposed that powerful currents operating at the base of the crust sweep oceanic materials—including great thicknesses of low-density siliceous sediments—beneath the continents, thus causing the isostatic rise of the continental blocks. Elsewhere, equally powerful currents wear off deep sialic roots or thin the sialic crust, causing depression below sea level of certain continental areas. Subcrustal erosion and deposition of sial was Gilluly's answer to the century-old enigma of isostatic adjustment. He wrote:

> If isostasy is to be maintained, powerful currents must return an equal mass of subcrustal materials to positions beneath the continents. . . .
>
> Perhaps this subcrustal flow from beneath the oceans carries the shelf sediments back into the continental mass. This may explain . . . the raised borders of the continents . . . the apparent lack of pre-Mesozoic continental shelves . . . and the underthrusting of continental borders along tectonically active coasts that is recorded in the deeps, the fault patterns, and the igneous activity there localized.

He went on to hazard the guess, based partly on the measurement of oceanic heat flow, that convection cells in the mantle may rise beneath the oceans and sink beneath the continents. At that time advocates of mantle convection favored currents rising beneath the continents, sinking beneath the oceans.

The Ocean Basins

When Alfred Wegener proposed his hypothesis of continental drift with blocks of sial moving through the sima, he believed that the ocean floors are mostly level plains. He knew they are not "smooth as a mud flat" under the pull of gravity as Dr. Bowie suggested they should be if the sima were a yielding layer. Yet he was convinced that, compared with the continents, they support relatively little variation in topography. As one line of evidence, Wegener cited the results of a hundred soundings made in preparation for the laying of a cable between the Pacific islands of Guam and Midway. Over a distance of 1,540 kilometers the most extreme differences in level amounted to only 747 meters, which he regarded as an astonishingly slight variation for so wide an area.

By the time he published the 1929 edition of his book, Wegener had come to view his two-level picture of the sial and sima as too simple. In his final crustal model he subdivided the ocean floors into a shallow basalt layer (sima) overlying a deeper dunite layer exposed at all depths below the 5,000-fathom line.

Wegener's knowledge of oceanic topography was apparently as adequate as most prevailing in his time, yet to a modern reader it seems surprisingly deficient in view of the active investigations that had been carried on by many nations.

Oceanographic research is generally reckoned as beginning with the voyage of H. M. S. *Challenger*, a 2,000-ton corvette with auxiliary steam power that sailed from Portsmouth, England, on December 21, 1872. During the following three and a half years the *Challenger* made detailed investigations in every ocean except the Arctic. At each station the scientists and technicians aboard took depth soundings and sampled the bottom sediments. Also they measured the temperature of the water and sampled its flora and fauna at regular intervals from surface to bottom, and they recorded the direction and velocity of the winds and currents. The vast accumulation of data collected on the *Challenger* filled 50 quarto volumes compiled under the direction of Sir John Murray.

Scientific expeditions under several flags soon followed, and by the beginning of the 20th century all of the major oceanic deeps had been discovered and charted, numerous submarine canyons were found incised into the continental shelves, and the broad configuration of the Mid-Atlantic Ridge was established. Still to remain unknown for the next 55 years was the full extent of the oceanic ridge system.

The underwater explorations carried on during and after World War II confirmed the evidence of two favored levels on the earth. Indeed, 50 percent of the earth's surface consists of broad abyssal plains lying between 4,000 and 5,000 meters below sea level. In addition to these smooth, stable areas the oceans conceal a rugged, world-encircling system of ridges and rises as well as fracture zones, deep trenches bordering island arcs, and tens of thousands of volcanic mountains.

With the Marianas Trench extending to 11,033 meters below sea level and Mauna Kea in Hawaii rising to a height of 4,205 meters above sea level, the ocean floors support a spectacular 15,238 meters (49,993 feet) of relief, which is about two times the continental span of 8,848 meters from sea level to the summit of Mount Everest. Mauna Kea is, in fact, the world's greatest mountain, with 10,203 meters of relief from base to summit.

Of the many features of the oceanic "landscape," the great ridges, the trenches, and the guyots may be singled out as of special interest for their implications with respect to ocean-floor tectonics.

Oceanic ridges

The global extent of the oceanic ridge system was first appreciated in the mid-1950s as a result of extensive echo sounding and seismic profiling in the oceans. With all branches, the ridge system is some 60,000 kilometers in length, 1,000 kilometers in average breadth, and up to 4,000 meters in relief.

The Mid-Atlantic Ridge is rugged in profile, with flanks broken into a succession of longitudinal uplands and valleys flanked by stepped plateaus. In 1953 members of the British *Discovery II* expedition found a deep trough in the crest of the Mid-Atlantic Ridge south of the Azores. Later, a similar trough or axial valley was found along numerous other ridge segments. The axial trough can be a spectacular feature. Along parts of the Mid-Atlantic Ridge the

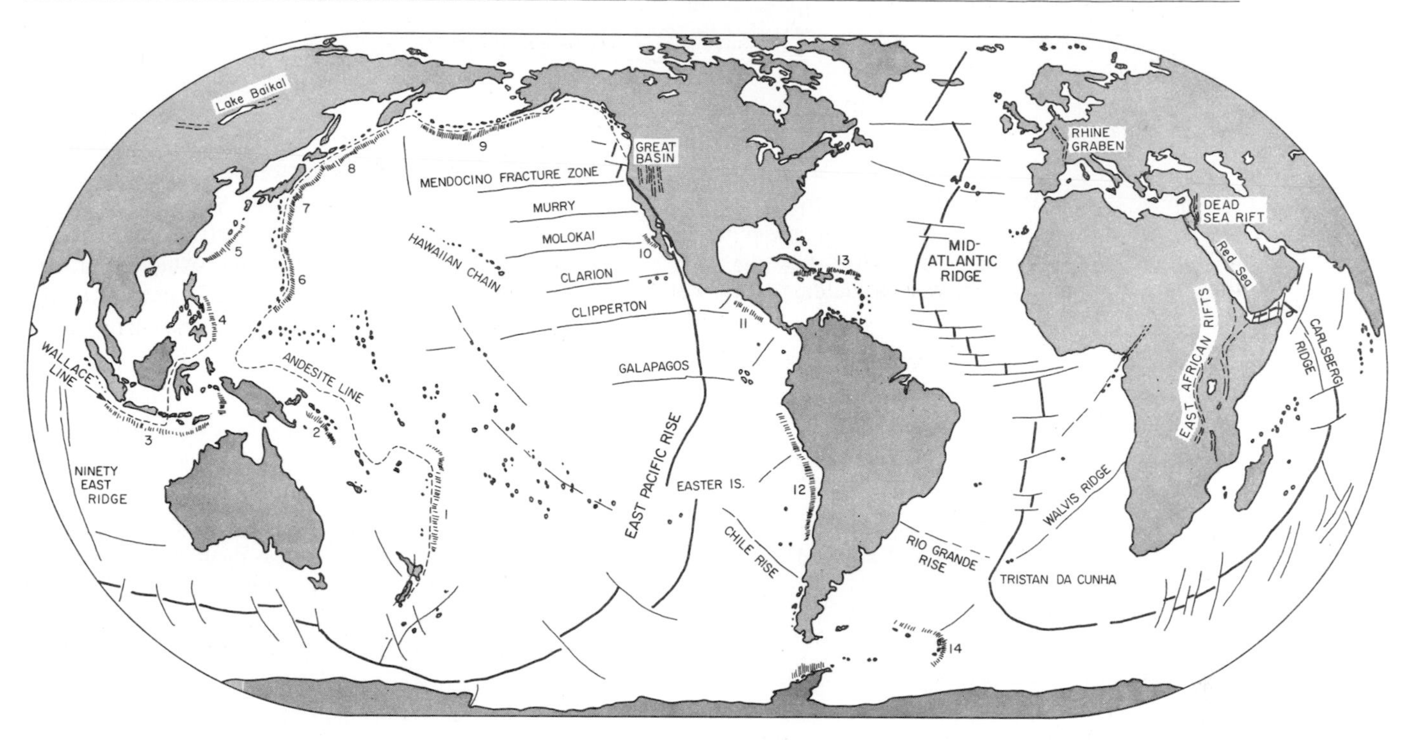

Figure 54. Worldwide distribution of continental rifts and of oceanic ridges, trenches, and fracture zones. The Andesite line, a petrological boundary, and the Wallace line, a biological boundary, are shown in the western Pacific. The numbered trenches are as follows: 1, Tonga-Kermadec; 2, New Britain; 3, Java; 4, Philippine; 5, Nansei-Shoto; 6, Marianas; 7, Japan; 8, Kurile-Kamchatka; 9, Aleutian; 10, Cedros trough; 11, Acapulco; 12, Peru-Chile; 13, Puerto Rico; 14, South Sandwich. (Compiled from many sources.)

valley floors are 12 to 50 kilometers broad and the walls are 600 to 2,000 meters high. However, the East Pacific Rise, which is a broad bulge rather than a steep-flanked ridge, lacks a crestal valley. Bruce Heezen, of Lamont Geological Observatory, was quick to point out a striking similarity in the topographic profiles of the oceanic ridge-rifts with the plateaus and rift valleys of East Africa.

A plan of the oceanic ridges is shown in Figure 54. Here several problematical aspects are immediately apparent. The ridges are fairly well centered in ocean basins with fractured coastlines, but not in the Pacific where the East Pacific Rise skirts the basin to the south and east. In two areas the ridges appear to strike inland and link up with continental rift or fault zones. One of these is the Afar depression of Ethiopia where the Sheba Ridge of the Gulf of Aden joins with the Red Sea and the African rift system. The other area is where the East Pacific Rise abuts the coast of Mexico and joins the San Andreas Fault zone of California. Most maps trace the junction through the Gulf of Baja California; others, less explicit, suggest a disturbed zone beginning farther south in Mexico.

The oceanic ridges are neither straight nor smoothly curving like most continental mountain ranges. They are sliced into short, offset segments along transverse fracture zones. The centering of the Mid-Atlantic Ridge between Brazil and West Africa, for example, involves at least 15 segments in only 15° of latitude, with an apparent total offset of 400 kilometers. The East Pacific Rise is offset by several large east-west fracture zones that cut the Pacific floor off-shore from the Americas. These Pacific fracture zones, discovered in the 1950s by H. W. Menard of Scripps Oceanographic Institution, are remarkably straight belts—thousands of kilometers long and from 100 to 200 kilometers wide—of ridge and trough topography with a relief of several kilometers. The Pacific fracture zones, like the ones cutting the Mid-Atlantic Ridge, strike toward the continents. Menard originally believed that some of them could be traced inland, but this is controversial. Linear magnetic anomalies trending north-south and offset by the fracture zones of the Pacific were measured in the late 1950s by A. D. Raff, R. G. Mason, and Victor Vacquier of Scripps Oceanographic Institution. After much searching, the matching anomalies across the Mendocino fracture zone were found. The total offset proved to be 1,160 kilometers.

Crustal motions of such a scale prove the reality of continental drift, said some investigators in the early 1960s. On the contrary, replied others, motions of this magnitude would slice continents to pieces and transpose them bit by bit rather than en bloc. No traces would remain of matching margins, structural provinces, or faunal zones. Until about 1965 no one doubted that these great fracture zones were all authentic strike-slip faults that had displaced slabs of the ocean floors and segments of the ridges in the sense suggested by the map pattern.

The midocean ridges are marked by shallow-focus earthquakes which account for about 5 percent of the total seismic energy released throughout the world. In fact, some of the ridge segments were "located" by their seismic activity before they were mapped by sounding or profiling techniques. Gravity determinations show that the ridges are largely compensated isostatically by a subjacent zone of appropriately low density. Throughout the 1950s and well into the 1960s many scientists regarded the ridge system as a permanent suture that has been the site of igneous and seismic activity from the beginning of geologic time.

In addition to the ridges that divide the oceans longitudinally there are a number of seismically inactive (aseismic) ridges or sills that trend across the ocean basins. Among the most impressive of these are the Lomonosov Ridge (2,000 to 3,000 meters in relief), which crosses the Arctic Ocean from the New Siberian Islands to Ellesmere Island; the Greenland-Faeroes Rise, which crosses the North Atlantic and includes Iceland; and the Walvis Ridge, which runs from South-West Africa to the Tristan da Cunha Island group in the South Atlantic where it joins with the Rio Grande Rise from the coast of Brazil. Some geologists saw these aseismic ridges, which extend from continent to continent, as confirmation of the idea of sunken land bridges or isthmian links.

Trenches and island arcs

Trenches stagger the imagination of landsmen attuned to the gentle relief of the continents. Trenches can be 3,200 kilometers long, 11 kilometers deep, and only 5 kilometers wide at the bottom. The Grand Canyon is 1.6 kilometers deep and about 320 kilometers long. To compare with the Tonga trench it would have to be six or seven times deeper and stretch from the Great Smoky Mountains of Tennessee to Flagstaff, Arizona.

There are some 20 trenches and troughs in the oceans; all lie close to land. The Peru-Chile trench, the Guatemala-Acapulco trench, and the Cedros trough border South and Central America. All other trenches parallel the convex margins of island arcs. Most of the trenches lie in the Pacific, but the Java trench is in the Indian Ocean and the Puerto Rico and South Sandwich trenches border the Antilles and South Sandwich islands of the Atlantic. Trenches do not occur along fractured coastlines.

Trenches are typically arcuate in plan—although some are straight for very long distances—and V-shaped in profile. Their bottoms often consist of a row of furrows separated by low saddles. Some trenches are partially filled with sediments, others contain virtually none (Figure 55). The side walls of trenches often are stepped with ridges or benches which act as sediment traps. In January 1960, Jacques Piccard and Don Walsh dived in the bathyscaphe *Trieste* to the bottom of the Marianas trench and, at a depth of 10,912 meters, found one fish, some tiny shrimplike crustaceans, and a film of soft, gray ooze.

Some 200 kilometers inside the concave curves of the trenches rise rows of volcanoes that are active or were recently active. Unlike the more tranquil dike and fissure eruptions of basalts along the ridges, the volcanoes of the island arcs vent explosively, producing lavas and tuffs that are predominantly andesitic. In some arcs the volcanoes stand on basaltic ocean floor; in others they are associated with belts of folded and partially metamorphosed sediments intruded by granites. A few islands have a double-arc structure with a range of folded mountains lying between the trench and the volcanoes. Island arcs are essentially strips of sial in the simatic ocean basins.

The enigma of islands characterized by andesitic volcanism in the same ocean with islands (such as the

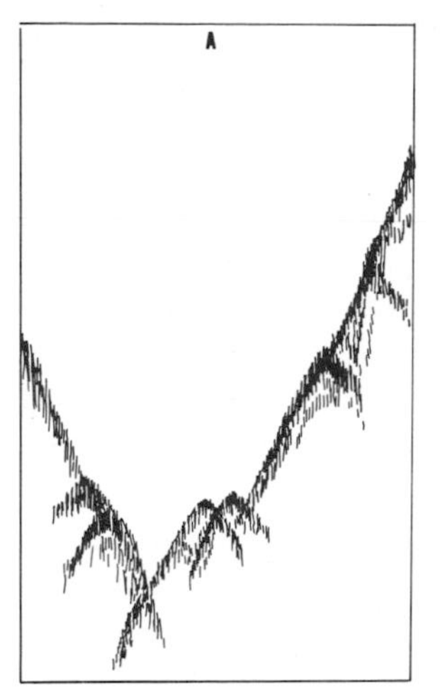

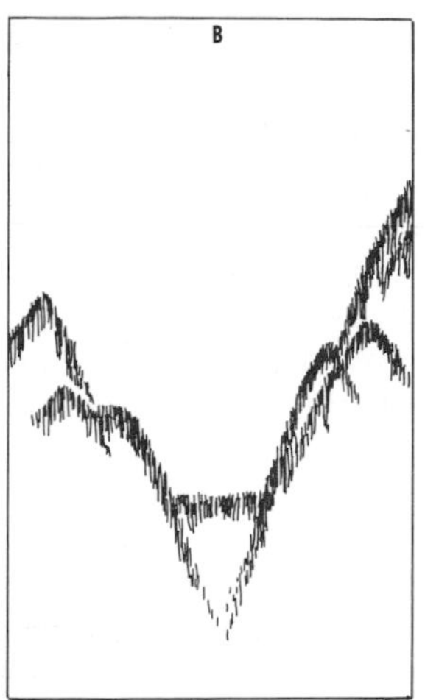

Figure 55. Echo-sounding profiles of the Acapulco trench. A: Near Acapulco the bottom tapers to a V at a depth of about 5,860 meters. B: Farther north, near Manzanillo, the trench has a flat bottom at a depth of 5,663 meters. (Sketches based on profiles recorded by Scripps Institution of Oceanography.)

Hawaiian group) built by basaltic eruptions was addressed in 1912 by Philip Marshall, who found that all of the andesitic volcanoes lie on the landward (continental) side of a line he drew through the southwestern Pacific archipelagoes. His famous "Andesite Line" has since been extended to skirt south of the Aleutians, east of the Kuriles and Japan, and southeast of the Bonin, Marianas, Fiji, and Tonga groups, as shown in Figure 54.

Although portions of the Andesite Line lie nearly 2,000 kilometers at sea, geologists jumped to the conclusion that it represents the "true" margin between the Pacific basin and continental Australasia. Many saw it as one of the most fundamentally important boundaries on the planet. By the early 1960s, however, it was apparent that most of the entrapped seas west of the andesitic arcs are floored with dense oceanic crust, and some of them, such as the Philippine Sea, are deeper than the main Pacific basin. Little justification remained for denying the uniqueness of the island arcs by defining them as a continental margin.

If the arcs are not continental in the traditional sense, they are sites characterized by the eruption of sial, a fact which led Marshall Kay, J. Tuzo Wilson, and others to conclude, in the 1950s, that the arcs are an important factor in continental accretion. In addition to sial, some of the island arcs contain sequences of mafic and ultramafic rocks, collectively named ophiolites, that are well known in the Alpine and other folded mountain ranges of the continents. Ophiolites consist of a layered succession of peridotites, dunites, gabbros, and basalts partially altered to serpentine or otherwise metamorphosed and capped by marine cherts and limestones. Typically, the basalts of an ophiolite sequence occur in lumpy aggregates called pillow lavas which form when volcanic flows congeal under water. (Although the oldest ophiolites are Ordovician, pillow lavas are common among the ancient rocks of the continental shields where they constitute direct evidence that the planet was at least partially covered by ocean water at an early date.) The intimate association of ultramafic rocks with marine sediments led to the idea that the ophiolite suite formed in the depths of eugeosynclines partly by the outpouring of magmas under deep water and partly by the incorporation of slices of the lower crust and upper mantle during orogenesis. The occurrence of the suite in island arcs emphasized the importance of the arcs for the information they could provide on mountain-building processes in general.

The trenches and island arcs mark zones of profound crustal instability. Wrote H. W. Menard in 1964: "The greatest depths, largest gravity anomalies, most extensive volcanism, most intense shallow seismicity, and almost all of the known deep focus earthquakes occur in island arcs and trenches."

The gravity anomalies were first discovered by Felix A. Vening Meinesz, who solved the problem of making determinations of gravity at sea by inventing a two-pendulum device to be used in a submarine. He began testing his device in 1923 on a voyage from Holland to the East Indies, and within two decades he had measured gravity in oceans around the world, thus opening an entirely new chapter in geodesy and geophysics. His most celebrated discovery was of long, arcuate strips marked by huge negative-gravity anomalies in the East and West Indies. These negative strips, thousands of kilometers long and 50 to 100 kilometers wide, parallel the trenches but are not centered over them. They follow submarine ridges or rows of small islands inside the concave curves of the trenches. Positive-gravity anomalies occur along the inner borders of the negative strips, and sometimes on

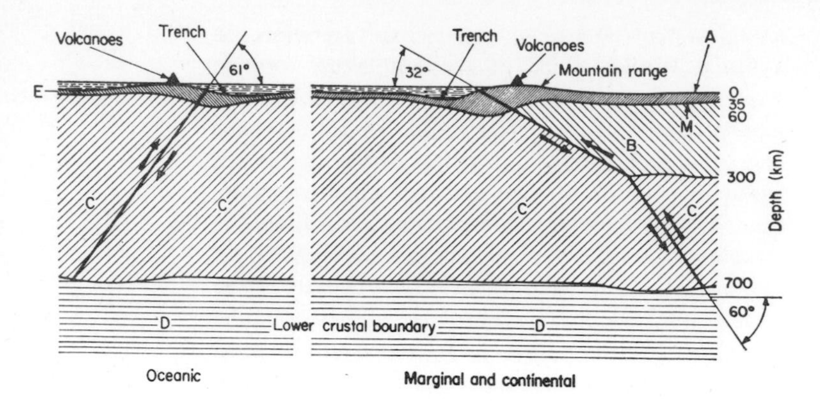

Figure 56. Cross sections published in 1955 by Hugo Benioff showing steeply dipping thrust planes separating continental crust from oceanic crust. (Used with permission of the Geological Society of America.)

the outer flanks as well. Vening Meinesz interpreted the negative strips as marking down-buckled furrows of sial projecting deep into the sima, a situation so unstable that isostatic adjustment would cancel it out very quickly if it were not dynamically maintained. He therefore proposed that the furrows of sial are held down by the compressive force of the descending limbs of convection currents in the mantle. As a result of his gravity studies, Vening Meinesz became one of the more dedicated proponents of mantle convection as the major factor in crustal tectonics.

The earthquakes of the Pacific margin are frequent and sometimes spectacularly destructive. Some of them trigger volcanism or tsunamis, ocean waves which radiate outward at speeds of 500 to 800 kilometers per hour and build to prodigious heights of up to 30 meters if they sweep across shallow bays or through constricted channels. Earthquake shocks originating at depths of a few kilometers were always well known on the Pacific margin, but in 1922 H. H. Turner in England reported that he had detected earthquakes propagated at depths of several hundred kilometers. His claim was discounted, partly because his evidence was insufficient and partly because of the widespread conviction that the earth has too little strength to build up stress differences at such depths. In 1930 Turner repeated his assertion, wondering out loud whether the intense seismicity of the Pacific margin indicates that the scar where the moon left the earth has even today not quite healed.

The evidence for deep-focus shocks continued to accumulate, and it brought on much speculation that the shocks are caused by sudden changes in state of deep-seated materials, such as the freezing of magma to crystals or the collapse of crystals to denser polymorphic phases. By 1936 Beno Gutenberg, then at the California Institute of Technology, was persuaded, on the basis of his own observations, that earthquake shocks are indeed propagated at depths of several hundred kilometers and that they are caused not by phase changes but by the fracturing and shearing of solid rocks.

Two years later, Gutenberg and C. L. Richter worked out a classification of earthquakes: normal (foci shallower than 60 km), intermediate (foci at 100–250 km), and deep (foci 400–700 km).* On the Pacific margin they found that the deeper the focus of an earthquake, the farther inland was its epicenter. From this they concluded that fault planes dip from the ocean floor beneath the land at angles of 30° to 60°. (The presence of thrust planes dipping beneath island arcs had been deduced in 1931 by Philip Lake solely on the islands' geometric form. Lake had concluded that the arcs mark the intersection of inclined planes with the spherical surface of the earth.)

In 1926 Gutenberg noted that earthquake waves appear to slow down as they pass through a layer about 100 to 200 kilometers deep, and he suggested the possibility of a glassy substratum of low rigidity in that region. He had precious little evidence, according to his colleagues, most of whom could detect no evidence at all; so, for 30 years his proposal was ignored. By 1960, however, a considerable body of data indicated the existence of a low-velocity channel at somewhat varying depths–120 to 275 kilometers under the continents, 50 to 240 kilometers under the oceans. The records of the Chilean earthquakes of 1960 confirmed the channel as worldwide. This feature, which is of the greatest tectonic significance, is now called the Gutenberg low-velocity channel.

The Pacific margin accounts for some 80 percent of the seismic energy released in earthquakes. The midocean ridges release about 5 percent, the Tethyan

*The classification now in use designates shocks as shallow (<70 km), intermediate (70–300 km) and deep-focus (300–720 km).

mountain belt about 15 percent, and shocks elsewhere account for 1 percent or less. Gutenberg's observations on the seismicity of the Pacific margin were refined in the 1950s by Hugo Benioff who showed that the fault zones are sometimes doubly or triply segmented as they plunge from the trenches beneath the continents or island arcs. Along the Chile-Peru trench, for example, the fault planes break in slope from about 32° to 60° at a depth of 300 kilometers. The landward margin is effectively decoupled from the oceanic crust (Figure 56). These dipping, seismically active planes are now called Benioff zones. The active volcanoes of the island arcs and the Pacific margin stand over that region where, at depths of 100 to 200 kilometers, the Benioff zones intersect the Gutenberg low-velocity channel.

Guyots and coral atolls

Submarine volcanoes that rise at least 1,000 meters above the ocean floor are called seamounts. Guyots are more special. They are volcanic mountains with flat summits submerged from 1 to 2 kilometers below sea level.

These features were first discovered on echo profiling records by Harry H. Hess while he was on duty in the Pacific in World War II. He recognized them as unique topographic forms and named them for Arnold Guyot, a 19th-century Swiss-born American geologist. The uniqueness of the guyots resides in their flat summits, which clearly suggest beveling by marine erosion at sea level followed by subsidence (Figure 57). Do guyots indicate that sea level was once 2 kilometers lower? Has the Pacific floor sunk by 2 kilometers? Or have both remained stable while the Pacific acquired a 2-kilometer thickness of sediments which has displaced the water upward by the same amount? Hess originally preferred the third explanation. He noted that guyots lack coral overgrowths, although some of them lie in warm, coral-bearing seas. He thought, therefore, that they must have been beveled by erosion in the Precambrian before the development of lime-secreting marine organisms.

Hess' guyots are a 20th-century version of the coral atoll problem, first puzzled over by Charles Darwin on the voyage of the *Beagle* in 1836. Coral polyps thrive only in the agitated surface waters of warm seas where the water temperature remains above 20° C. They build reefs on rock foundations—typically the eroded platforms around volcanic islands. In the western Pacific, however, Darwin found great rings (atolls) of coral enclosing wide, open lagoons or lagoons surrounding small remnants of volcanic peaks. He concluded that all atolls began as fringing reefs on volcanic islands but that the ocean floor had subsided thousands of meters, carrying the mountains below sea level while the industrious coral polyps continuously built their reefs upward, maintaining living colonies in the aerated surf zone (Figure 58). Darwin's interpretation, which encouraged the idea of sunken continents, seemed fully substantiated in 1952 when two borings were made through the coral of Eniwetok Atoll and a basaltic base was penetrated at depths of 1,384 and 1,500 meters. In the same year dredge samples from the summits of some of the Pacific guyots brought up shallow-water fossils of Cretaceous and early Tertiary age. The eroded platforms of Hess' guyots were therefore not Precambrian but at the oldest late Mesozoic. If they were beveled at sea level as recently as the Cretaceous, why did the guyots in tropical waters collect no crown of coral?*

Guyots are most abundant in the western Pacific, but they also occur in the Gulf of Alaska and in the Atlantic and Indian oceans. A row of them approaches the Aleutian trench, and one, the Kodiak guyot, appears to have "fallen in." It sits upright in the trench as though the floor had subsided beneath it. Once he appreciated the youth and the distribution of the guyots, Hess put his evidence together with other observations and developed a new theory of a mobile ocean floor.

The layering of the ocean floors

Experiments in measuring artificially induced seismic waves in shallow sea bottoms were begun by Maurice

*In recent years a number of arguments have been put forward that guyots are not eroded but are built under water as flat-topped volcanoes. Such an explanation, although plausible, fails to account for the shallow-water fossils on guyots and for the fact that relatively few oceanic volcanoes take this form.

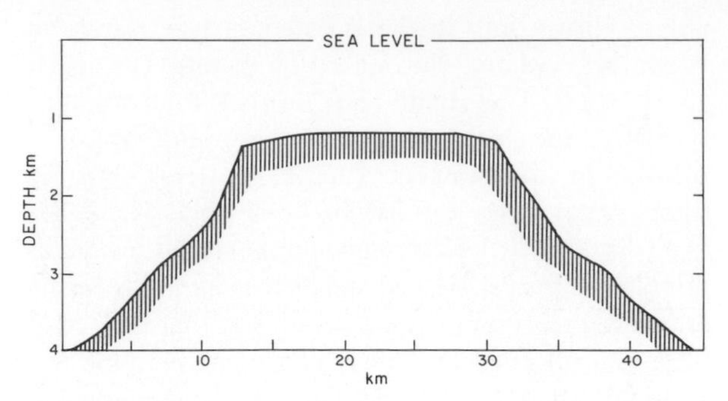

Figure 57. Echo-sounding profile of a guyot discovered by H. H. Hess in the Pacific Ocean south of Eniwetok atoll. The flat summit lies at a depth of 1,100 meters below sea level. (From *American Journal of Science*, volume 244, 1946; used with permission of the Journal.)

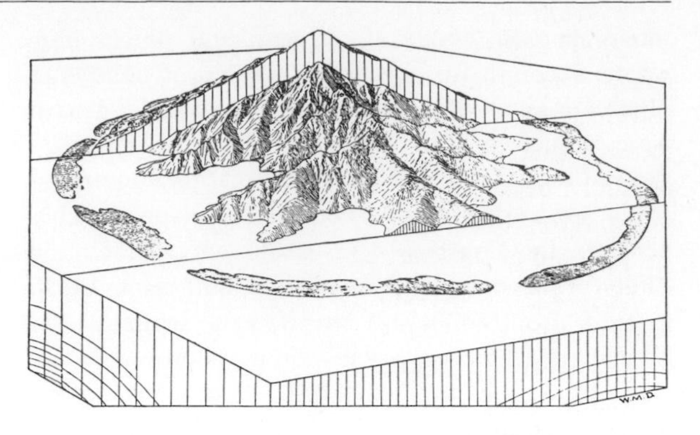

Figure 58. A triple-block diagram by William Morris Davis showing the development of a coral atoll during subsidence of the sea floor. Far section depicts a fringing reef on a volcanic island; middle section shows a barrier reef surrounding a partially submerged cone; front section shows an atoll enclosing an open lagoon. (From *The Coral Reef Problem*, 1928; used with permission of The American Geographical Society.)

Ewing in 1937. In succeeding years Dr. Ewing and numerous other scientists ran seismic reflection and refraction profiles of the ocean basins, clearly delineating their spectacular topography. The profiles also revealed a triple layering of the oceanic crust which, for proponents of the permanence of ocean basins, embodied a stunning surprise: the sedimentary cover on the ocean floors is very thin.

If ocean basins were indeed permanent features of the earth's crust, they must have slowly accumulated sediments throughout geologic time, even during the mysterious "missing" interval between the Precambrian and the Cambrian, which, on parts of the continents, was a period of profound erosion. One of the greatest adventures of marine science was intended to be the development of deep drilling techniques to bring up cores of the earth's entire sedimentary record and display the succession of fossil organisms that would bridge the gap between the primitive algae of the Precambrian and the teeming fauna of the 600-million-year-old Cambrian seas.* If the oceans were two billion years old, their sediments should be 3 kilometers thick, according to meticulous estimates of global erosion and deposition rates made by the Dutch geologist Phillip H. Kuenen in 1946. If the oceans were over four billion years old, as implied by Patterson's age of the earth, the sediments should be about 6 kilometers thick. In fact, however, the seismic profiles showed that the average thickness of unconsolidated sediments (layer 1) beyond the continental shelves is only a few hundred meters. Sediments are virtually absent from the midocean ridge crests, but they gradually thicken on the flanks until they attain a maximum of about 1.6 kilometers. In the Pacific the average thickness of layer 1 is about 300 meters; in the Atlantic and Indian oceans it is about 600 meters. At present rates of deposition, estimated at about 1 to 10 millimeters per 1,000 years, all of these sediments could have accumulated well within the past 200 million years.

Beneath the low-density sediments of layer 1, with seismic velocities of from 1.5 to 2.0 kilometers per second, Ewing identified two more layers above the Mohorovičić discontinuity (Figure 59). Layer 2 ranges in thickness from 1 to 4 kilometers, with seismic velocities of from 3.5 to 6.0 kilometers per second. Layer 3, which is from 4 to 5.5 kilometers thick, transmits velocities of from 6.4 to 6.8 kilometers per second. Layer 3 is the authentic bedrock of the ocean floor. Its thickness is uniform over vast areas of the oceans, and when first discovered it was believed to continue without interruption under the continents.

*Geological mythology has exaggerated the extent of the Precambrian-Cambrian unconformity, which in many places cannot be accurately located in the stratigraphic column—a fact which renders the absence of evolutionary life forms more mysterious than ever.

Layer 2 was, from the first, enigmatic. It could consist of compacted sediments, of low-density volcanics, or of an interlayering of both. Calculations showed that if layer 2 were wholly sedimentary it probably began to accumulate in the Silurian, about 400 million years ago. All hope was lost, therefore, for a record of the Precambrian-Cambrian interval.

Why are the oceanic sediments so thin? Is it because they are young? Proponents of continental drift had always argued that oceans with fractured coastlines are young; they never thought so of the Pacific. Indeed, a great age for the Pacific basin was one of the few uncontroversial issues among earth scientists. Yet the sedimentary record was implicit: all ocean floors are about the same age; all are new.

As oceanic dredging and drilling developed on a wide scale, the evidence from fossils confirmed the young age of the sediments and radiometric dating established ages of less than one million years for some of the ridge basalts. These findings were considered tentative until about 1965, by which time the evidence was unmistakable and its full import was becoming clear.

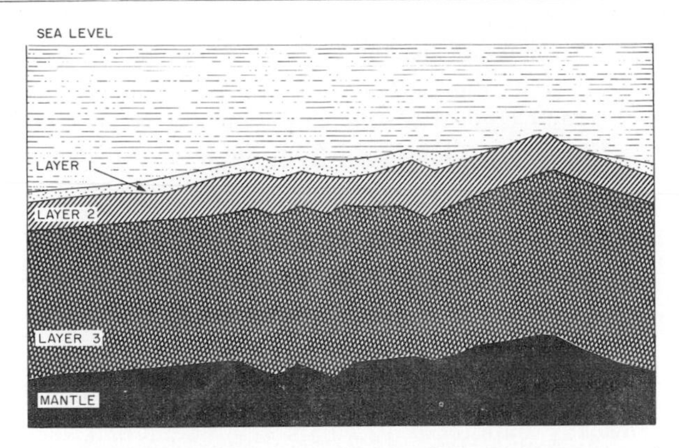

Figure 59. Schematic representation of the triple layering of the ocean floor overlying the Mohorovičić discontinuity.

Ocean waters and the atmosphere

The ocean basins have long been controversial. Almost equally mysterious are the ocean waters. Whence do they come? How long has the planet been awash with water and enveloped in a blanket of atmosphere? The oceans and the atmosphere are so dependent, one upon the other, that they may be considered as two phases of one system.

The belief in a primeval ocean covering the globe persisted from the time of Buffon to relatively recent years. Today it is generally assumed that the waters are not primeval but juvenile, having derived from the earth's interior. Believers in a high-temperature origin of the earth and those who believe in the low-temperature accretion of small particles both agree that all traces of any truly primeval gaseous atmosphere around the planet was swept away before the beginning of "geologic time."

In comparison with the sun, our atmosphere is severely depleted in the inert gases—neon, krypton, and most of the isotopes of argon. If it somehow lost its heavy gases, the atmosphere must also have lost its lighter ones, including water vapor, with a molecular weight of only 18. This situation seems to tell a tale of some cosmic hurricane that left the new planet barren. Since that early time every volcanic eruption has bled nitrogen, water, carbon monoxide, carbon dioxide, sulfur dioxide, and other gases to the surface where they have been precipitated into the ocean or held in the atmosphere by the powerful gravitational field. Over the past 4 to 4.5 billion years, then, the ocean waters have grown continuously. Or have they?

The evaporation cycle of water from ocean to cloud to rain to river and back to the ocean is well known. Some geologists believe in a more fundamental cycle that carries the water in sea sediments back into the upper mantle by way of geosynclines or underthrust continental margins where it feeds renewed volcanism. Three of the arguments supporting this theory were summarized in 1965 by Arthur Holmes: (1) Volcanic emanations are poorer in heavy water (D_2O) than is the sea. This should not be so if the emanations were juvenile waters arising from the mantle. It means, then, that volcanic waters are recycled seawater from which the D_2O was filtered out by fine pore spaces in marine sediments. (2) Recent calculations show that all of the sodium in seawater could have been introduced within the past 250 million years. (In 1898, John Joly had calculated 80 to 89 million years.) Where is the sodium that was introduced during the first 4 billion years of earth history? Holmes was convinced that the sodium, along with the water, is returned to crustal rocks via geosynclines where sodium plays an important role in granitization and the formation of andesites. (3) Calculations of the carbon cycle from volcanic carbon dioxide (CO_2) to marine limestone ($CaCO_3$) show that in 265 million years, from the Devonian to the Jurassic, 9.5×10^{15} tons of CO_2 were liberated from volcanoes along with 65 times as much H_2O. This is nearly half of all the water in the world's seas and oceans. At this rate of volcanism, all of the oceans would be overflowing in only 600 million years. "Obviously," wrote Holmes in 1965, "very little of the annual supply of 'volcanic' water can be juvenile, and any hypothesis based on the assumption that much of it *is* juvenile cannot be entertained."

In fact, however, that assumption is widely entertained. In 1951 William Rubey of the U.S. Geological Survey published a comprehensive analysis of data which indicated that juvenile water has been added to the oceans during geologic time as a by-product of the differentiation of sial from the mantle. Many earth scientists concur with his conclusion that ocean water has been slowly accumulating since the early Precambrian—although not necessarily at a uniform rate—rather than being ancient water cycled and recycled through the atmosphere, the hydrosphere, and the lithosphere. In either case, the sum total of our water and atmosphere was originally vented to the surface through volcanoes, apparent agents of destruction that have been the indispensable source of the life-supporting matter of the planet.

In the atmosphere, at least one component, Argon-40, has unquestionably been slowly accumulating over geologic time as a result of the radioactive decay of Potassium-40. The argon is too heavy to escape from the gravitational field and is too "noble" (unreactive) to combine with other elements and become fixed in compounds on the surface. The eternity and indestructability of argon has been illustrated by Harlow Shapley, the astronomer, in his book *Beyond the Observatory*:

> Now let us follow the career of one argon-rich breath—your next exhalation, let us suppose. We shall call it Breath X. It quickly spreads. Its argon, exhaled this morning, by nightfall is all over the neighborhood. In a week it is distributed all over the country; in a month, it is in all places where winds blow and gases diffuse. By the end of the year, the 3×10^{19} argon atoms of Breath X will be smoothly distributed throughout all the free air of the Earth. You will then be breathing some of those atoms again. . . .
>
> Your next breath will contain more than 400,000 of the argon atoms that Gandhi breathed in his long life. Argon atoms are here from the conversations at the Last Supper, from the arguments of the diplomats at Yalta, and from the recitations of the classic poets. . . . Our next breaths, yours and mine, will sample the snorts, sighs, bellows, shrieks, cheers, and spoken prayers of the prehistoric and historic past.

This saga of the argon atoms, Shapley points out, tells us of the smallness of units of matter, of the turbulence of the atmosphere, and of our intimate association with the past and future. It also tells of the pervasiveness of some atmospheric components, and of the fragility of "our healthful, gaseous envelope" which, he warns, we do not want "corrupted, polluted, poisoned."

Oceanic heat flow

The ocean basins yielded yet another surprise—this one for geophysicists—when a device was developed, about 1950, for measuring the rate of heat flow through the ocean floor. By that time few scientists seriously doubted that the oceanic crust is thin and largely free of sial. Some disagreement occurred over the predominant rock types—basalt or serpentinite—but both are markedly less radioactive than the continental granites and andesites.

Heat-flow measurements on the continents indicated that, aside from zones of active volcanism or hot springs, an average of from about 1.1 to 1.5 microcalories per square centimeter per second flows from the continental crust. All of this heat could easily be radiogenic, derived from the upper few kilometers of sial, with none at all contributed from the mantle.

In 1952 the first oceanic measurements were announced and discussed in two letters to *Nature* by Roger Revelle and A. E. Maxwell of Scripps Oceanographic Institution and Edward Bullard of Cambridge University. To the general consternation of geophysicists around the world, the results showed that the average heat flow from the ocean floors is the same as that from the continents. No explanations were immediately obvious, and new interpretations of the earth's thermal regime were clearly in order.

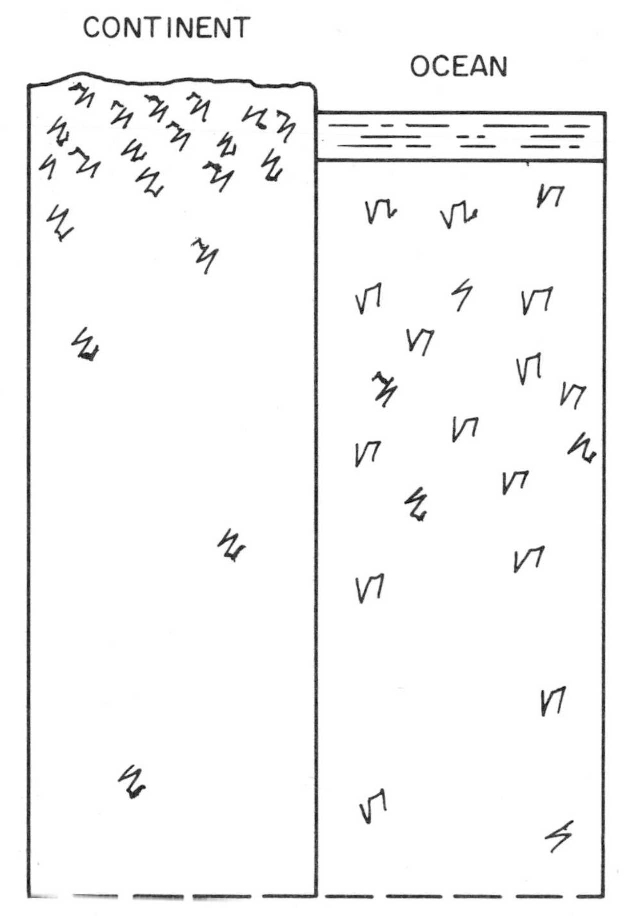

Figure 60. A diagram illustrating the principle that equal rates of heat-flow from continents and ocean floors imply differences in mantle composition extending to depths of several hundred kilometers. A marked concentration of heat-producing radioactive elements (jagged symbols) is found in the surface rocks of the continents; therefore, equal heat production is achieved only by counting all sources to a considerable depth. One can easily imagine the resulting contrast in surface heat-flow if a block of the continental crust (left) were to drift over the oceanic mantle (right). (Diagram is not drawn to scale.)

Protagonists of convection currents took the results as new evidence in their favor. Until then, convection currents were generally pictured as rising under the continents and sinking under the ocean basins. Given the new heat-flow values, such a pattern was seen to be impossible: it would tend to produce higher heat flow in continental than in oceanic crust aside from their radioactivities. With an admirable lack of embarrassment, the advocates of mantle convection simply reversed the direction of flow: the currents were now seen as rising under the oceans and sinking under the continents. Continental heat flow from radioactivity was balanced (fortuitously) by oceanic heat flow from convection.

On the opposite side of the argument, the new heat-flow data were interpreted as evidence for a solid, nonconvecting earth. All of the heat flowing from the interior was assumed to be radiogenic. The measured differences in radioactivity between continental and oceanic crustal rocks must therefore be compensated for by a reverse distribution in the upper mantle. Beneath the continents, the mantle must be severely depleted in radioactive elements and contributing nothing to heat flow; beneath the oceans, the feebly radioactive mantle must be contributing heat from such depths as to produce an overall equality with the continents (Figure 60).

Interpreted in this way, the new heat-flow measurements reinforced a growing impression that the mantle is a markedly inhomogeneous, chemically complex, earth shell. Seismic measurements had already shown density-layering in the mantle; and heat-flow measurements were adding major lateral inhomogeneities to the picture. Thermal convection, which is generally seen as a mixing process, appeared less likely than ever. One conclusion seemed certain: continental blocks have never migrated horizontally by slipping away from the subcontinental mantle and coming to rest over oceanic mantle. If they had, the heat-flow contrast should be exaggerated rather than averaged out. This new evidence against convection currents and continental drift was documented in explicit terms in 1965 by Gordon J. F. MacDonald of the University of California. In "The Deep Structure of the Continents," an article in the *Journal of Geophysical Research*, MacDonald concluded that, on the average, whenever rocky material has been located along an earth radius it will always remain in the vicinity of that radius. Here is a portrait of an earth that is solid indeed.

Paleomagnetism

That some rocks have a magnetic attraction for compasses and bits of iron is ancient knowledge. But the fact that rocks can themselves behave as compasses and carry a magnetic polarity aligned parallel to the earth's magnetic field was not discovered until 1849, when A. Delesse reported finding that property in recent lavas. A few years later M. Melloni experimented with bringing various rock specimens to red heat and then cooling them. He found that regardless of their orientation in his laboratory they acquired a magnetic moment aligned with the earth's field. Later in the century other investigators discovered that fired artifacts, both ancient and modern, have the same property. Cooling from high temperatures therefore seemed to be essential in producing the magnetic alignment.

In 1906 Bertrand Bruhnes in France discovered that some volcanic rocks are magnetized parallel to the earth's field but toward the opposite pole; in other words, they behave as south-seeking rather than north-seeking compasses. Bruhnes concluded that the geomagnetic field, at some recent time, had reversed its polarity. Other investigators soon confirmed his observation but no agreement developed as to its meaning.

Thus far, measurements had concentrated on the present magnetic field as shown by compass variations and the magnetic lineation of recently cooled rocks. In 1925 R. Chevallier demonstrated that the succession of lava flows at Mount Etna reflects past orientations of the earth's field. Five samples from the flow of 1699 were all aligned with the magnetic north as it had been recorded at that location in that year. Here, then, was authentic fossil magnetism frozen into the rock and preserved from the past. From that beginning Chevallier used the direction of rock magnetism in earlier and earlier flows to show how the field had varied at Mount Etna for the past 2,000 years.

The problem of normal and reversed magnetism in rocks was pursued by several investigators, including P. L. Mercanton in France and M. Matuyama in Japan. In 1929 Matuyama announced that of numerous samples of Cenozoic lavas from Japan and Manchuria those with reversed polarity were all early Quaternary (about one million years old) or older; younger rocks all had normal polarity. Matuyama's discovery was powerful evidence favoring Bruhnes' proposal of a pole reversal. Although Matuyama himself cautiously described the effects of the reversal as applying to the area he had sampled, it seemed clear that if reversals of the magnetic field occur at all they should be worldwide in scope and should affect all rocks of a given age.

About 1930 Mercanton suggested that, inasmuch as recently crystallized lavas all seem to indicate an approximate coincidence between the magnetic and the geographic poles, the study of paleomagnetism in older rocks should be useful in testing the possibilities of polar wandering and continental drift. We remember, however, that 1930 was just about the time that Wegener's hypothesis fell into general disrepute in most of Europe and America. In the clear hindsight we enjoy today we may see, impatiently, that Mercanton was right. But at that time the measurement of paleomagnetism was itself a "new" technique, carried on by relatively few investigators and regarded by the majority as unpromising. No sound theoretical basis had been developed to explain how rocks acquire their magnetic moment. Indeed, the origin of the geomagnetic field itself remained a mystery. "How can we test a theory of the continents on what-we-don't-know about magnetism? " asked a professor of geology in 1958. His predecessors must have been asking the same question in 1948 and 1938.

Remanence

Research on paleomagnetism was continued by a few investigators, mainly in Japan, France, and England. In time, three types of natural magnetization were distinguished: thermoremanent magnetization, depositional remanent magnetization, and chemical remanent magnetization. (The term remanent refers to the magnetization that remains in a substance after removal of the field that induced it.) All three types are produced in the iron-oxide minerals magnetite ($FeFe_2O_4$) and hematite (Fe_2O_3), both of which are common, if minor, constituents of igneous, sedimentary, and metamorphic rocks.

Thermoremanent magnetization is induced as a mineral cools from high temperatures. Each iron oxide has a separate temperature, called its Curie point, at which it begins to acquire ferromagnetic properties. Although the magnetic field of the earth is weak, the thermoremanent magnetism of iron oxides is very strong and very stable. Experimental attempts to reduce the strength or alter the alignment have confirmed this. Thermoremanent magnetization, therefore, is the most reliable indicator of the orientation of geologically ancient magnetic fields.

Depositional remanent magnetization derives from the fact that tiny grains of iron oxides will swivel in such a way as to align their magnetic moments with that of the earth's field as they settle through water and are deposited in bottom sediments. Although they correctly record the magnetic declination of their depositional site, they are unreliable indicators of the inclination. When tabular grains are deposited in horizontal beds they tend to lie parallel to the bedding planes; therefore, the magnetic inclination they show is from 10° to 20° too low. If they lie in sloping beds, such as the foreset layers of deltaic deposits, there is additional but unknown error. Furthermore, depositional remanent magnetism tends to be unstable. Samples have been known to alter in magnetic properties while sitting on laboratory shelves. Thus, as any given sediment sample may yield unreliable paleomagnetic data, measurements have to be made on groups of samples—preferably including ones of both sedimentary and igneous origin—in order to decipher the magnetic history of a predominantly sedimentary rock province.

Chemical remanent magnetism occurs in iron oxides that crystallize from solutions at temperatures well below their Curie points. This type of magnetization has proved to be fairly strong and stable. Iron oxides are a common cement in sediments such as red sandstones and shales, some of which are well-known horizon markers in the stratigraphic record. Caution must be used in interpreting their magnetization because the cement may form much later than the original sediments, or it may dissolve and be reprecipitated.

As paleomagnetic methods were being developed, the problem of reversed magnetism continued to puzzle investigators. Doubts about the significance of the observations—and on the paleomagnetic method itself—arose in 1952 when Louis Néel, in France, predicted the phenomenon of self-reversals in magnetic rocks and a self-reversal was observed in a volcanic pumice measured in a laboratory in Japan. The term self-reversal refers to the fact that some iron oxides acquire a magnetic inclination opposed to that of the earth's field. Self-reversed minerals crystallizing today are south-seeking rather than north-seeking. The problem of self-reversals clearly had to be solved before a convincing case could be made for reversals of the geomagnetic field as a whole.

Continental drift—polar wandering?

Paleomagnetic research did not come to a halt over the problem of self-reversals. By 1954 paleomagnetists at the laboratory of P. M. S. Blackett at the University of London announced the results of measurements on sediments from various parts of England: Triassic sandstones, Carboniferous coal measures, and the Devonian Old Red Sandstone. They found statistical evidence of three phenomena: (1) these sediments, which are 180 to 300 million years old, record two field reversals (one between the Devonian and Carboniferous and one within the Triassic) (2) the axis of the magnetic field in the Triassic was aligned in a direction about 37° east of present magnetic north; and (3) the magnetic inclination in the Triassic was only about 30°, whereas today it is about 65°.

The London group did not interpret these changes in terms of shifting magnetic poles. Instead, it proposed that while the geomagnetic field has remained stable in direction and inclination (if not in polarity) the landmass of England has rotated clockwise for 30° and drifted northward from a lower latitude. This would explain both the paleomagnetic record and the red color of the sandstones, which is generally (probably too generally) ascribed to deposition in warm, arid desert conditions. Here was an unequivocal reiteration, based on an entirely new type of evidence, of the theory of continental drift. At that time, however, the principles of continental accretion were preoccupying many geologists and heat-flow rates suggesting continental stability were being measured in the ocean floor. Few listened to the paleomagnetic news from London.

Blackett's research team extended its efforts to several other continents. Within three years the researchers found that the strong thermoremanent magnetism of the Deccan basalt flows of Late Cretaceous to Early Tertiary age in India showed a persistent upward inclination, indicating a cooling history in the southern hemisphere. This apparently meant that India has drifted northward for 7,000 kilometers within the last 80 million years. The paleomagnetic evidence thus supported paleoclimatic and fossil evidence—so graphically described by Alfred Wegener and others—that India was formerly a southern landmass, a fragment of Gondwanaland.

In the mid-1950s, however, a second school of English paleomagnetists, at the University of Newcastle-upon-Tyne, proposed a radically different interpretation of paleomagnetic results. Instead of resurrecting continental drift these scientists began by resurrecting polar wandering, but they soon discovered that they must encompass both.

The paleomagnetic method rests upon the assumption that the earth's magnetic field is and always has been predominantly dipolar. It is dipolar today, and measurements on Tertiary and Recent rocks show that it has been dipolar and close to the present orientation for the past 20 million years. The London scientists assumed that the axis has been dipolar and close to its present orientation throughout geologic time.

The Newcastle group, led by S. K. Runcorn, assumed that the earth's magnetic axis has always been predominantly dipolar but that the earth as a whole has changed position relative to the magnetic axis in the course of time. From replicate measurements on rocks of Europe and the British Isles, these paleomagnetists showed that the magnetic north pole can be traced for at least 21,000 kilometers over a long curving path from a site in western North America in the late Precambrian, about 1,000 million years ago, through the northern Pacific Ocean to northern Asia in the Mesozoic, about 150 million years ago, and finally to the Arctic in mid-Tertiary, about 20 million years ago. The curve they described is continuous, with no sudden skips or breaks. In 1955 Runcorn interpreted the results in terms of polar wandering without continental drift.

Turning their attention to North America, the Newcastle paleomagnetists found, over the next three years, that there, too, the rocks record a long, wandering curve followed by the north magnetic pole. Surprisingly, the North American curve was not the same as the European. The two curves converge at the present pole for late Tertiary rocks but they are systematically displaced by up to 30° of longitude for pre-Miocene rocks (Figure 61). In the search for an underlying cause of this displacement, one probability seemed manifest: while the poles have wandered the continents have drifted.

To test this probability, maps were reconstructed with the North American and the European polar-wandering curves superimposed. When this was done, the margins of the two continental masses came together, obliterating the North Atlantic Ocean. The new paleomagnetic maps from Newcastle-upon-Tyne bore an uncanny resemblance to the efforts of Alfred Wegener and his disciples.

Polar wandering reexamined

Polar wandering—an apparent motion of the magnetic and/or geographic poles with respect to the earth's surface—could be accomplished by any one of several different mechanisms (Figure 62). If the magnetic axis has always remained dipolar and closely associated with the spin axis, then the paleomagnetic and paleoclimatic evidence should indicate similar histories of polar wandering provided that the crust as a whole has migrated over the interior or that the body of the earth has undergone a plastic reorientation with respect to its axes, which have remained fixed in space. These are the types of motion that Alfred Wegener called superficial polar wandering.

If the magnetic axis can swivel independently from the spin axis, or if the magnetic field has been multipolar in the past, no correlation would be expected between the magnetic and climatic evidences of polar wandering. Finally, if astronomical polar wandering were to occur—if the spin axis, in concert with the magnetic axis, were to depart significantly from its angle of 23.5° from normal to the plane of the ecliptic—the sediments and fossils of the crust could record sensational climatic changes while paleomagnetic determinations would show no unusual variations at all. Astronomical polar wandering has long been rejected by most scientists for the reasons cited in the 18th century by Buffon:

periodic changes in the tilt of the axis are known but they are small and alternating. Current figures show that the axial tilt varies by about 2° over a period of 41,000 years. Alfred Wegener, who held a doctoral degree in astronomy, was one of the few investigators who seriously considered astronomical polar wandering a possibility. He thought that marked variations above and below the average angle of tilt may have occurred and resulted in worldwide fluctuations in the severity of climatic zoning. Wegener admitted that this hypothesis lacked proof.

The tilt of the spin axis has slowly increased during geologic time as the earth has yielded angular momentum to the moon. The increase probably amounts to about 13°, which leaves the initial 10° of tilt unaccounted for. The main result of the increase will have been a gradual intensification of the climatic contrasts between the earth's equator and poles. Such an enhancement of climatic zoning is not synonymous with polar wandering as the term is generally defined.

The ultimate cause of the tilt of the earth's axis is uncertain. Historically, philosophers unwilling to accept asymmetry as a primary condition have ascribed the tilt to a violent catastrophe such as the ripping of the moon from the earth or the moon's sudden capture from outer space. Neither of these events is widely favored today. The chemistry of the rocks returned from the moon's surface suggests that the moon accreted as a discrete body in the vicinity of the earth. If it did, then other large bodies were probably also present during the accretionary stage and caused the tilting of the earth either by near collision or direct impact. After that early date, however, it would be implausible to call upon a succession of cosmic catastrophes to explain the smooth polar wandering curves compiled by Köppen and Wegener and, more recently, by paleomagnetists.

At present, the earth's magnetic axis is inclined about 11° from the spin axis. The magnetic axis is not quite linear nor does it pass precisely through the center of the planet. The magnetic field fluctuates in alignment, intensity, and polarity, and this known instability has caused reluctance on the part of some scientists to take seriously the assumption, which is basic to the paleomagnetic method, of a permanently dipolar, geocentric axis. Paleomagnetists reply that the variations in the field average out over long periods of time and that measurements on rock

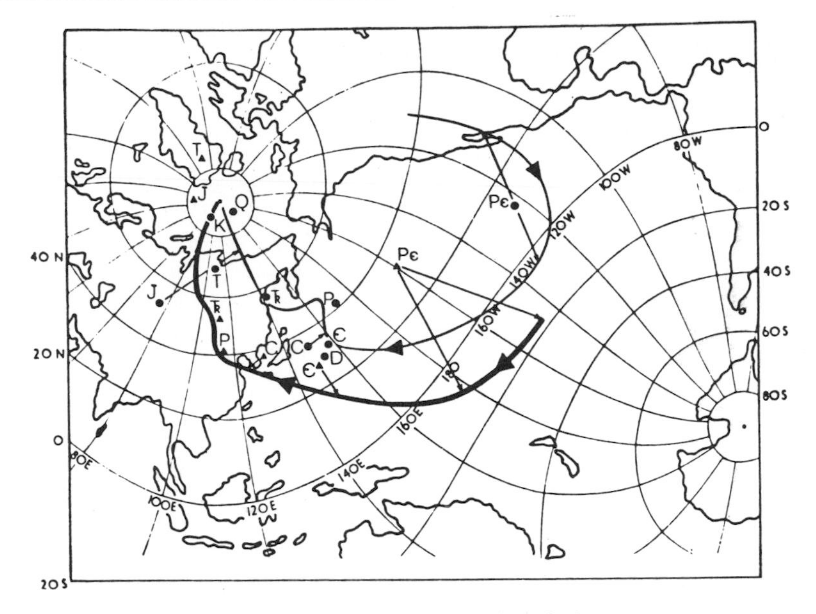

Figure 61. The wandering of the north magnetic pole from the Precambrian (Pc) to the present (Q) as indicated by pole positions measured on North American rocks (heavy black line) and Eurasian rocks (thin line). (From *A Symposium on Continental Drift*, 1965; used with permission of S. K. Runcorn and The Royal Society.)

samples representing periods of from 10,000 to 100,000 years allow accurate interpretations of past positions of an axis that has been predominantly dipolar and approximately coincident with the spin axis. The lingering doubts about the paleomagnetic method have emphasized the need for checking the results by an independent line of evidence, such as that for climatic polar wandering. Correlations have been made by E. Irving, K. M. Creer, and others who find remarkably good agreements between paleoclimatic and paleomagnetic polar-wandering curves.

How is polar wandering accomplished? If we reject the tilting of the earth's axis in space as a mechanism, we are left with the two possibilities cited above. The idea that the crust as a whole moves over the interior has had numerous advocates, including John Joly and, in recent years, George Bain of Amherst College. Alfred Wegener believed in crustal creep but doubted that it ever relates to any axis other than the rotation axis. He did not define westward or equatorward creep as polar wandering. Du Toit, in contrast, regarded the alternate poleward

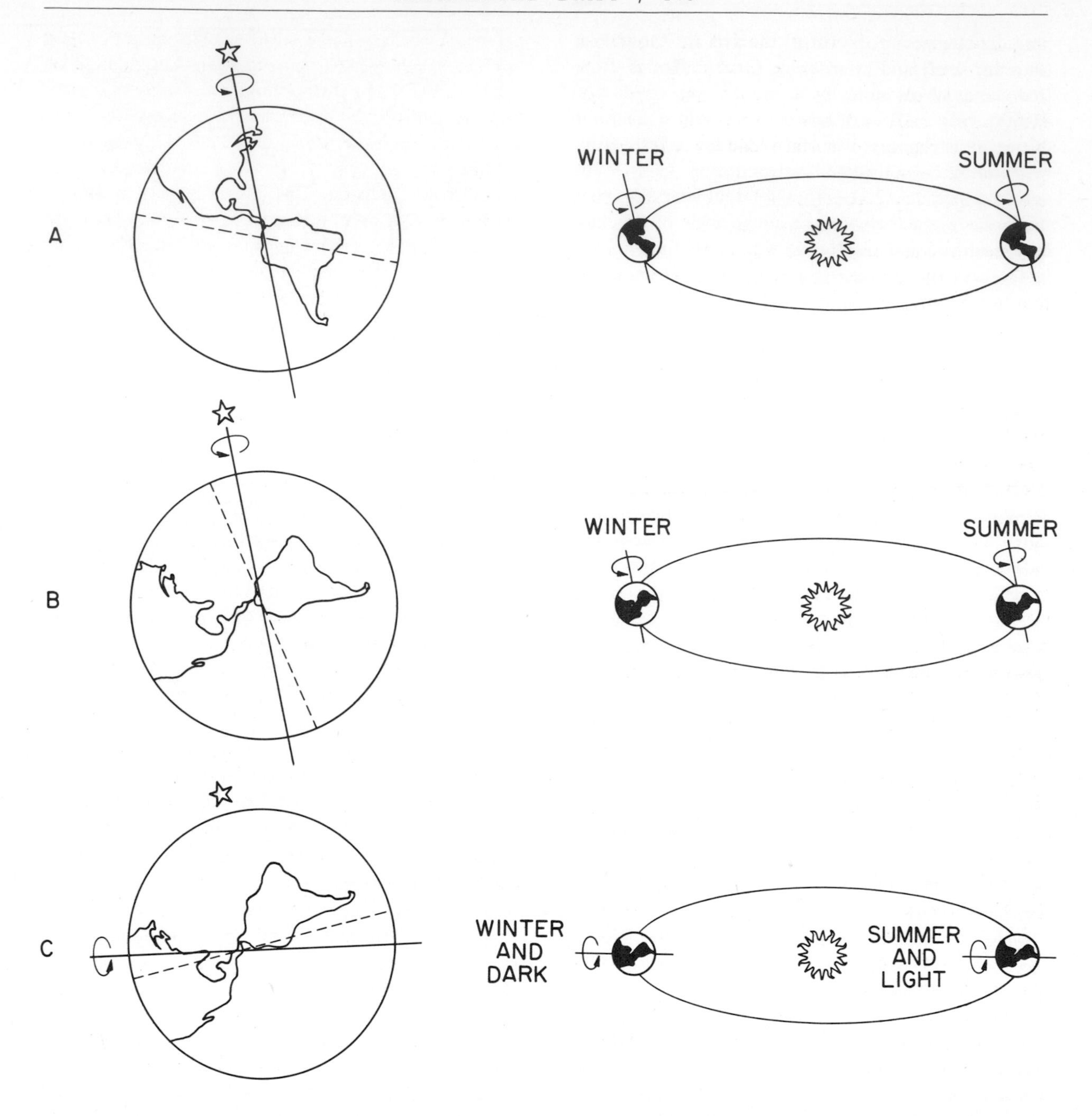

Figure 62. Diagrams illustrating four possible varieties of polar wandering. **A:** If the spin axis were to remain stable with respect to the fixed stars but the magnetic axis (dashed line) were to depart from it and wander about independently, the earth's climatic zones would remain unchanged but the rocks would record extensive magnetic polar wandering. **B:** If the spin and magnetic axes were to remain stable while (1) the earth's crust were to rotate, about some other axis, over the interior or (2) the crust and mantle were to be redistributed *en masse* with respect to the axes, the climatic zones and paleomagnetic record would show synchronous evidence of polar wandering. **C:** If the spin and magnetic axes were to remain close together and were to tilt about in space, the climatic zones could undergo spectacular changes but the magnetic poles would remain in Canada and Antarctica. Few scientists believe in this form of polar wandering.

and equatorward motions of Laurasia and Gondwana as a form of polar wandering. Bain argues that the crust as a whole does move around axes other than the rotation axis, and that such motion is generated by massive changes of surface load caused by global patterns of erosion and sedimentation. In 1963 he traced a detailed paleoclimatic history of the earth from the early Precambrian to the present in which continental areas have undergone marked changes in relation to the rotation axis by means of crustal polar wandering unaccompanied by any significant continental drift.

The remaining mechanism of polar wandering invokes the slow, plastic reorientation of the entire body of the planet, or of the crust and mantle, with respect to the spin and magnetic axes. This conception dates back at least to the 17th century and Robert Hooke, who believed that it would result in a wholesale exchange of land and sea. Many later investigators agreed, except for Lord Kelvin who calculated that the reorientation could occur with no perceptible change in distribution of land and sea. Alfred Wegener favored this mechanism of polar wandering and supposed that it has been accompanied by marine transgressions and regressions on the continents.

Calculations of the rate at which this type of reorientation, sometimes called global tumbling or toppling, can occur were published in 1955 by Thomas Gold, then at the Royal Greenwich Observatory. Gold concluded that major changes in the distribution of mass or density in the crust can result in rather rapid shifts of the earth's body around the rotation axis due to plastic flow at depth. One of his calculations showed that if a large continent such as South America were suddenly uplifted by 30 meters, the earth would topple over at a rate of 1° per thousand years, which would bring polar regions to the equator in less than 100,000 years. Gold regarded this rate as too high and suggested that it would be more realistic to postulate the rise of the continent by 3 meters in one million years and a large degree of toppling in several million years. He also suggested that toppling may be episodic, that the earth may become dynamically "trapped" in a given orientation until gross tectonic changes bring about an episode of compensatory polar wandering.

Until recently, the main argument against such reorientation involved the earth's disequilibrium figure. A perfect sphere might tumble around its axis; a rotating spheroid with an equatorial bulge should be more stable. The equatorial bulge of the earth is more pronounced by about 0.5 percent than it should be if the planet were rotating in a state of hydrostatic equilibrium. In 1960 W. H. Munk and Gordon J. F. MacDonald analyzed this problem and concurred with an earlier suggestion by Harold Jeffreys that the excess flattening originated at an early date when the earth rotated at a much faster rate and that the bulge was "frozen in" by the strength of the mantle. This implication that the interior is strong enough to maintain a disequilibrium bulge for long periods of geologic time became one of the most compelling geophysical arguments against global tumbling, mantle convection, and continental drift. Clearly, the mantle can be strong enough to resist long-term stress or it can be weak enough to readjust continually to superficial changes. It cannot be both.

In answer to this, S. K. Runcorn proposed that the disequilibrium aspects of the earth's figure are not static but are dynamically maintained by convection cells in the mantle. More recently, to anticipate our story once again, Peter Goldreich of the Massachusetts Institute of Technology and Alar Toomre of the University of California at Los Angeles showed that the bulge, though real, does not alter the planet's axial ratio significantly enough (beyond that expected of a randomly evolving spheroid) to justify an explanation in terms of a fossilized relic of a more rapid rate of rotation. Thus, the excess equatorial bulge has lost much persuasive force as evidence against polar wandering.

The geomagnetic dynamo

These speculations on polar wandering raise again the question of the nature and source of the earth's magnetic field. Astronomical explanations that sought to derive the field from solar radiation or other extraterrestrial sources were abandoned in the 19th century when Karl Friedrich Gauss showed that the spherical harmonics describe a field of internal origin. The presence of a weak external field has recently been demonstrated by space probes, but the predominant field is derived from the earth's interior.

Many people assume, intuitively, that the source

of the field is the earth's core of metallic iron. Yet the temperature of the core is far above the Curie point below which iron metal becomes magnetic. The core cannot, therefore, behave as a huge permanent magnet within the planet. The best alternative appears to involve motion in the liquid core, which, acting as a dynamo, sets up a fluctuating electromagnetic field. What drives the dynamo? How can it have remained self-sustaining throughout geologic time? Convection currents in the liquid have been proposed by Edward Bullard at Cambridge University and by Walter Elsasser at Yale as a source of dynamo power. Some geophysicists object that convection currents would carry upward enough heat to melt the base of the lower mantle. Turbulent motion, sustained by the precession of the earth and with the core and mantle rotating at slightly different rates, was suggested in 1968 by W. V. R. Malkus, of the University of California at Los Angeles, as an alternative to convection. The problem remains unsolved, but the precessional dynamo of Malkus is widely favored at present. If the geomagnetic dynamo is governed by the earth's precession, a close relationship between the magnetic and spin axes appears axiomatic, and it would seem that the evidence for magnetic and climatic polar wandering should always approximately coincide.

Second thoughts on paleomagnetism

In the 1950s Blackett's group in London had concluded that, since paleomagnetic measurements involve determinations of latitudes but not of longitudes, longitudinal motions cannot justifiably be read into the record. These scientists therefore replotted the worldwide paleomagnetic data in terms of continental changes in latitude in the absence of polar wandering. On their new map all continents (except Antarctica) appear to have moved northward and to have undergone some clockwise and counterclockwise rotation, a pattern recalling the "Fliesstheorie" of Gutenberg.

As paleomagnetic studies continued through the decade, voices of dissent began to be heard within the discipline. In 1961 F. F. Evison of New Zealand suggested that all continents are spreading outward under their own weight, in the manner of glacial ice caps, and that this motion, which is more rapid at the surface than in depth, deflects the magnetic alignments of rocks from their original positions, giving a spurious impression of changing polar orientations.

Other critics challenged the permanently dipolar character of the field and the statistical averaging of sample measurements as practiced by paleomagnetists.

The paleomagnetic researches of the 1950s proved disturbing enough to some geologists, specializing in other fields, to prompt them to seek alternative explanations for the evidence. Most, however, were totally unfamiliar with both the principles and the techniques involved and were quietly awaiting some degree of consensus among different schools of paleomagnetists before attempting to evaluate their claims. No doubt some believed that a consensus would never develop and they could comfortably dismiss the whole approach.

Geologists are traditionally attuned to an outdoor life, and many have always felt ill at ease when confronted with "black boxes" in the laboratory. Continental drift and polar wandering had been up for discussion before and had gone down to ignominious defeat. Those who would reopen these subjects would still do so at their peril.

In 1962 a volume edited by S. K. Runcorn summarized researches on paleomagnetism and other evidences of a mobile crust. Despite the evocative title, *Continental Drift*, Runcorn wrote in the preface: "This is not the time for a reappraisal of Wegener's work but it is hoped that this volume will stimulate a serious interest in a subject formerly considered by many earth scientists as already closed."

Intimations of Change

Despite the broad consensus against continental drift, a few investigators faced the peril and kept the question open. Arthur Holmes in England never doubted the validity of the geological evidence for drift or the efficacy of mantle convection as the driving force. Lester King, working in South Africa, carried forward the researches begun by du Toit, while a new advocate appeared in the person of S. Warren Carey, professor of geology at the University of Tasmania.

Remembering the claims and counterclaims regarding the matching of continental margins, Carey, in 1955, tested the fit of Africa and South America by very careful methods of projection. In deciding which isobath should be used, Carey commented: "All will agree without argument that the present coasts should be rejected and that some level on the slope from the continental shelves would give a more significant comparison." He chose the 2,000-meter isobath, which is half-way down the continental slopes, as the most logical site for the match. His results, published in a short article in *The Geological Magazine* entitled "Wegener's South America-Africa Assembly: Fit or Misfit? " demonstrated that the fit is excellent. Referring to Harold Jeffreys' statement (made in 1929 and reiterated in 1952) that "on a moment's examination of a globe this is seen to be really a misfit by almost 15°," Carey wrote: "If continental drift should be rejected, let it be rejected only on valid grounds." He concluded that the argument that South America and Africa do not fit "should never be used against it again."

Carey's out-of-hand dismissal of the present coastlines as the proper site for matching continents

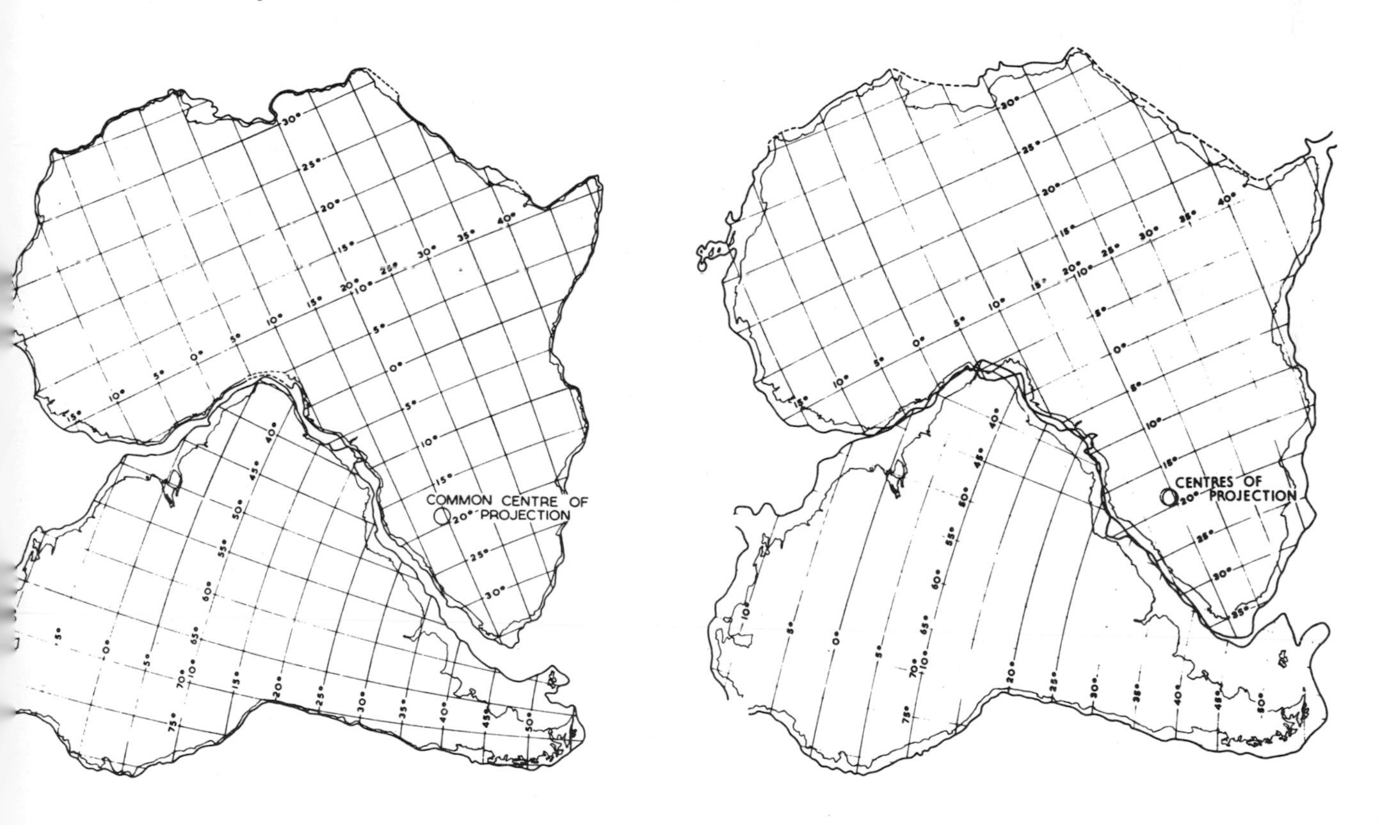

Figure 63. A comparison of the fit of South America and Africa at the 200-meter (100-fathom) isobath as proposed by Alfred Wegener (left) and at the 2,000-meter isobath as proposed by S. Warren Carey. (From *Continental Drift, A Symposium*, 1958; used with permission of S. Warren Carey.)

took on a life of its own, and in subsequent literature Carey often has been cited as the first man to show that continents fit better along their slopes than at their coastlines. As few readers delve back into original sources, the impression became widespread that Wegener and all other drift protagonists, presumably including du Toit, had followed the singularly witless procedure of matching continents at their present coastlines. Although du Toit erroneously made this charge against Wegener, Carey himself never made such a claim. In 1958 he published comparisons of Wegener's match at the 200-meter isobath with his own slightly better results (Figure 63) saying: "Wegener worked out the broad outlines of continental drift and the Paleozoic integration of the continents nearly 50 years ago, but my picture of the assembly does not differ greatly from his . . . I have even returned to Wegener's Pangaea."

Symposium in Tasmania

In 1956, at the University of Tasmania, Carey convened the first of the latter-day symposia on continental drift. It took place from March 6th to 9th, just 29 years and four months after the American symposium. Chester Longwell was the guest of honor. Six other Americans were present but the remaining overseas participants came from Africa, Brazil, India, and Australia, all fragments of old Gondwanaland.

Longwell reviewed some of the arguments he had presented when we last heard from him in his answer to du Toit in 1944. To these he added a summary of newer ideas on ice ages and a short commentary on S. K. Runcorn's recently published polar-wandering curves for Eurasia and North America. Longwell was persuaded of the reality of polar wandering, at least in the sense of the reorientation of the earth around its axes. He was not convinced that the available data were sufficiently exact to prove that either ice caps or paleomagnetic pole positions as determined for one continent were synchronous with other ice caps or pole positions determined for another continent. He therefore read the evidence as favoring polar wandering without continental drift. On that concept he pronounced himself still the friendly skeptic.

The one paper devoted to paleomagnetism was presented by E. Irving of the Australian National University. Irving reviewed the fundamentals of rock magnetism, described the polar-wandering curves for several continents, and showed alternative continental arrangements that could equally well accommodate the magnetic data. Such diagrams have rarely been seen in more recent years. Some solutions, for example, showed two or three of the southern continents superimposed, one atop the other. Irving conceded that, with rock magnetism as with other types of geophysical surveys, the inevitable ambiguities that arise will have to be settled by reference to geological evidence [!].

As the case for a permanently dipolar magnetic field was not yet resolved beyond all reasonable doubt, Irving concluded: "All that can be said at the present time, is that the balance of the evidence favors the idea that not only has the Earth's axis changed its position relatively to the Earth as a whole, but that also the continents have 'drifted' with respect to each other."

Lester King presented two papers—one proposed a new reconstruction of Laurasia and the other discussed the origin and significance of the great suboceanic ridges. King, a professor of geology at the University of Natal, had worked with Alexander du Toit and was an ardent supporter of the idea of continental drift. In 1951 he visited the United States and lectured at several universities, including Harvard and Columbia. At Harvard, King, a spellbinding speaker, ended his talk on erosion surfaces in southern Africa with the comment: "I am now going to show you what will undoubtedly be the most shockingly unorthodox hypothesis to be presented in many years to an audience at this institution." It was a slide depicting Gondwanaland reassembled. Dr. King was right. At least three decades had passed since such a map could have been displayed in that auditorium by any speaker maintaining a straight face. The audience did not stir. Confronted with such whimsy, no one was willing to be forced onto the defensive.

At Columbia, in contrast, some imaginative students arranged a debate between Lester King and Walter H. Bucher, then a professor of geology at that institution. Two more knowledgeable and articulate advocates—one for continental drift, the other for continental permanence—could scarcely be imagined.

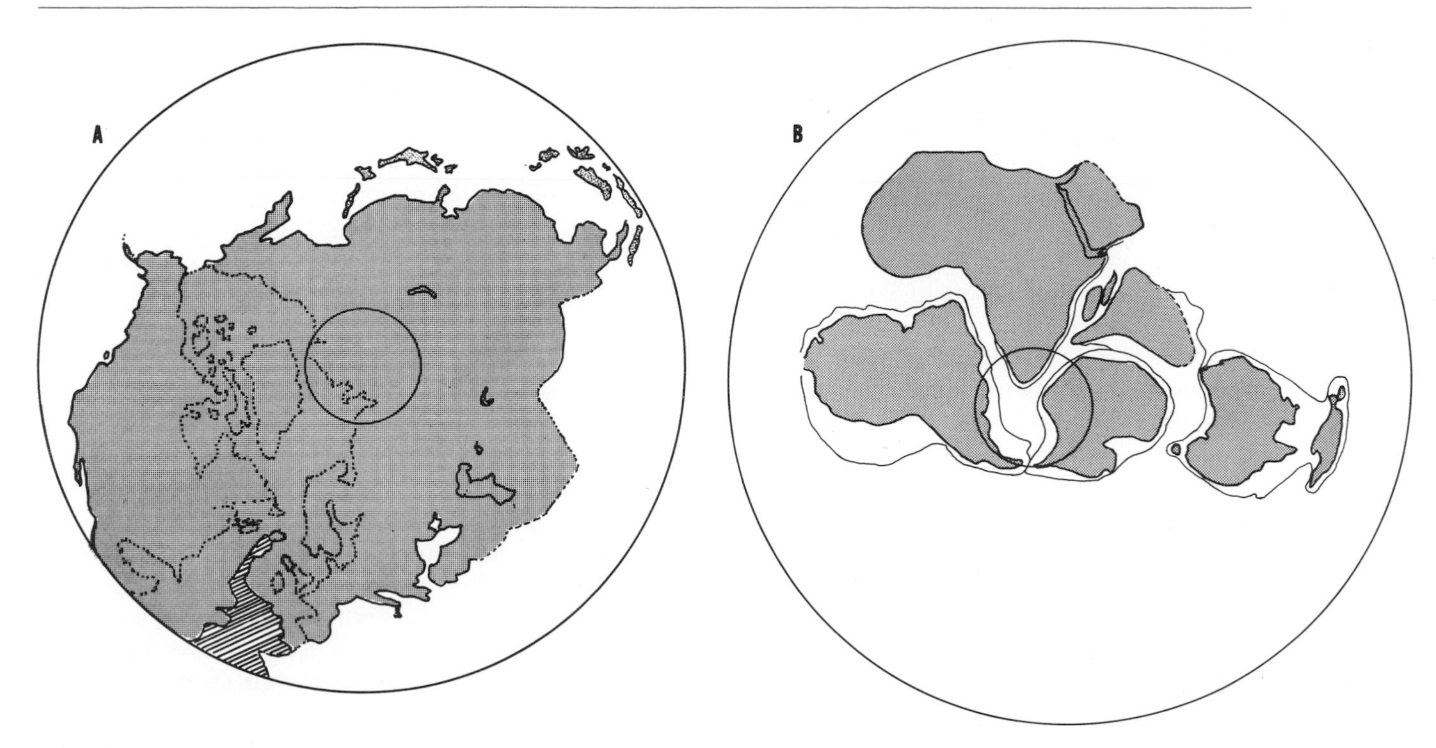

Figure 64. Laurasia and Gondwana. **A**: The landmass of Laurasia as it appeared in the late Jurassic, according to Lester King. This is a simplified version of King's original map which included Paleozoic and early Mesozoic orogenic belts and was dedicated "In memory of Alfred Wegener, author of the hypothesis of Continental Drift, who lies buried in the Greenland Ice Cap at the very heart of Laurasia." **B**: King's reconstruction of Gondwana in the late Paleozoic. The south pole lies at latitude 78° south, longitude 30° west. (From *Continental Drift, a Symposium*; used with permission of the author and of S. W. Carey, convener and publisher.)

Reports on the outcome vary. One version alleges that, in an unofficial vote, the students favored King ten to one—an early manifestation of student independence.

At the Tasmanian symposium, King emphasized the importance of the oceanic ridges, the global extent of which was then newly discovered. Taking the ridges into account he reconstructed two polar continents of almost equal area: Laurasia with 76.4 million square kilometers, centered at 83° N, 65° E; and Gondwanaland with 72.5 million square kilometers, centered at 78° S, 30° W (Figure 64). He dated the breakup of these continents as beginning in the late Jurassic because marine strata older than Jurassic are absent from the margins of the Atlantic and Indian oceans. A schematic cross section depicting Gondwana strata on three continents is shown in Figure 65. Continental drift ceased, according to King, sometime in the Tertiary, as he could discern no visible evidence of motion since the Pliocene.

King concluded that oceanic ridges are of three types. One type is essentially sialic and marked by islands such as Madagascar and the Seychelles, and these ridges have migrated laterally since they were formed. A second type, including the Hawaiian group, are built of Tertiary volcanics and are fixed in position over deep fracture zones. The third type, which includes the Mid-Atlantic Ridge, are partly continental, partly volcanic, and King thought that such ridges mark the lines of rifting where continental fragmentation has taken place.

Alan H. Voisey, of the University of New England, New South Wales, introduced a memorable description (without illustrative diagrams) of the matching of the east coast of Australia with the east coast of North America. Not only did the two match in outline but also in stratigraphy, volcanism, and age. Voisey explained that he had changed sides within

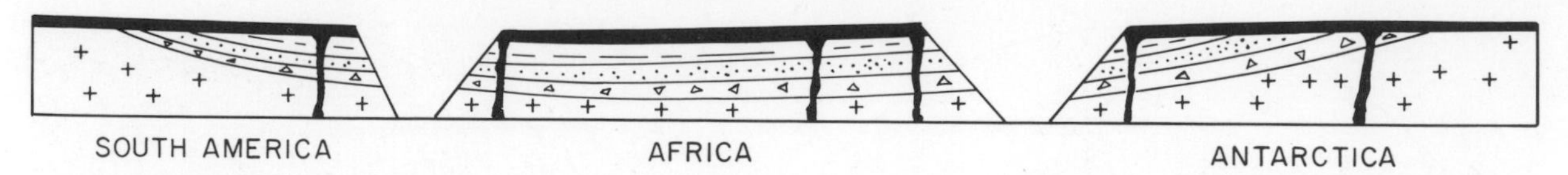

Figure 65. A sketch of the Gondwana strata on three continents. (From *The Morphology of the Earth* by Lester King; used with permission of the author and of Oliver and Boyd, Ltd., Edinburgh.)

the past five years, from being a drifter to an antidrifter. He was now ready to take a firm stand with the continental blocks. He believed in continental accretion, accomplished through cycles of erosion, sedimentation, and orogenesis that are simultaneous throughout the world. As a result, almost any continental margin should record a history much like that of any other. By placing Cape York in eastern Australia against Florida he found striking similarities throughout the stratigraphic section. Yet, as an argument for continental drift, this match would, of course, be absurd.

Voisey predicted that the final answer for or against drift would come from a study of the ocean floors. He still saw reason to believe that sedimentary cores from some areas might preserve a more complete record of earth history than any found on the continents.

Half of the symposium volume, published in 1958, was devoted to a paper by S. W. Carey on "The Tectonic Approach to Continental Drift." After he had matched Africa to South America, Carey tried fitting all of the other continental fragments but met with yawning gaps or crude misfits at every turn. Suddenly the thought occurred to him that he was reconstructing Pangaea on too large a globe—the earth must have expanded since the formation of Pangaea. To test this possibility, Carey developed a model of an expanding planet originally covered by a uniform crust of volcanic greenstones. His early earth, which he thought might have consisted of closely packed nuclear materials, had about one-half the diameter, one-fourth the surface area, and eight times the density of the present planet. It had a much greater gravitational pull and rate of rotation. Carey postulated that expansion, possibly due to heating or to a decrease in the gravitational constant, has resulted in the development of our familiar, diffuse electron-bound atoms and is continuing at an accelerating rate.

Carey interpreted geological structures on the continents in terms of dilation, rifting, relaxation, and creep of crustal materials in response to an expanding interior. He pictured large plates or slabs of the crust as maintaining enough rigidity to rift and move apart with little or no deformation except along the zones of weakness where mountain-building has occurred. He showed how the Atlantic and Arctic oceans might have opened and widened while the Americas, Antarctica, and Australia all swung clockwise some 20° to 30° "like a gate about the pivot of the Alaskan orocline" (Figure 66). He defined oroclines as mountain belts marked by pronounced angles or horseshoe bends. He identified several sites on the continents where pivoting at an orocline appears to have ripped apart the rigid portions of landmasses and opened new oceans. Carey predicted that lineations in the Atlantic floor should be found marking the course of the continental separation.

In his ardent championship of continental drift Carey was in advance of the times. In 1966 he visited the United States and spoke at a number of universities with such persuasiveness that, much to the discomfort of geological faculties, he aroused the interest of many graduate students in continental drift, an idea totally new to them. Some of Carey's structural terms—including rhombochasm and sphenochasm to designate ruptures in the crust that are rectangular or triangular, respectively—have come into wide use. However, his general view of continental displacement lost much of its credibility because it was associated with the idea of earth expansion.

The theory of the expanding earth was championed in 1933 by O. C. Hilgenberg, in Germany, whose work is remembered mainly for an illustration depicting a series of globes, three of which are shown in Figure 67. Otherwise, Hilgenberg's ideas were ignored because he postulated an increase in the

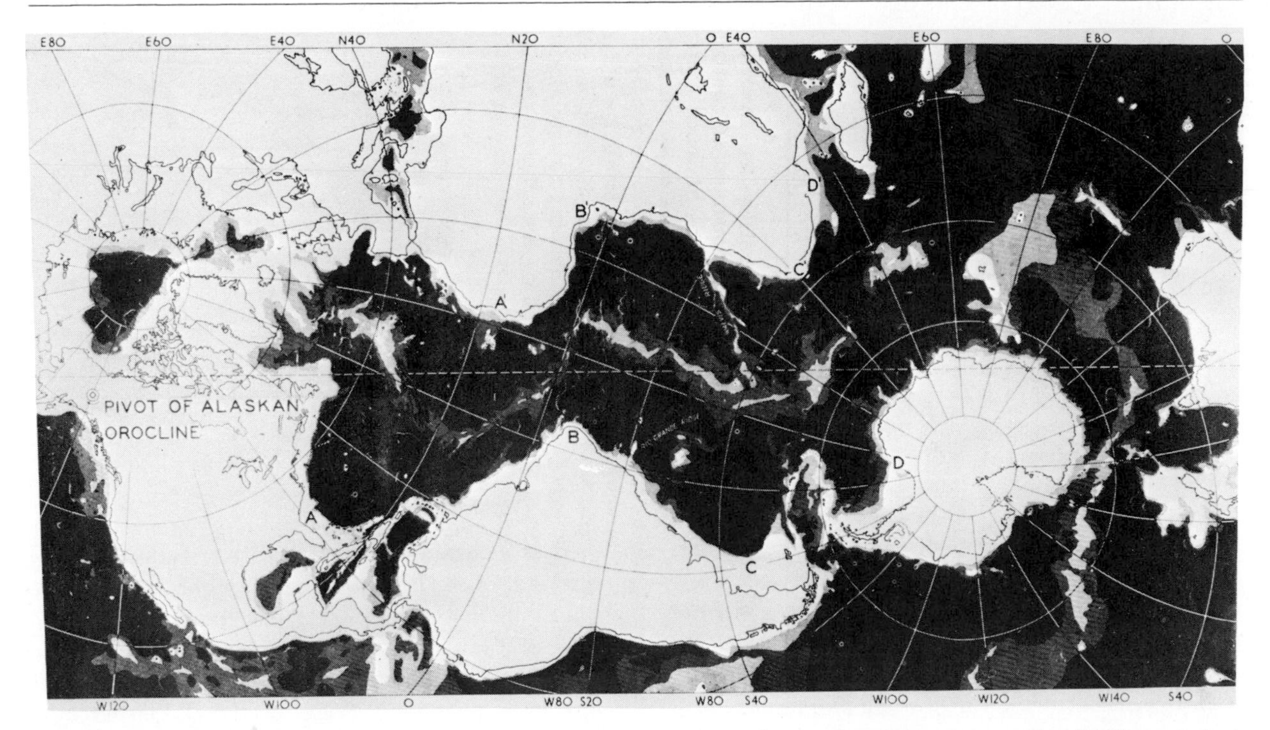

Figure 66. The Atlantic Ocean opening as a deep rhombochasm while the earth expands and the western strip of sialic crust pivots around the Alaskan orocline. (From *Continental Drift, A Symposium*, 1958; used with permission of S. Warren Carey.)

earth's mass as well as its volume. In 1935, J. K. E. Halm, a South African astronomer, proposed the idea, later adopted by Carey, that the earth is expanding because it originally consisted in part of densely packed atomic matter. Halm defended his hypothesis by reference to the stars called white dwarfs which have densities ranging up to 100,000 times that of the earth and can be explained only as compressed aggregates of atomic nuclei. However, Halm's primeval earth had five-sixths the diameter and twice the density of the present planet and therefore was neither as small nor as densely packed as that of Carey. Expansion was favored again in 1955 by L. Egyed, director of the Geophysical Institute of Budapest. Egyed also believed that the interior consists of close-packed nuclei, and he attributed expansion in part to a decrease in the gravitational constant. In 1958 Bruce C. Heezen of the Lamont Geological Observatory advocated expansion because the rifted character of the oceanic ridge system seemed to demand such an explanation.

The idea of earth expansion never attracted many followers partly because an expansion of the planet due to phase changes in atomic matter or a decrease in the gravitational constant should also occur in the sun, altering appreciably the intensity of solar radiation. But the earth's fossil and climatic record shows no signs of any progressive change in the influx of solar heat. Bruce Heezen abandoned the idea of expansion in 1966 when he calculated that it would require a radial expansion of 4 to 8 millimeters per year for the past 200 million years to account for the width of the Atlantic Ocean alone. An upper limit of 0.8 millimeters per year for the possible expansion since the Devonian was calculated in 1963 by J. W. Wells of Cornell University. Wells counted the daily growth rings on living corals and on fossil ones of Devonian and Upper Carboniferous age. He found that modern corals produce about 360 growth rings per year, which suggests that they skip a day on rare occasions of no sunshine. Devonian corals 370 million years old produced about 400 rings per year; Upper Carboniferous ones 280 million years old about 387 rings per year. Taken as daily timekeepers, these corals show that the turning of the earth from sunrise to sunrise has been gradually slowing down at a rate

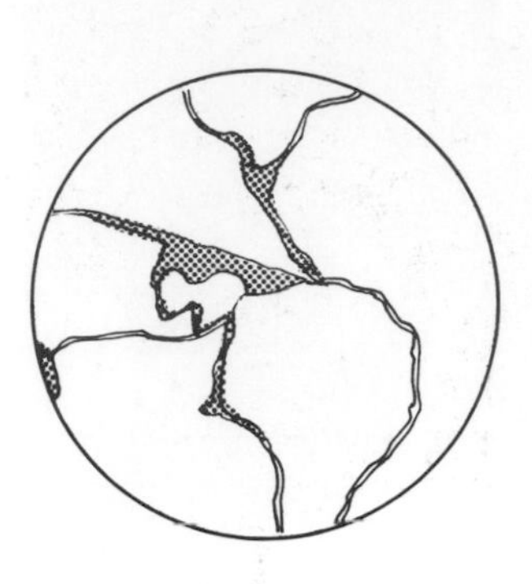

Figure 67. Three stages of crustal rifting on an expanding earth. (After O. C. Hilgenberg, 1933.)

of about 2.2 seconds per 100,000 years, so that the days are longer but fewer of them are crowded into each year. The diminishing rate of rotation is ascribed mainly to the frictional drag, from the moon's tidal attraction, of waters on the floors of shallow-shelf seas. Expansion of the earth would produce the same effect because the conservation of angular momentum requires the slowing down of an expanding, rotating body. However, the actual rate of change, indicated by corals in the past 400 million years, has been too slow by more than an order of magnitude to account for the area of the oceans. This appears to rule out earth expansion as a significant factor in global tectonics.

Ocean-floor spreading: an essay in geopoetry

A proposal that, in retrospect, generally is taken as opening a new epoch in ideas about earth science was made in 1960 by Harry H. Hess at Princeton University. Formal publication came two years later in a short article entitled "History of the Ocean Basins" in a volume honoring the petrologist A. F. Buddington that was published by the Geological Society of America.

"I shall consider this paper an essay in geopoetry," wrote Hess. "In order not to travel any further into the realm of fantasy than is absolutely necessary I shall hold as closely as possible to a uniformitarian approach; even so, at least one great catastrophe will be required early in the Earth's history."

Hess assumed that the earth was formed by the accretion of small particles having a bulk composition similar to that of the sun. While the accreting cloud was still fairly dispersed, vast amounts of gases, including hydrogen, neon, carbon, and water, were lost by evaporation. Some of these volatiles were trapped within the earth during the final stages of accretion, but the surface of the primitive planet was devoid of any ocean or atmosphere. Immediately after its formation the earth was heated and melted by the decay of short-lived radionuclides until one large convection cell formed. In one-half of an overturn this cell promoted the sinking of heavy metal to form the earth's core and the rising of sialic material to cover that half of the globe that lay over the rising limb. The other half never acquired a crust of primitive sial. Thus, Hess saw the earth's bilateral asymmetry—one-half continental, one-half oceanic—as having formed at the beginning of geologic time. This was his "great catastrophe": core formation and crustal asymmetry established by a process that operated only once. After core formation, a single global convection cell would be geometrically impossible. The currents would divide into smaller cells circulating within the confines of the mantle. The changing patterns of these cells would control the subsequent distribution of continents and oceans.

A single, primeval, planet-wide convection cell had been proposed by Harold Jeffreys and also by F. A. Vening Meinesz, from whom Hess borrowed the illustration reproduced here as Figure 68. Convection currents in the mantle were favored by numerous scientists, among whom Hess listed Arthur Holmes,

David Griggs, and Jean Verhoogen. It puzzled Hess, therefore, that convection was still considered a radical hypothesis unacceptable to most geologists and geophysicists. To him it was the essential key to earth history.

Hess calculated that about one-half of the continental sial and one-third of the ocean water were extruded from the interior in the initial convective overturn. The remaining sial and water have been derived from the mantle over the past four aeons. Hess believed that the mantle is predominantly peridotite and that the bedrock of the ocean floors (layer 3) is largely serpentinite, formed by the hydration of the peridotite at low temperatures. This aspect of the thesis resembled Alfred Wegener's idea that the ocean floors are surface exposures of mantle rock. Hess proposed that the oceanic ridges are formed over the rising limbs of convection cells and that when these limbs reach the surface they spread laterally to either side in the manner of opposing conveyor belts. Heated waters, brought up with the rising limbs, reach a level where the temperature is below 500° C. In this zone the waters react with the peridotite to form serpentinite, which rides horizontally away as a low-density rind on the upper mantle and slowly becomes blanketed with marine sediments. Where the spreading ocean floors plunge downward again, along disturbed zones such as the Pacific margin, the rind is reheated, the water is expelled into the oceans, and the serpentinite is transformed back into peridotite. ("The great advantage of serpentine–it is disposable," said Hess.) The unconsolidated sea-sediments ride down into the "jaw crusher" between the ocean floor and continents or island arcs, then they are heated, partially metamorphosed, and welded to the landmasses. Thus, by one and the same agency (mantle convection) Hess explained the youth of the ocean floors, the growing volume of ocean water, the origin of ridges and trenches, and continental accretion.

Hess also explained continental drift. He assumed that where the limbs of new convection cells rise under continents the sialic crust is split open and the fragments are carried passively to either side at equal rates, which he calculated at about 1 centimeter per year. This uniformity of spreading preserves the median position of ridges while continental rift valleys widen to new oceans. If, however, the limbs of a new convection cell rise beneath an ocean there is no reason for the ridge to be centered. This, according to Hess, explains the peripheral position of the East Pacific Rise.

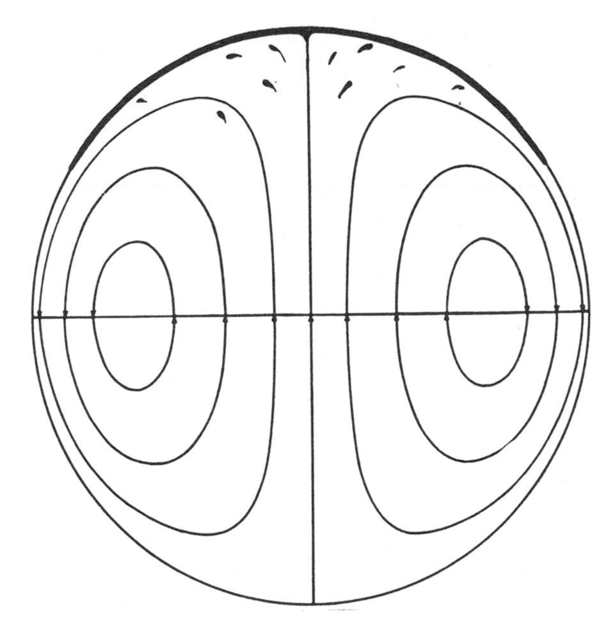

Figure 68. The single, primeval, planet-wide convection cell which, according to Hess, extruded sial (heavy black layer) over the rising limb and established the asymmetrical distribution of the earth's sial and sima. (Used with permission of the Geological Society of America.)

In speculating on the early history of oceanic ridges, Hess recalled the guyots he had discovered 15 years earlier in the Pacific, with their eroded platforms lying 2 kilometers below sea level and their enigmatic lack of coral. He concluded that the guyots bear witness to the existence in the Mesozoic of a ridge that trended southeasterly from the Marianas toward Chile. Volcanic islands formed on the ridge were beveled at sea level and then carried downslope toward the northeast and southwest by the spreading ocean floor. When the pattern of convection changed, the ridge subsided, leaving no trace except the deeply submerged Darwin Rise, which Hess regarded as the only known fossil ocean ridge. He supposed that the lifetime of a convection cell, and therefore of a ridge, would be about 200–300 million years. The lack of coral on the guyots he now attributed to the cold temperatures of the northwestern Pacific waters when the Permocarboniferous north pole lay in Siberia.

A spreading ocean floor, with new material

constantly added at the ridge and the old floor constantly destroyed, would account not only for the thinness and youth of sea-sediments but also for the surprisingly small number of volcanic seamounts in the oceans. Some 10,000 of these were known in all oceans. Hess believed there should be ten times as many. Seamounts (and guyots) too, he reasoned, must ride down into the jaw crusher and be eliminated. He suggested that the volcanic islands and seamounts should be progressively older with increasing distance from the oceanic ridges.

Hess did not dwell on the fact, implicit in his model, that, in all of geologic time, continents and oceans have never migrated far enough to disturb the global asymmetry that was established at the beginning. He did, however, provide new insights on many problems, including the evolution of ocean water. Shallow seas have always overlapped the continents, but this would occur, he said, regardless of the volume of ocean water. Continents and oceans are balanced isostatically. Therefore, however shallow or deep the oceans may be, the continents will always be eroded until they stand just above sea level. As ocean water has grown in volume by escaping from the mantle to the serpentinite of the ocean floor and thence to the open ocean, the continents have kept pace by a compensatory rise.

Hess' new model of earth tectonics was championed and given the name sea-floor spreading by Robert Dietz, then at the U.S. Coast and Geodetic Survey, who substituted the eclogite-basalt phase change for the peridotite-serpentine transformation as an explanation for oceanic bedrock. Otherwise, the idea gained few immediate supporters. After more than a century of adherence to the doctrine of permanence of ocean basins, and several decades of determined opposition to the idea of moving continents, earth scientists were not yet prepared to contemplate the wholesale creep of the ocean floors.

Magnetic lineations on oceanic ridge flanks

In 1963, the year after Hess published his article on ocean-floor spreading, magnetic anomalies occurring in long, linear strips along the midocean ridge flanks were discussed by F. J. Vine and D. H. Matthews of Cambridge University in a short article in *Nature*. Such anomalies had been measured by various investigators in the North Atlantic, Antarctic, and Indian oceans and they all followed much the same pattern: a pronounced central anomaly associated with the median valley of the ridge, narrow strips over the rugged mountains on either side of the valley, and broad anomalies over the foothills. Vine and Matthews suggested two possible explanations for these anomalies: (1) they record differences in the strength of magnetization in alternate slabs of bedrock and are in part topographically controlled, or (2) they register the effects of magnetic pole reversals in the basalt layer of the spreading sea floors. To Vine and Matthews, the second alternative seemed the more plausible.

When the Vine and Matthews paper (slightly less than three pages long) was published it no doubt sparked the interest of a few scientists who were studying paleomagnetism and the ocean floors, but by and large it went unnoticed. Reversals of the earth's magnetic field still seemed implausible to the majority of scientists. So did ocean-floor spreading. A combination of the two presumably doubled the improbability.

The London symposium

On March 19 and 20, 1964, a symposium on continental drift was sponsored by the Royal Society in London. Although it was not the first international symposium on the subject since the famous American one of 1926, the symposium in London was pivotal—it changed opinions. Up until then, most earth scientists had rested easy in the assumption that continental drift had been effectively killed and that the search for truth led elsewhere. Perhaps the majority continued to feel that way, but some were beset by serious doubts. Patrick Hurley, who had helped develop evidence for continental accretion, probably expressed a common feeling with the comment: "I went to London orthodox; I came home a drifter."

What happened in London that was so portentous? Today nothing appears new or startling in the symposium volume, published in 1965. There was no emphasis on the idea of sea-floor spreading and no discussion at all of linear magnetic anomalies along

oceanic ridge flanks. Yet, the aggregate of results that had been slowly accumulating over the past decade was compelling. Few could contemplate the discoveries in paleomagnetism or oceanography and suppress a mounting excitement along with a suspicion that the theories of permanence and of continental accretion might need revision.

The presentation that fired many imaginations was a map showing the fit of the Atlantic continents that was prepared by Sir Edward Bullard of Cambridge University and his colleagues J. E. Everett and A. G. Smith. Their continental reconstruction (Figure 69) was achieved by a computer (not a "modern computer" of the type for whom Alfred Wegener had transcribed the Alphonsine Tables but a mechanical EDSAC-2) that had been programmed to find the best fit of the continental margins. The first fit to be tried and error-tested by the least squares method was that between Africa and South America. The next step tested the matching of Europe, Greenland, and North America. The results, according to the authors, exceeded all expectations. In both hemispheres the Atlantic continents matched along the 500-fathom (1,000 meter) contour of the continental slope with minor gaps and overlaps of only about one degree (30-90 km). Finally, an attempt was made to match the northern to the southern continents en bloc, but here the results were much less impressive.

Of the matches across the North and South Atlantic, Bullard and his colleagues proposed two possible explanations: either the fit is meaningless and on a par with the chance resemblance of the coast of Italy to a boot, or it is meaningful, and the continents, once united, were separated by the formation of the Atlantic Ocean.

If the continental reconstructions of Alfred Wegener, Alexander du Toit, Lester King, and S. Warren Carey failed to impress geologists, we may well ask why the map presented at the 1964 symposium by Bullard, Everett, and Smith was suddenly and widely taken as significant. Cynics have ascribed its acceptance to the fact that it was drawn by a computer rather than by a scientist. As another very important factor we can simply suggest that the time was ripe.

The fit of the Atlantic continents as displayed on the Bullard map could, in the 1960s, be tested by geochronological methods. The earlier attempts of Wegener and others were often in doubt because of the lack of precise dating of lithologic units. At the London meeting J. A. Miller of the University of Cambridge presented evidence that rock provinces, separated into six age groups on the basis of both the K/Ar and the Sr/Rb dating methods, could be matched across the North Atlantic on a reconstruction similar to that of Bullard, Everett, and Smith. P. M. Hurley resolved to test the matching of dated provinces between Africa and South America.

After returning from London, Hurley arranged a collaborative effort between the geochronology laboratories at Massachusetts Institute of Technology and the University of São Paulo, Brazil. Samples from well-documented sites were solicited from geologists working in Brazil, Venezuela, and western Africa. European geochronologists had already established the location and trend of a sharp boundary in Ghana separating two Precambrian rock provinces, one more than 2,000 million years and the other about 600 million years old. This boundary strikes southwest past Accra until it is truncated by the Atlantic coastline. Using Bullard's reconstruction of Africa and South America, Hurley and his colleagues predicted that the boundary would be found again in the vicinity of São Luis in northeastern Brazil.

At São Luis the Precambrian basement is largely covered by younger Paleozoic strata, but numerous drill holes had been put down during explorations for oil, and, west of the city, the ancient formations are exposed in a few small "windows" eroded through the younger formations. After basement samples were finally acquired and dated by the K/Ar and Sr/Rb methods, the predictions were borne out in splendid fashion: the rocks fell into the same two age groups separated by a contact precisely in line with the one in Ghana (Figure 70). This result, announced in 1967, proved crucial in persuading many geologists of the probability of continental drift. In subsequent work the same laboratories correlated several more rock provinces between Africa and South America that match in petrology and ore deposits as well as in age.

Transform faults

One of the problems discussed by several participants at the London symposium involved transcurrent or strike-slip faults, which are essentially vertical frac-

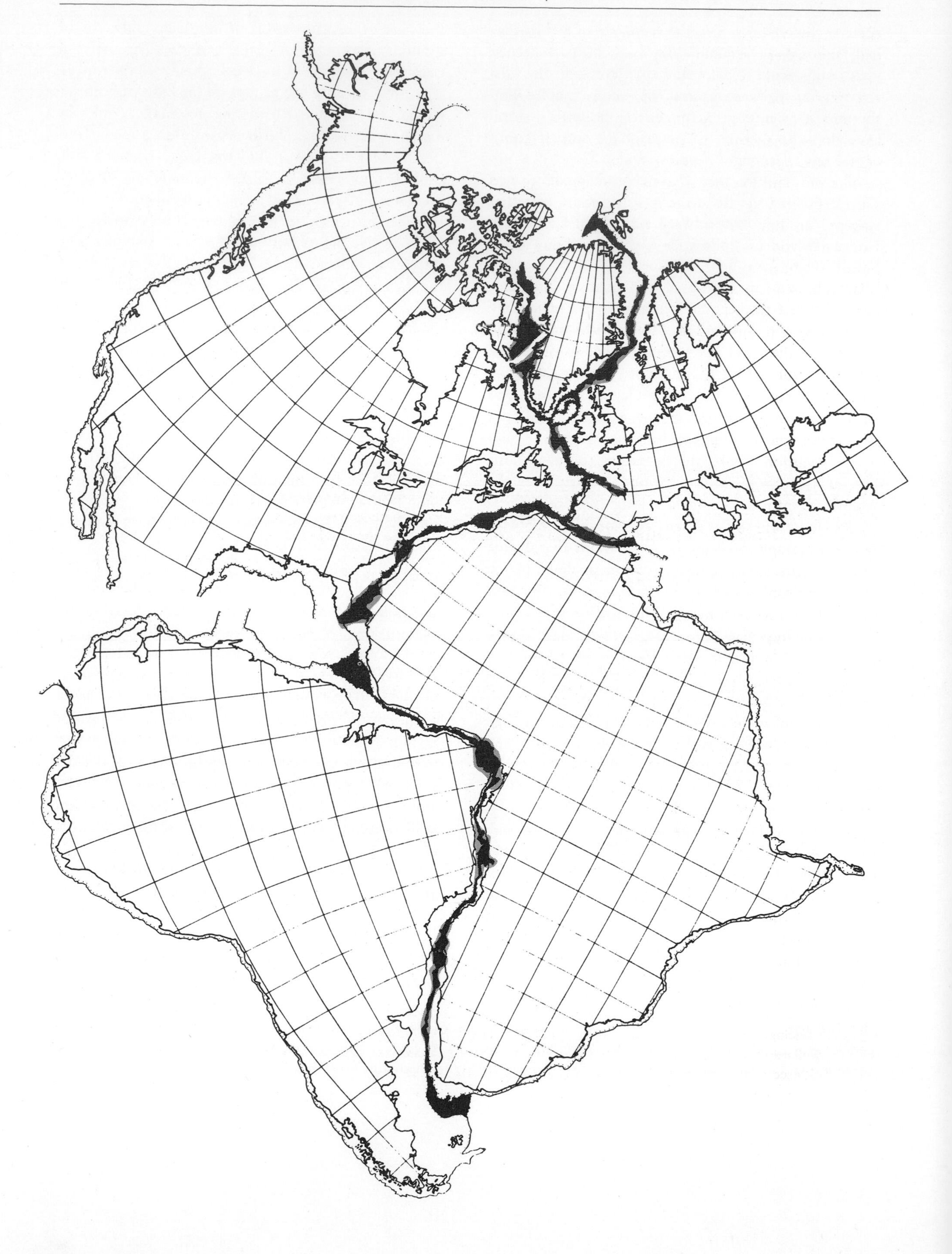

tures along which one wall has apparently sheared past the other for very long horizontal distances. Small-scale strike-slip faults with displacements of a few meters or even a few kilometers are familiar features of structural geology. But numerous faults having displacements of hundreds or thousands of kilometers were now known. These were traced in California, Turkey, Japan, and New Zealand, and they had also been discovered, by means of magnetic surveys, on the eastern floor of the Pacific Ocean. Relative motions of this scale took on a disturbingly global significance.

One mysterious aspect of these great faults arose from the fact that they end abruptly in undisturbed crust. How is it possible, geometrically, for a fault to develop in the crust, for one rigid wall to shear past the other for thousands of kilometers, and for the fault trace to suddenly vanish? Theoretically, at least, there seems no reason why a transcurrent fault should not continue right around the globe.

The solution to the problem of transcurrent faults was provided by J. Tuzo Wilson in 1965 when he introduced into geology the new structural concept of the transform fault. When we last spoke of Wilson he was developing evidence for the marginal accretion of continents along permanent fracture zones in the mantle. Wilson's imagination was captured by Hess' proposal of ocean-floor spreading, and he devised a test for the hypothesis. At the London symposium he presented a map showing that the islands of the Atlantic Ocean, dated by both fossiliferous and radiometric methods, are younger in the vicinity of the ridge and older near both coastlines, as though they were indeed created at the ridge and conveyed toward both shores. His data (disputed by H. W. Menard) were the main evidence for ocean-floor spreading presented at London.

A year after the London symposium Wilson published in *Nature* a short article describing transform faults. The name "transform" refers to the fact that, unlike transcurrent faults where the shear motion continues indefinitely, the motion on these faults ends abruptly where it is transformed into a zone of extension or compression.

Figure 70. The matching of age provinces between Africa and South America. Dark gray areas are at least 2,000 million years old; stippled areas, more than 600 million years old. The heavy line marks the dated contact that extends from the vicinity of Accra, Ghana, to that of São Luis, Brazil. (Compiled from data of P. M. Hurley. Used with permission.)

Transform faults are a natural consequence of ocean-floor spreading, as shown in Figure 71. Here, along a ridge-to-ridge transform fault, all slabs move directly away from the ridge crests where a new ocean floor is generated. As plate A shears past plate B the relative motion along the fault plane is directly opposite to that envisioned in classical transcurrent faults. And the faults come to an end at or near the ridge crests.

Given ridge-to-ridge transform faults of this geometry, the crests need never move closer together or farther apart as a result of sea-floor spreading. Why, then, are the ridge crests displaced into short offset segments? Most investigators today regard the offsets

Figure 69 (facing page). The fit of the Atlantic continents at the 500-fathom (1,000 meter) isobath, as prepared by a computer for E. C. Bullard, J. E. Everett, and A. G. Smith. Areas of overlap, light shading; gaps, black. (From *A Symposium on Continental Drift*, 1965; used with permission of Sir Edward Bullard and The Royal Society.)

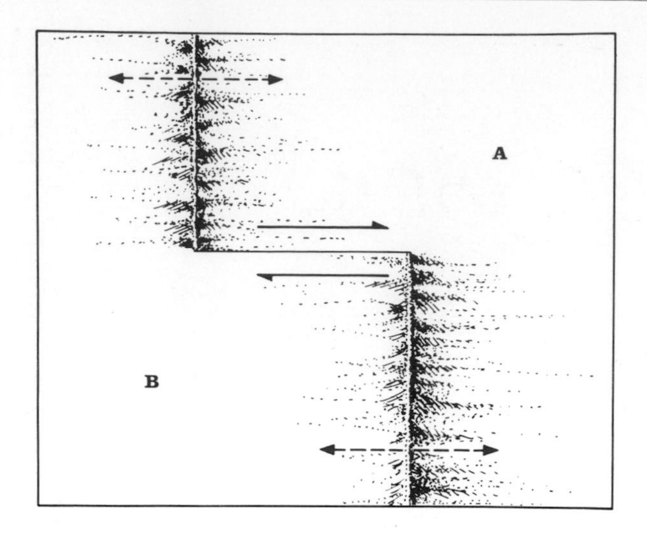

Figure 71. A transform fault connecting two segments of a spreading ridge. While the ridge segments remain fixed, plates A and B shear past each other in the direction indicated by the arrows.

as primary features arising from imperfections in the original ridge-forming process. This explanation appears reasonable enough until one recalls that Hess and others envisioned the ridges as governed by gigantic convection cells in the mantle. Convection is a means of heat transfer and as such is not likely to end abruptly at closely spaced fault planes and to pick up again hundreds of kilometers to either side. Transform faulting is a structural necessity in ocean-floor spreading. The causes of ridge formation and spreading are still not understood.

Proof of ocean-floor spreading?

Two of the papers presented at the Geological Society of America meeting at San Francisco in November 1966 proved fundamental to the new interpretation of the earth. The first, "Geomagnetic Reversals: A Practical Tool for Global Stratigraphic Correlations," by Alan Cox, Richard Doell, and Brent Dalrymple of the U.S. Geological Survey, presented conclusive evidence that volcanic rocks and marine sediments record reversals of the magnetic field that are worldwide in effect. The authors had correlated paleomagnetic measurements with K/Ar age determinations on the same samples and established the timing of nine pole reversals that have occurred within the past four million years.

The second paper, by F. J. Vine, was entitled "Proof of Ocean-Floor Spreading? " Despite the question mark in the title, Vine entertained no doubt that ocean-floor spreading is a worldwide phenomenon demonstrated by the pattern of magnetic anomalies on the flanks of the midocean ridges. He stated his case in forthright terms:

> It is suggested that the whole history of the ocean basins in terms of ocean-floor spreading is "frozen in" in the oceanic crust. Variations in the intensity and polarity of the earth's magnetic field are considered to be recorded in the remanent magnetism of the igneous rocks as they solidify and cool through the Curie temperature at the crest of an oceanic ridge and subsequently spread away from it at a steady rate.

"I have only one objection!" shouted a California professor of planetary physics at a lecture on ocean-floor spreading in December 1966. "It violates every principle of geophysics!"

But the floodgates were open upstream, and by January 1967 nearly 70 abstracts on sea-floor spreading were submitted for presentation at the Washington meeting of the American Geophysical Union to be held in April. Scientists all over the world examined their data in a new context and saw the pieces falling into place. Four sessions of the four-day meeting in Washington were devoted to sea-floor spreading. Additional papers on the subject dominated sessions on tectonics, seismology, petrology, geochemistry, and paleomagnetism. A crowd overflowed the largest available auditorium to hear Harry Hess recount the history of the concept of sea-floor spreading.

The doctrine of the permanence of continents and ocean basins, born in America in November 1846, died in America in April 1967.

The Moving Lithosphere

The evidence for sea-floor spreading that proved so universally persuasive rested upon the correlation of magnetic anomalies on the flanks of the oceanic ridges with a time scale for reversals of the earth's magnetic field.* This time scale, in turn, was worked out by measuring the remanent magnetism in samples of young lavas from many parts of the world and determining their age by the potassium-argon method. Thus were established two epochs of predominantly normal (north-seeking) polarity alternating with two epochs of predominantly reverse (south-seeking) polarity within the past 4.5 million years. Each of the main epochs included short-term events (lasting less than about 100,000 years) when the field was opposite to that predominating for the epoch. The short events were named for localities; the main epochs for the pioneers of geomagnetic research—Brunhes, Matuyama, Gauss, and Gilbert.

The linear anomaly belts along oceanic ridge segments were matched with periods of normal and reversed polarity—at first, not by magnetic or age determinations on deep sea samples but by testing various estimated rates of spreading until a reasonable fit was obtained with the widths of the belts.

Correlating the magnetic and age data measured on continental samples with the oceanic anomaly belts, Vine, in 1966, presented his case for sea-floor spreading and cited the following estimated rates of motion, in centimeters, at several of the more active ridge segments:

Locality	*Spreading rate (cm per year)*	*Total separation (cm per year)*
Reykjanes Ridge, south of Iceland	0.95	1.8
Mid-Atlantic Ridge (lat. 38° S)	1.5	3.0
Carlsberg Ridge (lat. 5° N, Indian Ocean)	1.5	3.0
Red Sea floor	1.0	2.0
Juan de Fuca Ridge, south of Vancouver Island	2.9	5.8
East Pacific Rise (lat. 51° S)	4.4	8.8

A year later, in 1967. Neil D. Opdyke and his colleagues at Lamont Geological Observatory reported that sedimentary cores from the deep oceans also record a succession of pole reversals that can be precisely dated by their microfossil content. In cores from the Antarctic Ocean they established several more short reversal events within the past 4.5 million years. Their data, and also those from various igneous rocks of the continents, indicate that pole reversals are accomplished within a short time span of a few hundred years to no more than about 5,000 years.

*The phenomenon of self-reversals within rocks had been shown to be a peculiarity of certain titanium-bearing iron oxides and of no overall importance in the paleomagnetic method.

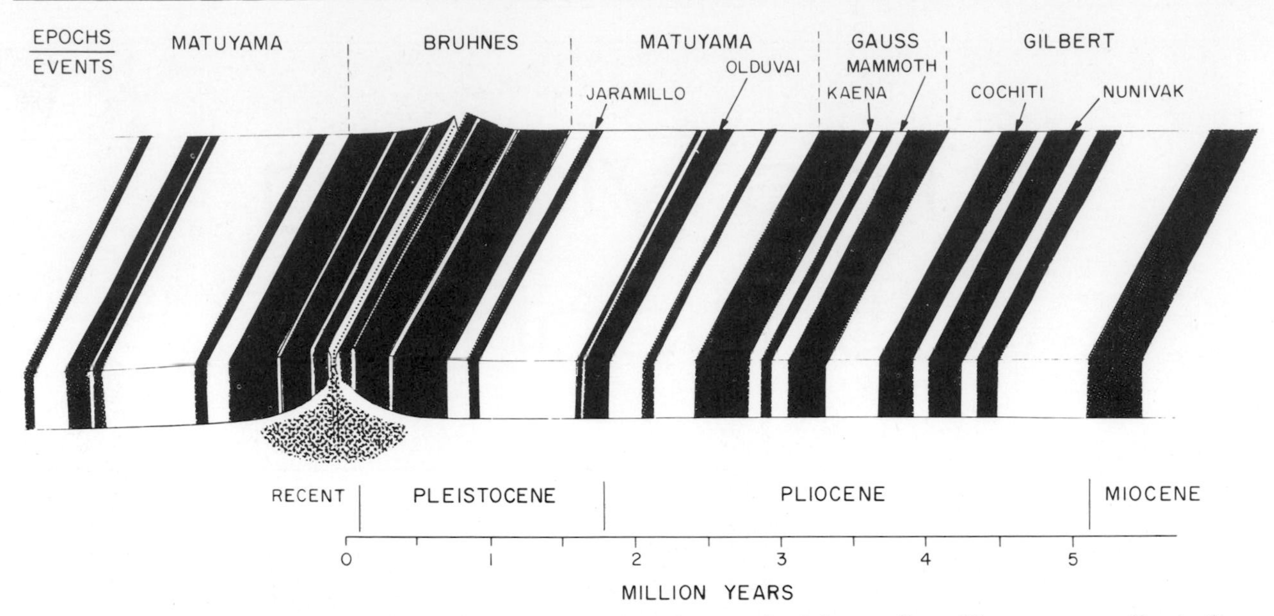

Figure 72. The formation of linear belts recording the magnetic pole reversals of the past five million years, according to the principle of sea-floor spreading. Dark stripes represent normal polarity; light stripes, reverse polarity. (Compiled from several sources).

Figure 72 illustrates the succession of magnetized belts that should occur in a sea floor that has been spreading at a uniform rate for the past 5 million years.

Further confirmation of sea-floor spreading came from the earthquakes generated at the fault planes offsetting the ridge crests. After J. Tuzo Wilson introduced the concept of transform faults in 1965 it was widely assumed, without proof, that the fault zones crosscutting the oceanic ridges were of that variety. The actual identity of these planes as transform faults was established by seismological evidence developed in 1967 by Lynn Sykes of Lamont Geological Observatory. Sykes found that crustal motion is limited mainly to the fault planes connecting the ridge segments and that, along these faults, the sense of first motion in earthquakes is in the direction of shear appropriate to transform faulting. Thus, evidence from seismology, paleomagnetism, and geochronology combined in support of sea-floor spreading.

Once the probability was established that sea-floor spreading has occurred in the last 4.5 million years, geologists did not hesitate to extrapolate backward many more millions of years and postulate a long history of spreading based on worldwide measurements of the oceanic magnetic anomaly belts. The map reproduced in Figure 73 was worked out in 1968 by J. R. Heirtzler and his colleagues (also at Lamont Geological Observatory, one of the most active centers of research on sea-floor spreading). In this map the information provided by the magnetic anomalies has been translated into estimated distances of spreading during intervals of 10 million years; therefore, the map depicts the estimated age of the ocean floors on either side of the ridges. Where the magnetic data was most complete, anomaly belts gave evidence of at least 171 reversals of the geomagnetic field during the past 76 million years—which goes back to the end of the Cretaceous.

The "lines of equal age" on the map indicate different spreading rates in different oceans and across separate ridge segments in the same ocean. Spreading appears to have been continuous throughout the Tertiary in some areas, episodic in others, and dormant for long periods in still others. The highest rates, ranging up to 10 centimeters per year (total separation), appear to have occurred over the East Pacific Rise. In the Atlantic, spreading has apparently been more continuous in the south than in the north, but the average rate of separation of about 2 centimeters per year is of the right order to account

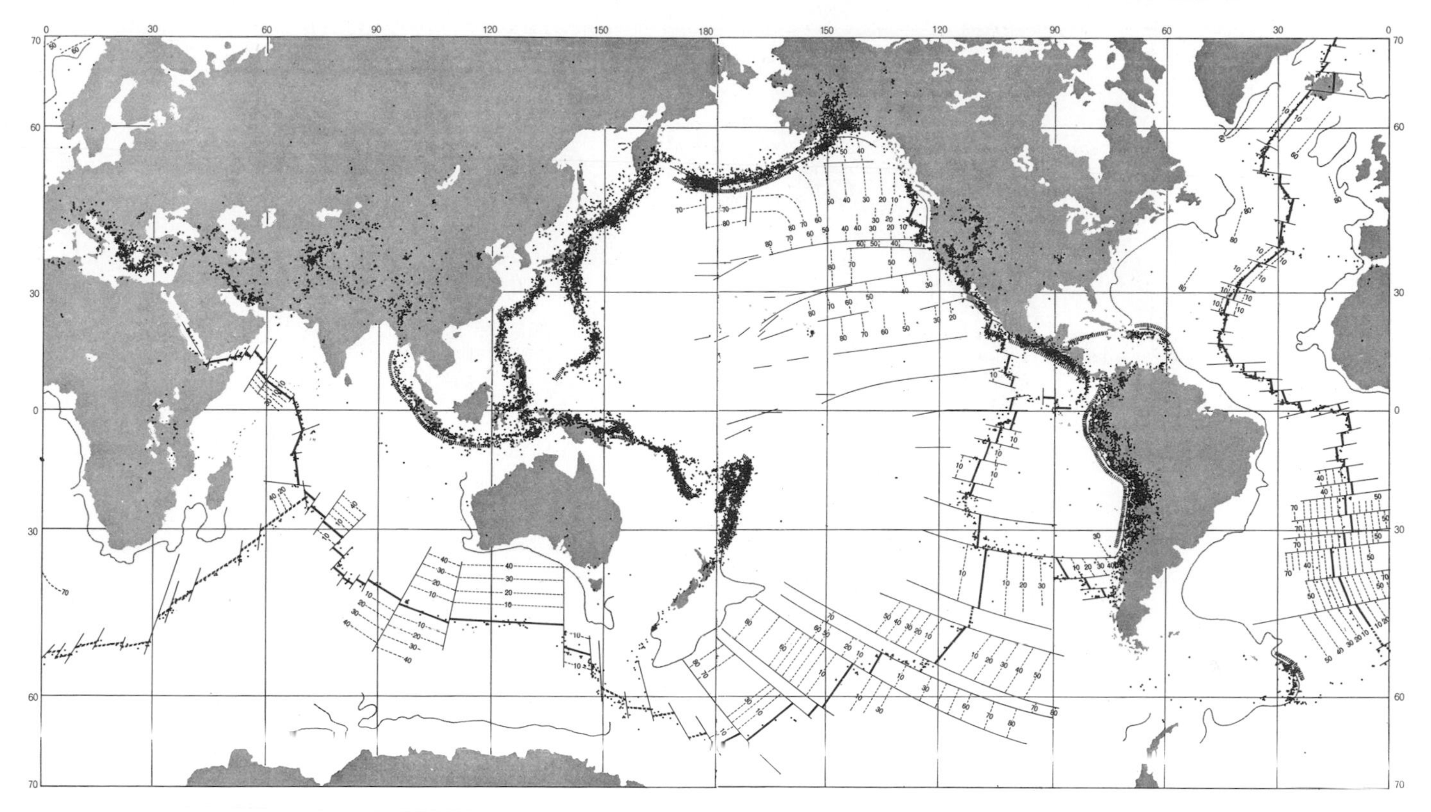

Figure 73. A plan of the ocean floors illustrating the configuration of midocean ridges (heavy black lines), transform faults and transverse fracture zones (gray lines), deep trenches (hatching), earthquake epicenters (gray dots), and "isochrons" or lines of equal age estimated from magnetic data (dashed lines labeled with numbers indicating the proposed age in millions of years). (From "Sea-Floor Spreading" by J. R. Heirtzler. Copyright 1968 by Scientific American, Inc. All rights reserved.)

for the separation of the Americas from Euro-Africa since the Jurassic, about 180 million years ago.

The concept of sea-floor spreading broke with static-earth geophysics and implied mobility on a scale never before seriously contemplated even by Alfred Wegener, who had limited major displacements to the continental blocks. Sea-floor spreading was quickly seen, however, to engender its own special inconsistencies. An immediate dilemma was posed by the global pattern of the oceanic ridges and the manner in which they encircle some continental fragments, such as Africa and Antarctica. If a new ocean floor, generated at the ridge crests, is moving slowly to either side, carrying or pushing the passive continental blocks, then where is Africa drifting? Is it drifting eastward from the Atlantic ridge, westward from the Carlsberg Ridge, or is it compressed between two descending slabs of ocean floor? Africa is not rimmed by deep trenches, folded mountain ranges, or any other sign of convergence. Indeed, the rift valleys of East Africa are commonly described as a zone of tension fracturing. Thus, Africa would appear surrounded and also split by rising limbs of convection cells with no intervening zones of downward motion. Antarctica, too, is surrounded by ridges with no evidence of crustal resorption along its margins.

The occurrences of two or more ridges with no intermediate zones of convergence prompted the conclusion that the oceanic ridges themselves must migrate away from each other, maintaining, somehow, their connection with heated materials at depth and preserving the symmetry of their magnetic anomaly belts. Ridge migration had not been envisioned by Hess in the original hypothesis of sea-floor spreading.

On the Pacific margin, where convergence is

Figure 74. Schematic representation of the Great Magnetic Bight in the floor of the North Pacific Ocean. The alternate belts of normal and reversed polarity increase in age toward the southwest; thus, the younger, more northerly belts appear to have preceded older ones into a trench. (After Pitman and Hayes, 1968; used with permission of W. C. Pitman, III, and the American Geophysical Union.)

assumed, totally different types of problems arise. Here, along the coasts of South and Central America, are the deep trenches and active Benioff zones where the sea floor and its cover of sediments should be riding down into the jaw crusher. Yet, long strips of these trenches are empty of sediments while others are partially filled with an orderly succession of horizontal layers. There is no hint of a crumpled, sinking mass or of any jaw crusher.

Farther north, between Baja California and Oregon, there is no trench or active Benioff zone at all. The East Pacific Rise appears to strike inland and disappear beneath the continent. This has encouraged much speculation that the anomalous crustal characteristics of the western United States—the tensional character of the Great Basin, the uplift of the Colorado Plateau, the unusually high rates of heat flow in some areas—may result from the overriding of the oceanic rise by the westward-drifting continent. But could a continent in fact ride up over a ridge of the type envisioned by Hess? Could the North American continent, in particular, pushed by the distant, slowly spreading Mid-Atlantic Ridge, overwhelm the rapidly spreading East Pacific Rise? Does the northward motion of coastal California along the San Andreas Fault zone conform with such geometry? Or does it indicate that the East Pacific Rise is discontinuous and that the San Andreas is a transform fault connecting the ridge crest south of the Gulf of Baja California with the Juan de Fuca Ridge south of Vancouver? Whatever the actual situation, the simple picture of orderly convection currents carrying passive continents *away* from spreading ridges somehow failed in western North America.

Surprise, if not dismay, greeted the discovery in 1968 that the magnetic anomaly belts in the Gulf of Alaska (Figure 74) describe a pronounced bend of 80° and that northward-trending strips of ocean floor appear to plunge into the Aleutian trench with *younger* strips preceding older ones!

All of these problems indicated that the straightforward conveyor-belt model of sea-floor spreading in the form originally proposed by Hess was far too simple. Hess had explained crustal tectonics in terms of deep, rolling, mantle-wide convection cells which would be unlikely to (1) develop two or three adjacent sets of rising limbs which migrate horizontally; (2) break up longitudinally into short, offset segments, each of which ends abruptly against a wall of colder material; or (3) push a continent over a rapidly spreading ridge.

As a guide to a new interpretation of the earth, Hess' concept was invaluable. Second thoughts were in order on the dynamics involved.

The New Global Tectonics

By the end of 1968 the concept of sea-floor spreading had been revised and incorporated into a far more fundamental and all-embracing theory of the earth called the new global tectonics, or plate tectonics. The first paper to use the term "plate" in this connection was published in 1967 by D. P. McKenzie and R. L. Parker, then at the University of California at San Diego.

Plate tectonics rests upon the concept that the earth's crust is a mosaic of large plates—akin, on a mega-scale, to ice floes or paving stones—which move as rigid units and suffer deformation only at their margins with neighboring plates. The moving plates are not the continents, as postulated by Alfred Wegener, nor are they the individual ocean floors. Most of them include continental and oceanic crust coupled together to form pieces in a new global jigsaw puzzle. The pieces of this puzzle are outlined by the earth's seismically active zones, illustrated in spectacular fashion on the map of earthquake epicenters compiled in 1968 from data collected by the Environmental Science Services Administration and the U. S. Coast and Geodetic Survey (Figure 75). Here it can be seen that the narrow belts delineating the oceanic ridges, the Pacific margin, and the Tethyan mountain belt surround large areas that are free of all but a few sporadic events. In 1968 W. Jason Morgan of Princeton University constructed a

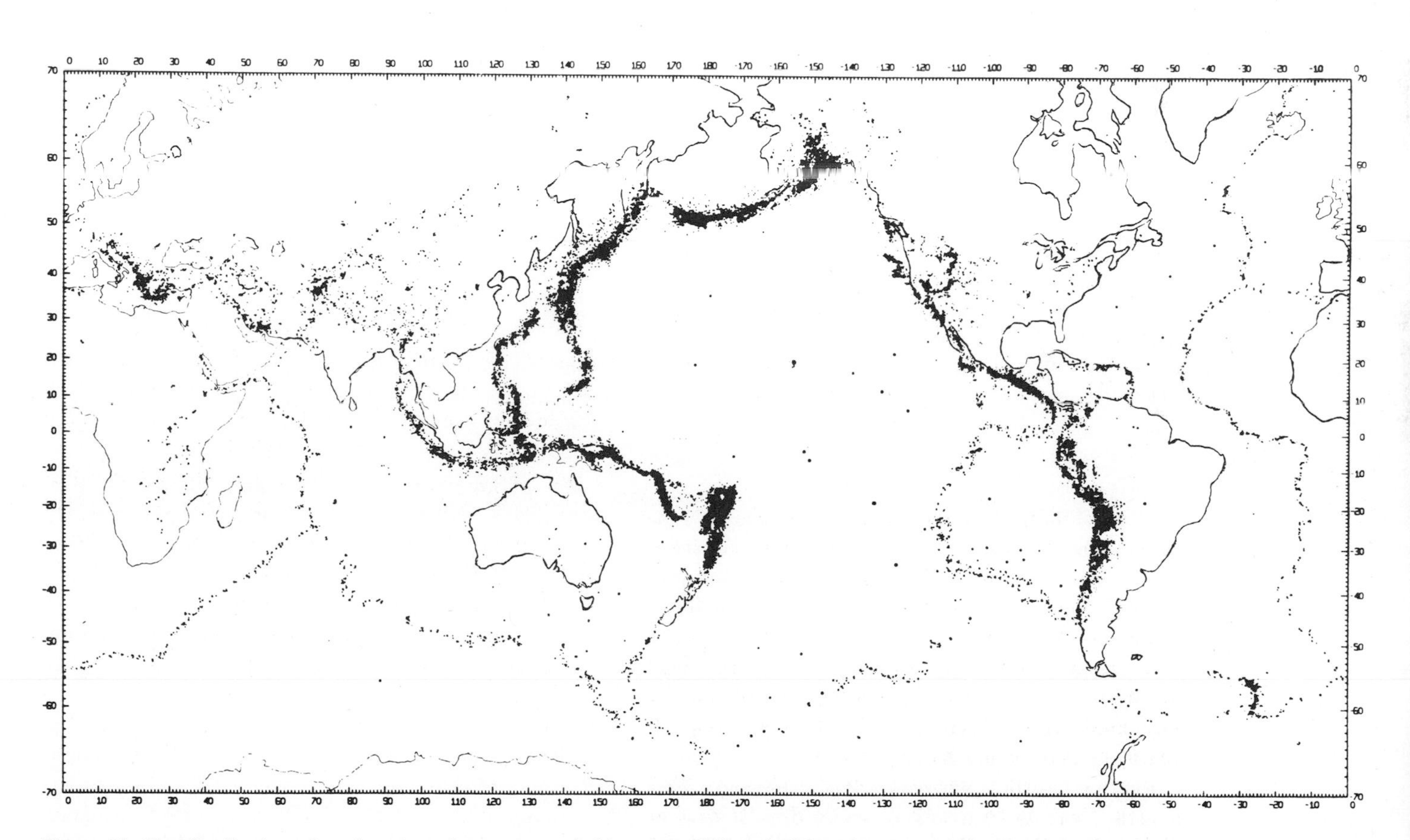

Figure 75. The distribution of earthquake epicenters recorded between 1961 and 1967. This map, which clearly delineates the earth's narrow, linear zones of most intense seismic activity, was crucial in influencing numerous earth scientists to begin viewing seriously the theory of plate tectonics. (Compiled by Muawi Barazangi and James Dorman; used with permission of the authors and the Seismological Society of America.)

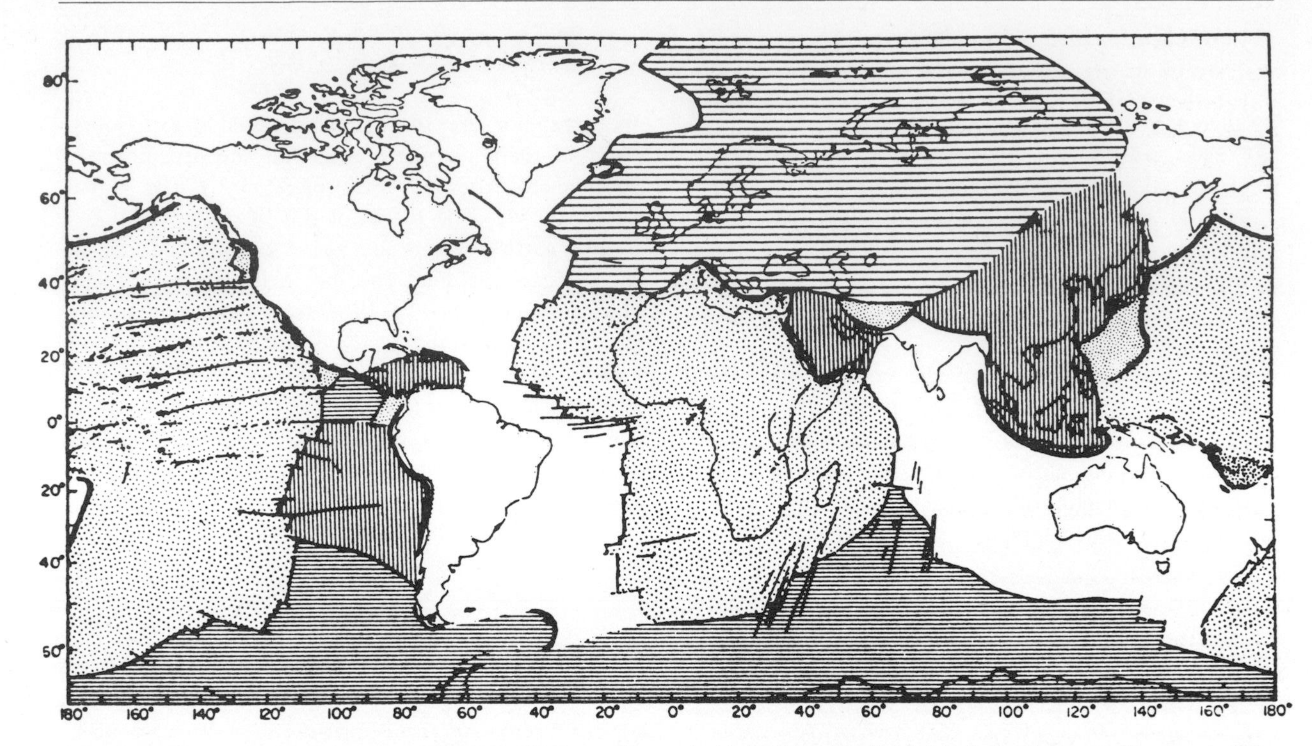

Figure 76. The array of rigid plates outlined by W. Jason Morgan in 1968. One plate (white) includes the ocean floor west of the Mid-Atlantic Ridge, both American continents, and northwestern Siberia. India occupies the same plate with Australia, but the rest of continental Eurasia is distributed among four plates. (Used with permission of the author and the American Geophysical Union.)

world map (Figure 76) in which the aseismic areas were designated as six large plates and twelve subplates. Since that time other subdivisions have been proposed by different scientists addressing specific problems.

One of the first questions to be answered is how rigid plates, which are in contact with one another and appear to be firmly wedged in place, can move over a sphere of constant volume. The crustal plates can move only if they are underlain by a pronounced zone of weakness and if, in addition, they are bounded by three kinds of surfaces: a surface where new area is generated (ridges), a surface where old area is destroyed (trenches or folded mountain ranges), and surfaces where horizontal shearing occurs between neighboring plates (transform faults, or megashears). In terms of a theorem of spherical geometry worked out by Leonhard Euler in the 18th century, the motion of each plate relative to another plate is completely described by a single rotation about an axis that passes through the center of the sphere. In an ideal case, transform faults bounding any one plate should be parallel to each other and should describe latitudes, or small circles, around the axis of rotation. The ridge axis should describe meridians on the same sphere, as shown in Figure 77.

In actuality, the plate margins rarely conform to this ideal shape. With due care, however, it is possible to locate a probable axis of rotation to account for the relative motion of pairs of individual plates, and in 1968 this was attempted in a unified global synthesis by Xavier Le Pichon, then visiting at Lamont Geological Observatory. On the surface of a sphere of fixed diameter it is evident that no sector (plate) can move without responding to, or evoking a response from, some or all other sectors. This axiom is implicit in Le Pichon's analysis, whereby—given only the data for the oceanic ridges together with the geographical outlines of the plates—he was able to calculate motion vectors for each plate boundary.

Le Pichon's data were reinforced by a global analysis of seismic first-motion vectors: the "new global tectonics" of Bryan Isaacs, Jack Oliver, and Lynn Sykes, again of Lamont Geological Observatory. All of Le Pichon's directions of relative plate motions were confirmed by seismological analysis, and, with the development of a theory (by James Brune and others) of the amount of movement at an earthquake focus, Le Pichon's rates of motion were also confirmed—within wide limits but still to an impressive degree.

To accomodate plate theory, the outer shells of the earth have been redefined in terms of their strength rather than their chemical composition and—with full honors accorded to Reginald A. Daly—are once again designated as the lithosphere, the asthenosphere, and the mesosphere. The lithosphere, a strong and rigid layer about 100 kilometers thick (estimates range from 70 to 125 km), includes the crust and the uppermost mantle. The Mohorovičić discontinuity lies within the lithosphere but is of no special dynamic significance to the theory. The asthenosphere is a weak layer, unable to resist stress over long periods. It extends from the base of the lithosphere to a depth of some 700 kilometers. The mesosphere is the lower mantle from 700 to 3,900 kilometers. It may or may not have strength, but it is regarded by many (although by no means all) geophysicists as inert with respect to surficial tectonic processes.

The upper part of the asthenosphere is equated with the Gutenberg low-velocity channel where high-frequency shear waves are strongly attenuated, proving low rigidity. This, then, is the layer upon which the plates ("paving blocks") of the lithosphere slowly slide and rotate, shearing, jostling, and underthrusting one another.

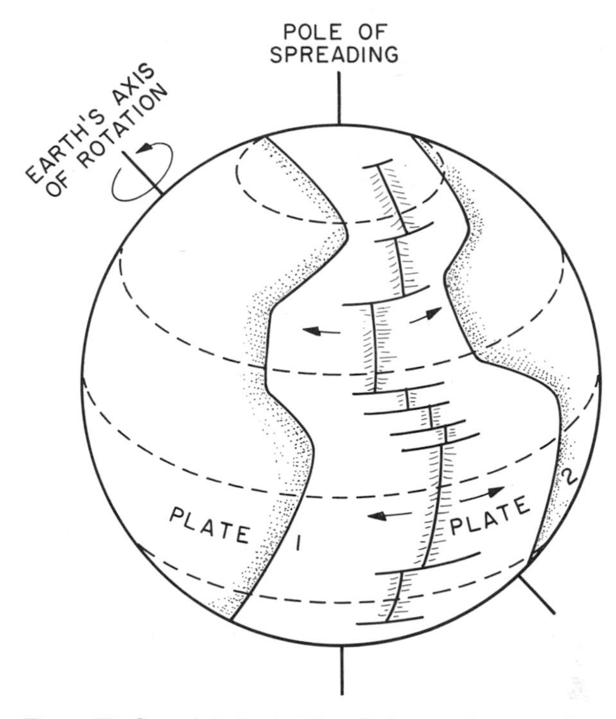

Figure 77. Geometrical principles of plate motion according to the theorem of Euler. The ridge crests strike parallel to the axis of spreading and the transform faults describe latitudes around the same axis.

Consequences of Plate Motions

What are the geological consequences of plate motions as generally conceived at present? Without attempting to encompass the details, we may construct the following outlines of the broad principles.

Zones of divergence

Oceanic ridges.—The oceanic ridges are zones of tension where new lithosphere (oceanic sima) is added to the plate margins. According to estimates available early in 1972, the oceanic ridges generate fresh basalt at a rate of about 50 billion tons per year. The plate-forming process requires the repeated opening of fissures along the ridge axis and the injection of hot material which, as it cools, acquires thermoremanent magnetism recording the polarity of the geomagnetic field. Each new fissure will presumably open along the hottest, weakest portion (the center) of the material injected previously. This "halving" of each central intrusion generates the

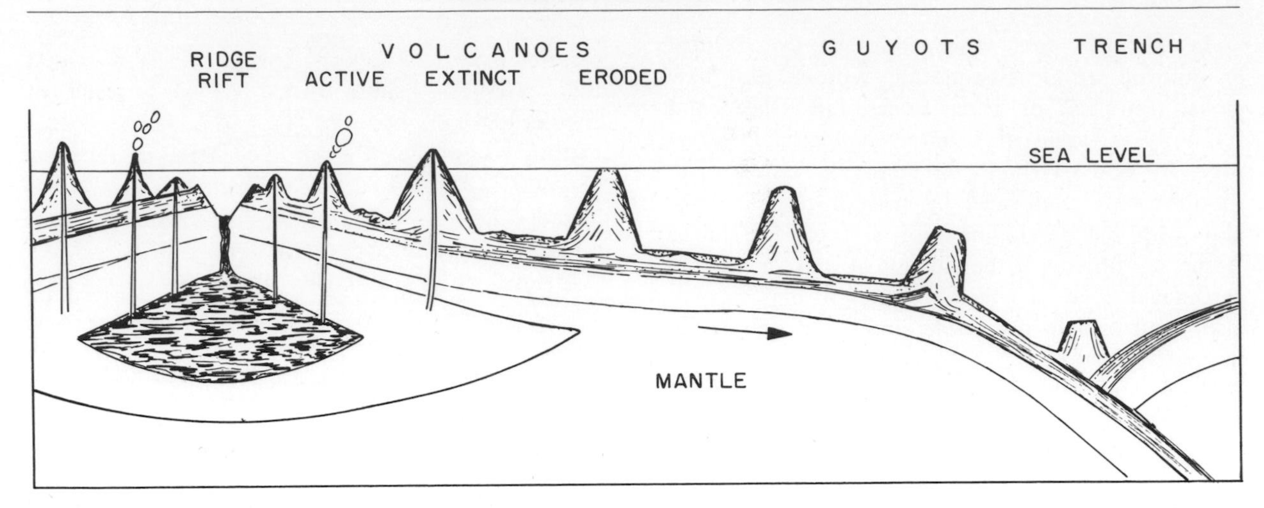

Figure 78. A composite diagram illustrating the growth, erosion, submersion, and destruction of volcanoes on a spreading sea floor. A guyot with a tilted platform sits on the lip of the Tonga trench; the Kodiak guyot sits upright in the bottom of the Aleutian trench. (Composite from several sources, especially Menard, 1969.)

symmetry in the pattern of magnetic strips on the ridge flanks. According to early versions of plate theory, the symmetry should be preserved regardless of whether both plates move directly away from the crest at equal rates or whether only one plate moves while the other remains stationary. In the latter case, the ridge must itself migrate laterally, following, at half speed, the retreating margin of the moving plate and continuing to split along the center of each successive intrusion. More recently, plate theory has been modified to accomodate the mounting evidence of asymmetry in the spreading of ridges.

The axial rift valleys are a striking and enigmatic feature of some oceanic ridges. No valley exists along the crest of the rapidly spreading East Pacific Rise, where the plates simply move apart down the sloping flanks. Where spreading is slower, as from the Atlantic and Indian ocean ridges, the plate margins slip upward for hundreds of meters, forming precipitous scarps on either side of a narrow valley, before migrating down the flanks. The reason for this phenomenon is not clear.

The seismicity associated with the ridges shows that relative motion between rigid plates occurs at shallow depths of less than 30 kilometers and is limited to the ridge axes and to the transform fault planes connecting the offset axes. Elsewhere there are no stress differences sufficient to rupture the oceanic lithosphere.

The nature of ocean-floor volcanism poses numerous problems, some of which are difficult to reconcile with plate theory. The majority of large volcanic islands and submerged seamounts are found in the midst of the oceanic plates rather than on their margins. After a careful study of this problem H. W. Menard proposed, in 1969, that many basaltic volcanoes begin to develop at the ridge crests as small cones and continue to grow larger as they ride away on the moving sea floor (Figure 78). He estimated that at least ten million years of eruptive activity are required for a volcano to build up to sea level. It may then persist as an island for another ten to twenty million years.

How do migrating volcanoes, fed by narrow pipes or fissures, maintain contact with a source of magma? Menard suggests that either the magma chambers are shallow enough to move along within the plate or that the volcanic conduits continue to tap a widespread source layer at depth. In either case the summit of a migrating oceanic volcano will be built not only at a later date but also in a geographical location different from that where its base was initiated.

Numerous oceanic islands are old enough for their lava flows to record a succession of magnetic pole reversals, but some seamounts are uniformly magnetized and must therefore have developed within a single magnetic event. Such seamounts are useful guides to the position of the magnetic pole at the time they formed as well as to the rates of submarine volcanic cone-building.

Volcanoes may not always be born on ridge crests. Certain of them develop in the center of oceanic plates, and H. W. Menard cites the Hawaiian group as a probable example. That group forms part of a long chain of islands, atolls, and seamounts extending for 2,500 kilometers, from the large island of Hawaii on the southeast to the Emperor seamounts at the northwest. Active volcanism occurs only on Hawaii and Maui, the two youngest and most southerly of the chain. The remaining islands of the group consist of extinct cones which are progressively older and more deeply dissected until, northwest of Kauai, they die out to a row of eroded platforms. Still farther to the northwest lie the coral atolls of the Midway group and beyond these is the submerged chain of Emperor seamounts. This configuration suggests that a moving plate has passed slowly northwestward over a deep, permanent "hot spot" which has continuously erupted lavas and built one island after another along a zone of weakness in the plate. After moving beyond the zone of active eruptions, each island has succumbed to erosion and then has been submerged as it rides downward toward the trenches of the northwestern Pacific. An alternative explanation that has been offered is that a prolific magma source migrated southeastward beneath a fixed ocean floor, leaving in its wake a row of islands. The tectonic characteristics of the Pacific margin and the configuration of other long chains of Pacific islands favor the former explanation.

If the ocean floor is new crustal material, generated at the ridge axes and accreted to the trailing edges of spreading plates, what types of rocks make up the three layers identified by seismic refraction profiling? Reviewing this problem in 1971, John F. Dewey and John M. Bird of the University of New York described the superficial layer 1 as a thin sequence of chert, silts, clays, and limestones; layer 2 as extrusive basalts occurring mainly as pillow lavas cut by minor intrusions; and layer 3 as predominantly gabbros. Many of the basalts and gabbros have been partially metamorphosed to serpentinite, amphibolite, or a sodium-rich basaltic rock called spilite. Rocks dredged from the depths of oceanic fracture zones include ultramafic types, such as peridotites and dunites, that probably represent the upper mantle underlying layer 3. Dewey and Bird suggest that primitive mantle rock rises as a result of decompression at the ridge axes and undergoes partial melting to produce a gabbroic magma and a residue of dense, ultramafic crystalline material. The gabbro intrudes the axial valley as a dike that is incorporated into layer 3 while swarms of thinner dikes feed the extrusions of pillow basalts that make up layer 2. Some of these pillow lavas carry concentrations of copper or manganese-rich minerals, whereas the ultramafic rocks at depth are associated with chromite ores.

If that description of the ocean floor—with its sequence of lime and silica-rich sediments overlying basaltic pillow lavas and mafic and ultramafic crystallines—has a familiar ring, it is because it duplicates the description of the ophiolite suite of rocks which, as we noted earlier, occurs in many of the folded mountain ranges of the continents and island arcs. When Harry Hess pointed out the similarity in 1962, ophiolites were believed to be portions of the lower crust or upper mantle emplaced in the deepest portions of eugeosynclines. Although their mode of origin was controversial, ophiolites were universally believed to be of fundamental importance in understanding the petrological aspects of the transformation of geosynclines into mountain ranges. According to plate theory, ophiolites are, quite simply, samples of oceanic lithosphere, and they originate not in the deepest furrows of the earth's crust but on the elevated ridges and horizontal plains of the ocean basins. From the ridge axes, the ophiolite sequence is rafted horizontally away to the trenches where, by means of faulting and compression, slices of it are incorporated by tectonic processes with the sedimentary and metamorphic rocks of future mountain ranges. Their presence in these ranges is therefore a petrological "accident" resulting from a juxtaposition by tectonic processes of rocks that originated in two different environments.

Continental rifts.—Continental rift zones, exemplified by the magnificent valleys of eastern Africa, the Rhine graben of Germany, and the Baikal rift of Siberia, appear to be tensional structures having certain features in common with oceanic rifts. The conclusion that the two are identical, however, may be premature. The continental rifts occur in regions of broad domal uplift, are associated with shallow seismicity, and have a higher than average rate of heat flow (but no longitudinal magnetic anomalies) along the valley floors. Geomorphological evidence indicates that the walls of the rifts have pushed upward on either side of the valleys, but these walls clearly expose ancient continental rocks and certainly are not generated by the injection and subsequent uplift of new material from depth in the manner advocated for the oceanic rifts (Figure 79). The nature of continental rift valley floors beneath their cover of sediments or volcanic lavas and tuffs remains controversial. According to plate theory, the rift valleys should be fissures filling at depth with injected oceanic basalts. In fact, eruptive rocks occur sporadically within the continental rifts, and, where found, the basalts are significantly more alkaline than the oceanic tholeiites. Lavas highly enriched in sodium or calcium silicates or carbonates (pantellerites and carbonatites) are characteristic of the rift valleys. Petrologists remain in disagreement on whether the rift volcanics are contaminated by remelted sialic crust.

The rates of separation across continental rifts are difficult to measure, but they must be slower by one or more orders of magnitude than those across the oceanic rifts. In Kenya and Uganda, for example, where the valleys are 40 to 60 kilometers wide, rifting is believed to have begun in the Miocene some 20 million years ago. In this region separation has occurred at a rate of roughly 0.25 centimeters per year. Farther south, rifting may have begun as early as the Permian, about 250 million years ago. If so, separation there has proceeded at only 0.02 centimeters per year. Meanwhile, beginning in the Jurassic, about 180 million years ago, the South Atlantic grew to an ocean some 5,000 kilometers wide at a rate of about 2 to 3 centimeters per year. The East African rift system is often spoken of as an ocean of the future; more appropriately it might be called the ocean that failed.

The Red Sea, in contrast, appears to have started as a continental rift that is a potential ocean. A central valley along the axis of the Red Sea floor exposes oceanic basalt. Within some of the deepest depressions of this valley are pools of hot, metalliferous brines suggestive of ore deposits in the making. The Red Sea and Gulf of Aden are widening at an estimated rate of 2 centimeters per year as Arabia

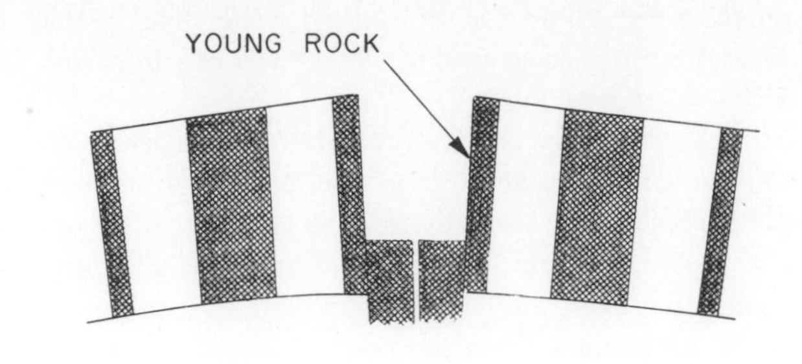

Figure 79. Upper sketch: An oceanic rift where new, magnetically polarized sima is added to the spreading sea floor. Lower sketch: A continental rift with walls of ancient crystalline sial. The nature of the material under the valley floor is highly speculative.

moves northeastward away from Africa.

The Great Basin of western North America is included in many lists of continental rifts because it is a zone of uplift and tension associated with alkaline volcanism (recently extinct). The Great Basin, however, is an immense area extending for some 950 kilometers from Idaho to Arizona and 700 kilometers from the eastern scarp of the Sierra Nevada Range of California to the western scarp of the Wasatch Range of Utah. It embraces well over 100 longitudinal valleys with interior drainage separated by block-faulted mountain ranges. In terms of plate tectonics it has been variously interpreted as a region of new rifting, as a portion of the continent that has overridden a segment of the East Pacific Rise, and as a broad, soft zone of deformation between the Pacific plate and the main body of the American plate. In no case can the Great Basin be regarded as a site of simple rifting and spreading of the type envisioned for the breakup of Gondwanaland, when continents, frozen into rigid lithospheric plates, split abruptly along fractures clean enough to be used to fit them neatly back together again. Some geologists of long field experience in East Africa argue that none of today's rift valleys are of the type that broke Gondwanaland. In their view, therefore, the present continental rifts are not the key to understanding those of the past.

Zones of convergence

Old lithosphere is reduced in area or removed from the surface along zones where two moving plates converge. Somewhat different results are predicted depending on whether both plate margins are oceanic, one is oceanic and one continental, or both are continental.

If both margins are oceanic, one will override the other and the lower one will bend downward and thrust or sink into the asthenosphere along a Benioff zone. A trench marks the site of underthrusting; a volcanic island arc arises on the margin of the overriding plate. If one plate margin is continental, its buoyant sial will always override an oncoming oceanic plate. Once again a trench will form at the line of underthrusting, and the continental margin will be compressed into folded mountains backed by a row of volcanoes. If both plate margins are sialic, the profound sinking of either one is unlikely. Crustal shortening will be accomplished mainly by folding and overthrusting unaccompanied by large-scale volcanism.

The descent of the lithosphere into the asthenosphere is now commonly called subduction, a term introduced into the literature in 1951 by Andre Amstutz of Switzerland and largely neglected until the advent of plate tectonics. Conversely, any upthrusting of slices of deep-seated crust or mantle materials which may occur during plate convergence is called obduction.

Oceanic plates.—A descending slab of oceanic lithosphere with its cover of marine sediments may (1) plunge as a cold, heavy plate all the way down to the mesosphere; (2) break into segments that move downward separately; (3) react to the increased temperatures and pressures at depth, undergo phase changes or partial melting, and be resorbed into the mantle; or (4) melt wholly or partially and contribute to the volcanism directly above the slab. Any of these processes may occur, but various combinations of them seem to be the general rule. The trenches which mark the zones of convergence should, in Hess' terminology, feed materials into the jaw crusher where plates meet under compression. However, the lack of sediments in some trenches and the horizontal layering in others suggest that trenches are not zones of compression. Some seismic evidence suggests that near the surface the trenches occupy zones of tension, and that only at depths approaching 100 kilometers do slabs of lithosphere shear past each other along compressional thrust planes. Clearly, much remains to be explained about the mechanics of subduction.

Although plates of oceanic lithosphere are expected to migrate directly away from the ridge axes at an angle of 90°, no theoretical constraints are placed on the angles at which plates plunge into trenches. Depending on their shape and rate and direction of motion the plates may enter trenches at right angles, oblique angles, or with a large component of strike-slip motion parallel to the trench axis. Estimates of the rates of plate convergence are made solely by reference to the release of seismic energy along these zones. The rates presently computed range up to 15 centimeters per year, which is about one-third higher than the maximum estimated for rates of spreading at the ridges.

An interesting situation arises when the leading edge of an oceanic plate is consumed at a trench more rapidly than its trailing edge is generated at a ridge. Presently the entire plate will disappear into the asthenosphere, pulling the axis of the ridge toward the trench and thus threatening to combine a zone of generation with one of consumption (Figure 80). In terms of convection, a rising limb approaches until it merges with a descending limb. The patent impossibility of such a thermal regime constitutes one of the strongest arguments against cellular convection as a driving force in plate motion. If an oceanic ridge is seen not as a zone of convective upwelling but rather as a zone of thin, weak lithosphere where magma rises like water between ice floes, then no exceptionally dramatic consequences are predicted in a ridge-trench convergence. Spreading simply ceases. The loss of almost an entire oceanic plate, called the Farallon plate, beneath western North America and the overriding of the former ridge of that plate by the

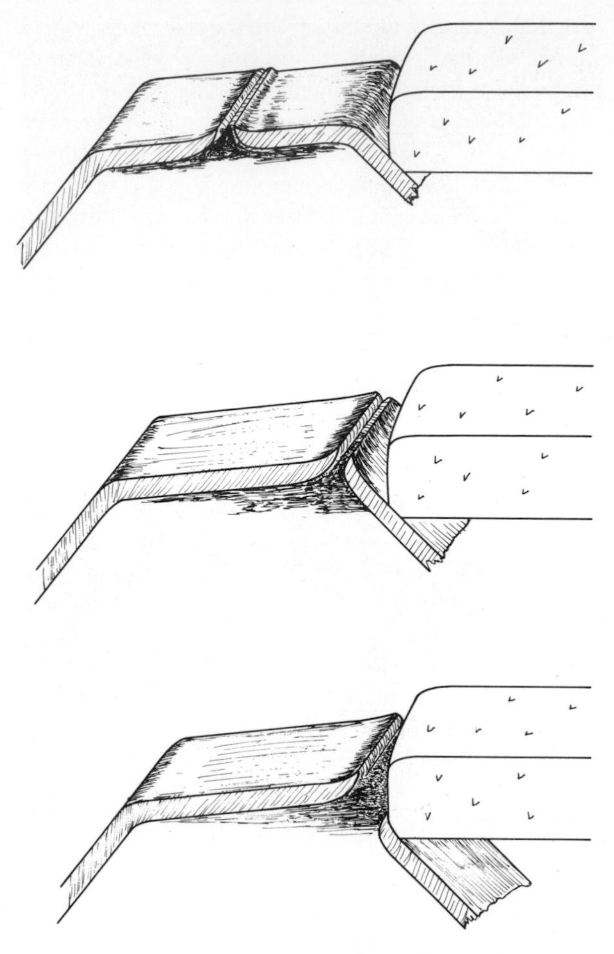

Figure 80. The coalescence of a ridge with a trench. In the upper two sketches a plate of oceanic lithosphere is consumed in a trench faster than it is generated at a ridge. The ridge migrates toward the trench until the two meet (lower sketch) and sea-floor spreading ceases.

edge of the continent was postulated by McKenzie and Parker in 1967 to explain the asymmetrical magnetic pattern of the Pacific floor and the sense of motion along the San Andreas Fault zone.

Since that time the history of the northeastern Pacific floor has been investigated in detail on the basis of the magnetic pattern, the array of fracture zones, and the distribution and ages of bottom sediments. The results are somewhat contradictory. In order to explain the Great Magnetic Bight (Figure 74), Walter C. Pitman III and Dennis E. Hayes of Lamont Geological Observatory proposed in 1968 that in the late Cretaceous three spreading ridges met at a triple junction as shown in the upper left panel of Figure 81. In a much simplified version, the other small panels of the figure depict the main developments up to the Pliocene. The figure illustrates how a simple pattern of spreading can lead to a result of great complexity. The structural features involved are four plates, four ridges, and two trenches. Two of the four plates remain stationary but all four ridges migrate trenchward while they add symmetrical bands of normal and reversely magnetized basalt to the plate margins. The present configuration of ridge remnants, fracture zones, and ocean floor ages inferred from the magnetic pattern are shown in the figure's bottom panel which is based on a map published by Hayes and Pitman in 1970. These authors postulate that two separate epochs of spreading and one change in vector of motion were required to produce this pattern.

In 1970 Tanya Atwater, at the Scripps Oceanographic Institution, proposed that the subduction of the Farallon plate (plate III in Figure 81) was

Figure 81 (facing page). The complex pattern of age belts inferred from magnetic anomalies in the North Pacific Ocean. The upper four panels illustrate the principles of how the Great Magnetic Bight could have been generated. Upper left panel shows late Cretaceous time before the onset of spreading. Four ridges are present, three of which meet in a triple junction. Plates I and IV remain stationary while plates II and III migrate northward and eastward toward trenches. All four ridges also migrate trenchward, and three of them change in length as they create symmetrical belts of normal and reversed polarity on the plate margins. In panel at lower left, Plate II and the two ridges bounding it have been subducted into the northern trench. In panel at lower right, which depicts Pliocene time, Plate III and the small northerly ridge segment have been consumed and the last vestiges of the ridge between Plates I and III are disappearing into the eastern trench. The continental margin is shown in its present form solely for purposes of reference. (Upper four pannels based on diagrams in *Journal of Geophysical Research*, volume 73. 1968, page 6576; used with permission of W. C. Pitman, III, and the American Geophysical Union. Bottom panel simplified from *The Geological Society of America, Memoir* 126, 1970, plate 2; used with permission of Dennis E. Hayes and the Geological Society of America.)

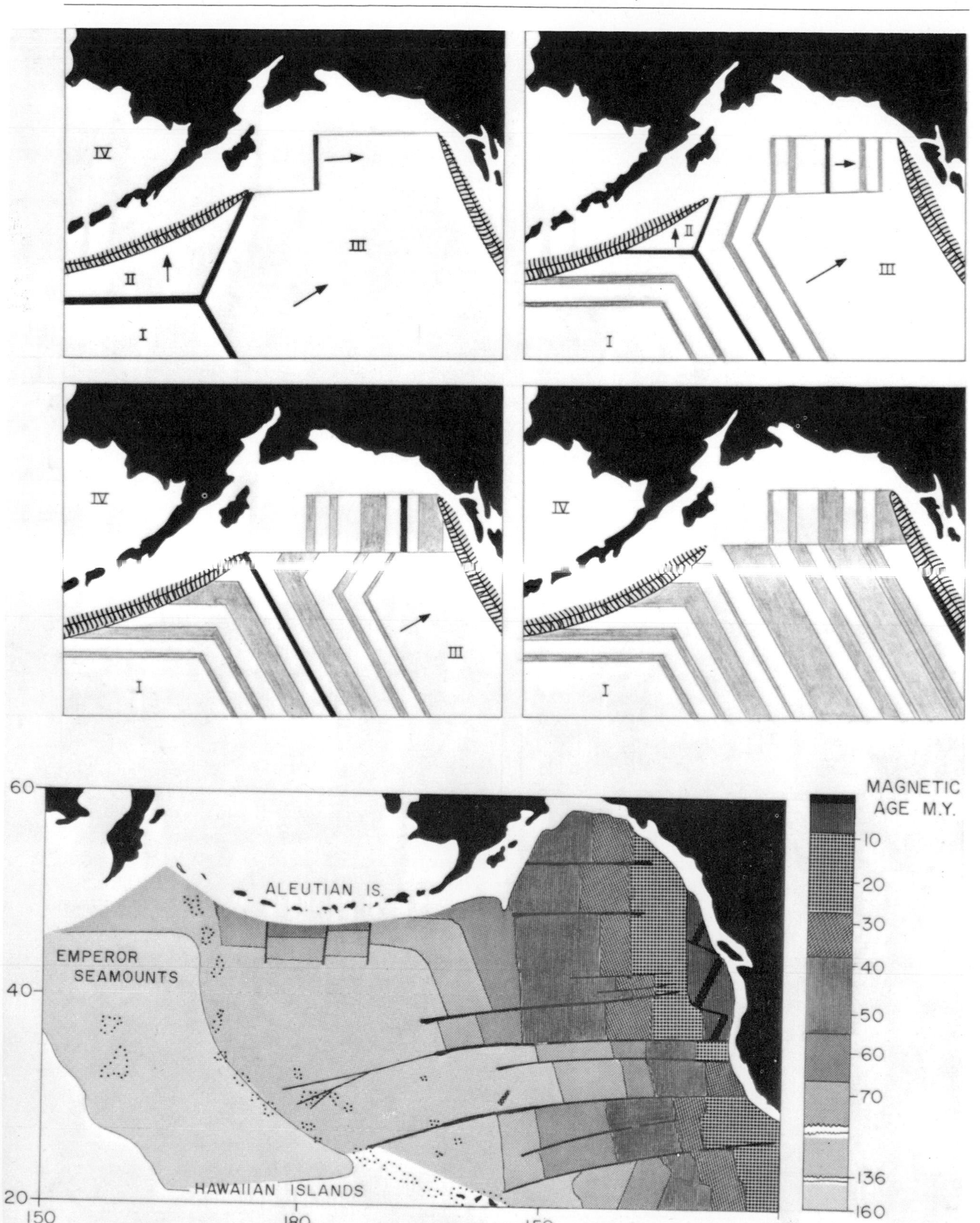
IV
II
I
III
IV
II
I
III
IV
I
III
IV
I
60
40
20
150
180
150
120
ALEUTIAN IS.
EMPEROR SEAMOUNTS
HAWAIIAN ISLANDS
MAGNETIC AGE M.Y.
10
20
30
40
50
60
70
136
160

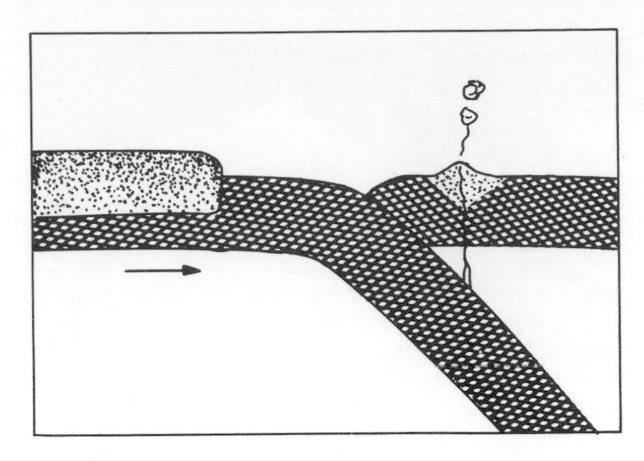

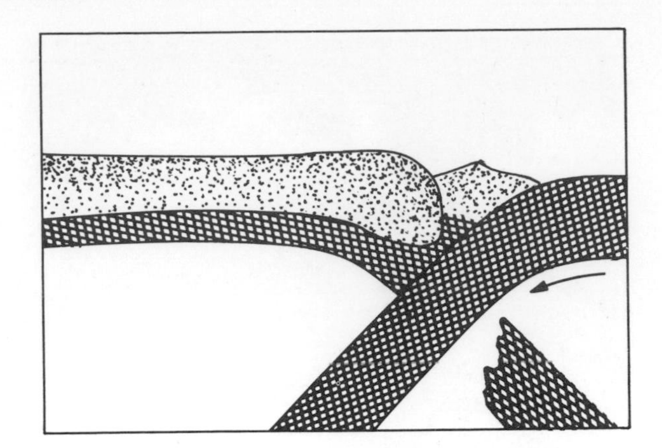

Figure 82. A collision of a continent with an island arc followed by trench-flipping.

substantially complete about 30 million years ago. At that time the edge of the sialic North American plate overrode the trench and became attached to the opposite Pacific plate (plate I in Figure 81), which was slipping rapidly northward relative to the continental margin. The attached strip of sial (coastal California) is still shearing northward along the San Andreas Fault zone past the rest of North America at a rate of about 6 centimeters per year. Reviewing the evidence again, in 1972, Tanya Atwater pointed out that the magnetic record preserved in the anomaly belts of the Pacific Ocean floor and in the paleopole positions of certain Pacific seamounts indicates that the Pacific plate has moved northward with respect to the North American plate by some 50° to 60° of latitude since the Cretaceous. On the other hand, the drilling results of the Joint Oceanographic Institutions Deep Earth Sampling (JOIDES) program show a thick band of sediments lying parallel to and displaced northward by only about 4° from the present equator. This distribution of sediment suggests that the Pacific floor has moved very little from the position it occupied in the Cretaceous. The contradiction between these two lines of evidence is yet to be resolved. As one possible solution to the dilemma, Atwater suggested that the earth's magnetic and spin axes may not have been closely coincident in the Cretaceous and early Tertiary.

When spreading ceases for any reason all the plates of the world will feel the change in rate and vector of motion. As lithosphere must be conserved, new trenches must be formed or "trench-flipping" may occur. Trench-flipping is a geometrical reorientation of the dip-slope of a Benioff zone. If, for example, a block of sial embedded in an oceanic plate arrives at a trench, its buoyancy prevents it from plunging downward and it therefore comes to rest over the trench. What has been the active Benioff zone will now be dipping away from rather than beneath the block of sial. In time this inclined subduction zone is expected to reverse its angle of dip, as shown in Figure 82. Trench-flipping, which looks simple enough on paper, is perhaps one of the more difficult concepts to envision as taking place in the real earth.

The long oblique plunge of the lithospheric plates results in internal stress as well as shearing stress and gives rise to the observed seismicity along the inclined Benioff zones down to depths of 700 kilometers. Abnormally low heat flow would be predicted in the area immediately above each sinking slab of cold lithosphere. In fact, however, most of the world's active volcanoes stand over these slabs. This unexpected thermal anomaly is now generally ascribed either to frictional heat generated by plate motion or to strain melting within the slabs.

The volcanic lavas erupted in zones of convergence are, as we have seen, predominantly andesites. The seismicity associated with the eruptions indicates that the andesites rise from depths of about 100 to 300 kilometers where the Benioff zones intersect the Gutenberg low-velocity channel. The origin of andesites and their petrological relationship, if any, to deep-seated granitic rocks have long been controversial. In 1965 Arthur Holmes pointed to the occur-

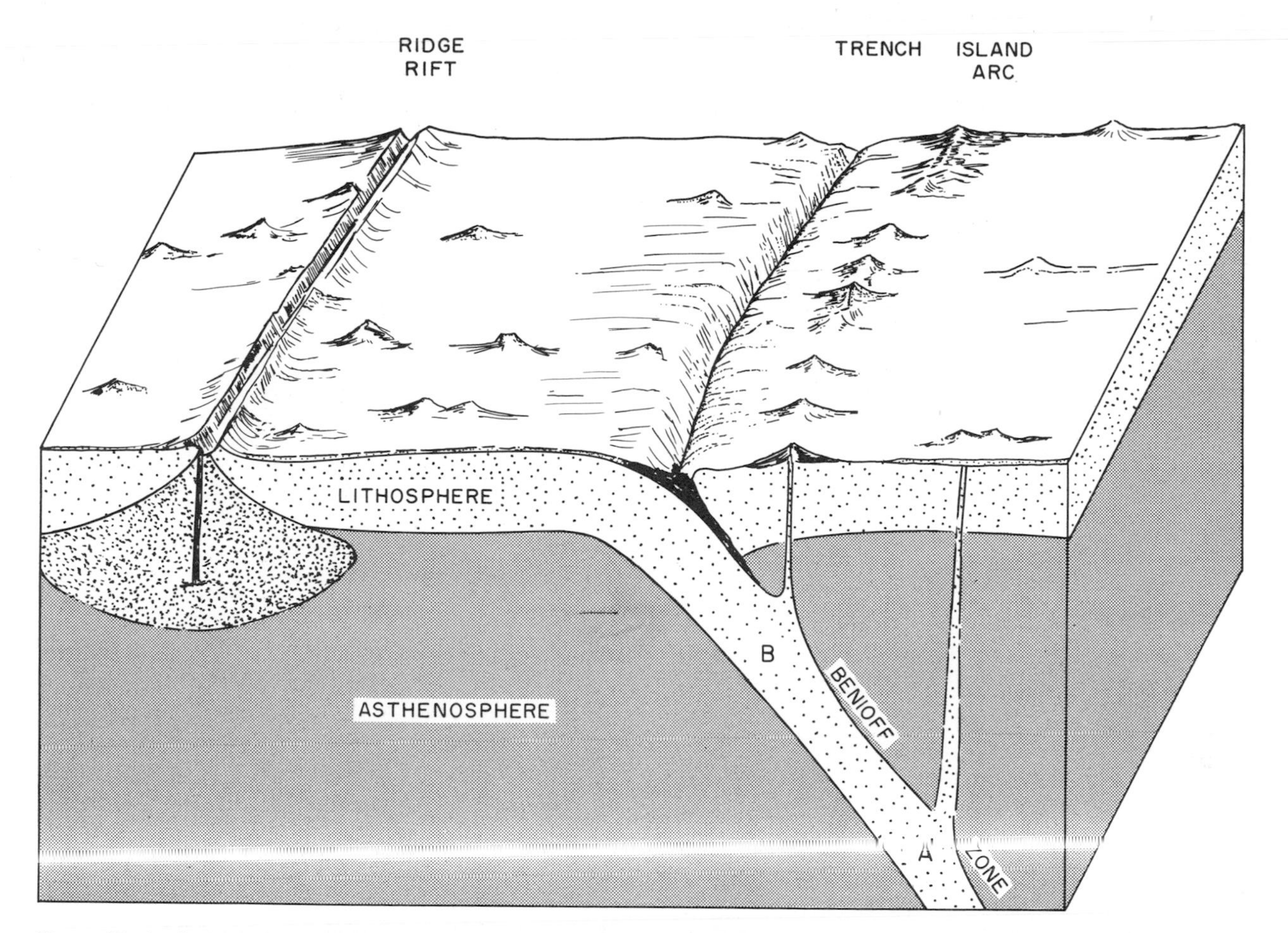

Figure 83. A schematic diagram of the generation of basaltic lithosphere at an oceanic ridge and of andesitic islands over a Benioff zone. The andesites originating near **A** contain more potash than those originating at **B**. The andesitic magmas may result from partial melting of the descending slab (as shown) or from a process of mantle differentiation in which composition of the product varies as a function of depth.

rence of andesitic volcanoes along coastlines and concluded that andesites are generated where sediments soaked in seawater are thrust downward and admixed with mantle materials. His statement admitted of little doubt: "It seems now to be well established that most andesites and dacites are hybrid lavas which result from mixing and reaction between basaltic magma and sialic rocks."

Today, William R. Dickinson of Stanford University rejects this idea of sialic contamination on several grounds, including the lack of enhanced values of radiogenic strontium-87 which should be contributed by assimilated masses of preexisting siliceous crust. He concludes that andesites are generated either by partial melting of peridotites in the low-velocity zone above the slabs of descending lithosphere or by partial melting of eclogites within the descending slabs. In either case the heat source for melting is provided by seismic energy released by motion of the lithosphere.

Dickinson carries his argument a step further and defines granites and andesites as cogenetic rocks that differ mainly in that they occupy different levels within the island arc complexes. A factor which led him to this conclusion was his discovery of compositional gradients that trend across the arc structures. For one example, he finds that the potash (K_2O) content of the andesitic volcanics increases as a function of increasing depth to the Benioff zone, and this trend holds regardless of whether a line of volcanoes rests on oceanic crust or on continental crust (Figure 83). A similar variation has been found along a line transverse to the trend of granite batholiths of western North America. Dickinson

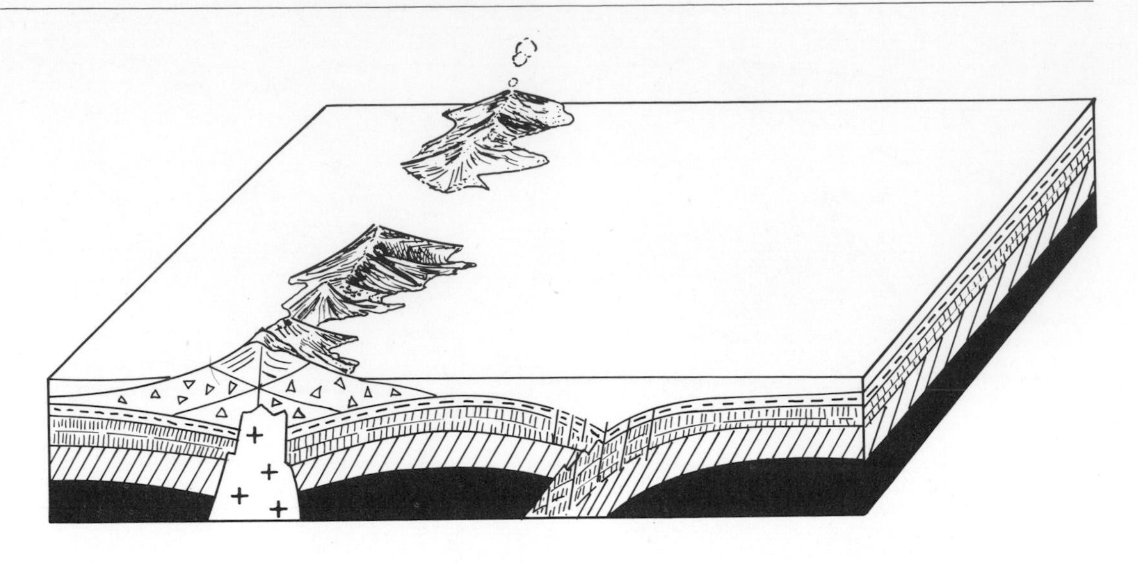

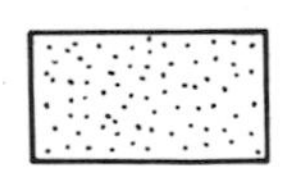
TURBIDITES
CLASTIC SEDIMENTS
AND VOLCANIC DETRITUS

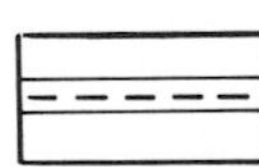
LAYER I
SEDIMENTS AND VOLCANIC DETRITUS

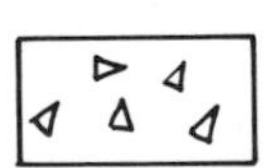
ANDESITIC
VOLCANICS

LAYER 2
BASALTS

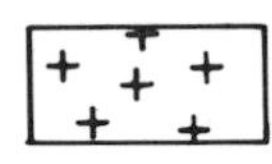
GRANITIC
INTRUSIVES

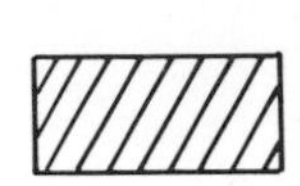
LAYER 3
ULTRAMAFIC IGNEOUS AND
METAMORPHIC ROCKS

MANTLE

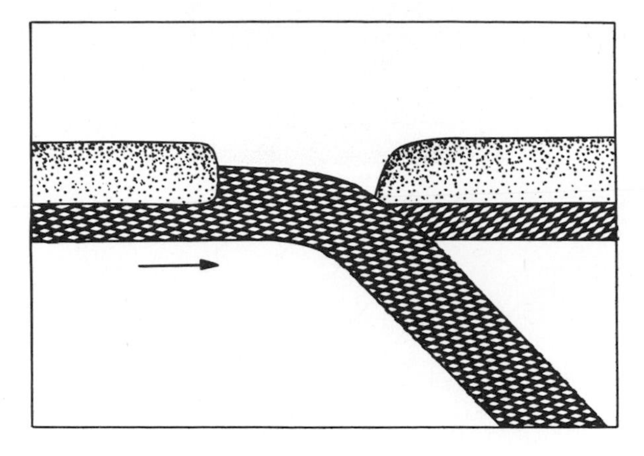

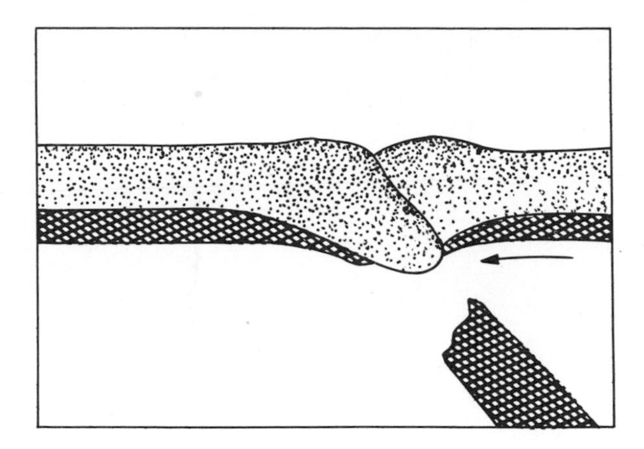

Figure 85. A continent-continent collision. The closing of the ocean is followed by thickening of the sial and uplift of a mountain range marking the suture.

suggests that such compositional gradients can be used to search out the island arcs built over the "paleoseismic" subduction zones of the past.

Zones of convergence have replaced geosynclines as the proposed antecedents of folded mountain ranges and the agents of continental accretion. Wrote William R. Dickinson in 1970: "This change in thinking requires only that observational data on orogenic rock masses be passed through a new theoretical filter." Once past this filter we see that the most fundamental aspect of the change involves the descent of a slab of spreading sea floor as the cause of all the tectonic and petrologic characteristics of these ranges—the folding, thrusting, and metamorphism of continental shelf sediments; the andesitic volcanism and granitic intrusion, which are generated by the process of subduction; and the emplacement of ophiolites, which began to form at the ridges. The thickening of low-density crust by the compression and underthrusting of sediments and siliceous igneous rocks leads to the subsequent uplift of the light materials into mountain ranges. The new theory, illustrated in Figure 84, seems to clarify many ancient problems. However, if a sinking slab of lithosphere is indeed the key to orogenic phenomena we should perhaps contemplate with renewed respect Reginald A. Daly's illustration of continental sliding (Figure 39) published in 1926.

Continental plates.—If two plates with sialic margins converge in a continent-continent or a continent-island arc collision, the results will include (1) the closing of an oceanic strip and deactivation of a Benioff zone; (2) the welding together of two sialic masses; (3) the crumpling of mountain ranges incorporating coastal sediments and slices of oceanic lithosphere with all the structural and petrologic complexities of orogenesis except for active volcanism; and (4) thickening and some possible underthrusting of sial resulting in buoyant uplift of mountains and plateaus (Figure 85).

The prototype of continent-continent collisions is that of India with Asia which resulted in the ranges of the Himalayas and the uplift of the Tibetan plateau. The lack of volcanism at this suture is ascribed to the absence of an underthrusting oceanic plate. However, most advocates of plate theory picture India as having moved northward for a very long distance on the back of a subducting plate (as shown in the left panel of Figure 85). Thus, it would seem that a long history of volcanism should have preceded the collision of India with Asia.

Figure 84 (facing page). The evolution of island arcs. Upper block shows a volcanic arc forming over the Benioff zone where one slab of oceanic lithosphere is thrusting beneath another. Lower block depicts a later stage when underthrusting has ceased. The former trench with its load of sediments and volcanic detritus has been compressed and uplifted to form an outer sedimentary arc parallel to the extinct volcanic arc. (Based on cross sections by A. H. Mitchell and H. G. Reading, *Journal of Geology*, volume 79, 1971; used with permission of A. H. Mitchell and University of Chicago Press.)

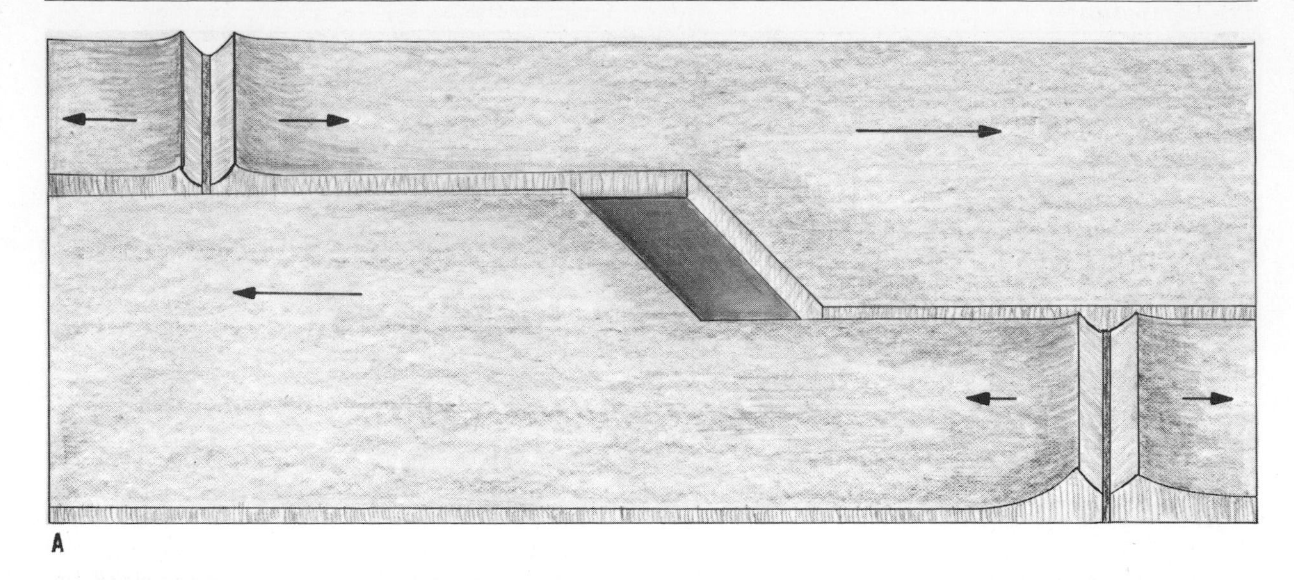

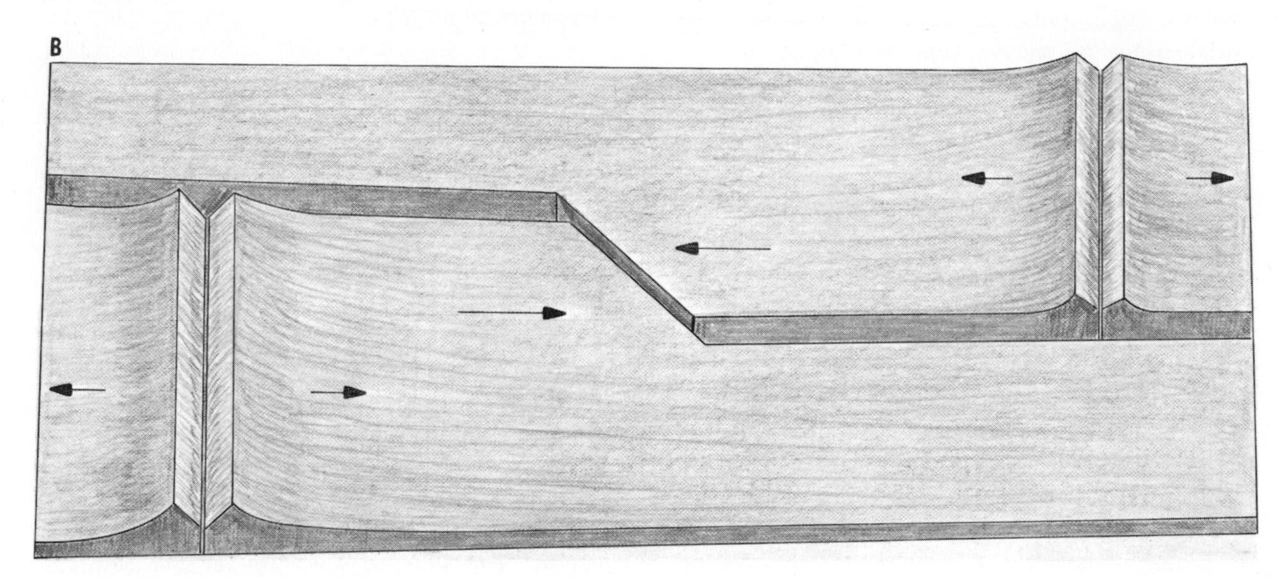

Figure 86. Megashears. **A**: Deep rhombic chasm opening where a transform fault is offset at an angle. Such an origin has been suggested for the basin of the Dead Sea. **B**: The walls of an offset locking under pressure. Motion will probably cease until the buildup of sufficient stress causes rupture and a high-magnitude earthquake. The San Andreas Fault zone has this configuration in the vicinity of Los Angeles.

Megashears

The third type of plate boundary is characterized neither by addition nor destruction of the lithosphere but solely by faulting, which may involve much crushing and fracturing as plates shear past one another. Some of the faults are linear for long distances, others are offset by abrupt angles. If plate motion pulls the walls of offsets away from each other, deep rents open in the lithosphere (Figure 86A); if it locks them together under pressure (Figure 86B) the result will be a temporary halt to spreading and the buildup of stress until rupture occurs, heralding an earthquake.

Triple junctions

Few plates are rectangular in shape, and plate boundaries do not always meet at right angles. Interesting consequences arise when three plates meet at so-called triple junctions. Analyzing this problem in 1969, D. P. McKenzie found that there are 16 possible combinations of plate margins—ridges, trenches, and transform faults—that can converge at triple junctions. Adding to the complexity of the situation is the fact that any of the three contiguous plates at the junction may be stable or moving at a rate different from the others. Perhaps the easiest type of triple junction to visualize is the meeting of three spreading ridges that we reviewed in Figure 84. A triple junction active at the present time is postulated at Bab el Mandeb, where the Red Sea meets the Gulf of Aden and the Afar rift of Ethiopia.

The Mechanism—and Other Perplexities

Wegener's hypothesis had been shipwrecked earlier on the problem of an ultimate cause, a mechanism of sufficient force to achieve continental drift. In the fourth edition of his book Wegener conceded the inadequacy of his proposed mechanisms, stating that the Newton of the drift theory had not yet appeared. But that absence caused Wegener no alarm. Wegener believed that the clear, unambiguous evidence for continental drift could be seen on every side, with or without an adequate causal force. Today we stand rather close to where Wegener stood. The clear evidence of plate motions may be seen at every hand; no adequate driving mechanism has been agreed upon. The Newton of the theory is still absent. This time, however, the substantial weight of geophysical opinion supports the theory.

The concept of plate tectonics has brought with it a new resolution of the formerly conflicting pieces of evidence with respect to the strength of the earth's interior. Earthquake waves are propagated to depths of 700 kilometers, but this fact can no longer be taken to mean that the whole earth possesses great strength to such depths. Only along the Benioff zones where slabs of sinking lithosphere plunge all the way to the mesosphere are deep-focus earthquakes recorded. Elsewhere, at depths below about 100 kilometers, the asthenosphere is sufficiently weak and yielding to allow for large-scale slippage of the overlying plates without the accumulation and release of strain which generates seismic waves.

"So far as I know, no one has yet suggested a model for the generation of plate motion that is acceptable to anyone else," wrote James Gilluly, in 1971, in a succinct appraisal of a situation that is likely to persist for a long time. The mechanisms that have been proposed fall into three broad categories: those driven by dynamic motion in the asthenosphere, those deriving mainly from gravity, and those impelled by global deformations. Ultimately, the force of gravity is an important factor in all conceptions of crustal motion.

Motion in the asthenosphere results from thermal or chemical instability and may assume the form of cellular convection, penetrative convection, or random flow. The idea of cellular convection, which was opposed by many proponents of solid earth geophysics, enjoyed a burst of popularity with the revival of interest in continental drift. H. H. Hess proposed it in 1960 as the explanation for sea-floor spreading. At about the same time, S. K. Runcorn worked out a theory of convectional cooling in which the continuous growth of the earth's core, due to differentiation of the mantle and the sinking of metallic particles, has forced changes in pattern from one large cell to several smaller cells (Figure 87). Runcorn suggested that the shift from four to five cells took place about 200 million years ago and was the event which initiated the present episode of continental drift.

Not all geophysicists agree that the earth's core has grown significantly over geologic time, or that cellular convection has persisted within the mantle. In 1965, Francis Birch of Harvard University speculated that formation of the core was a catastrophic event which was completed very early—perhaps as early as 4.5 billion years ago. Indeed, he suggested that core-

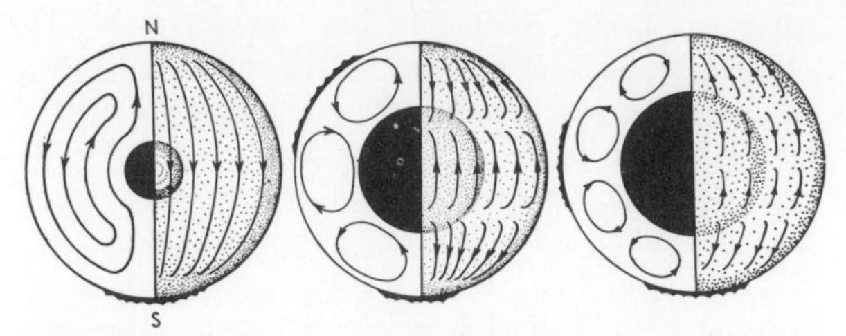

Figure 87. S. K. Runcorn's illustration of the increase in number and decrease in size of convection cells in the mantle that would result from growth of the earth's core. Continents are carried to sites where limbs converge. (Used with permission of S. K. Runcorn and *Nature*.)

mantle separation may be the event which marks the beginning of geologic time. More recently, Sidney Clarke, Karl Turekian, and Lawrence Grossman of Yale have argued that the core accreted directly from the planetary nebula and that, in all subsequent time, it has never had a convective exchange of material with the upper mantle. Aside from problems of the core-mantle relationship, the major discontinuity between the upper and lower mantles has prompted many investigators to regard the lower mantle as tectonically inert and to search for causes of crustal instability solely within the upper mantle. The upper mantle, however, is now known to be a complex domain with major vertical and lateral variations in density, temperature, and seismic properties. Such differences must be both a cause and a result of chemical and thermal instability and they surely lead to crustal deformation. Do they lead to mantle currents that are cellular and symmetrical?

Some advocates of plate tectonics, looking at the configuration of plate margins with a critical eye, have found it well nigh impossible to imagine in three dimensions an array of convection cells with rising limbs under the ridges. Not only are the ridges irregular in plan and offset into short segments but the global pattern is dominated by rising limbs on one side of the world and sinking ones on the other—bounding the margin of the Pacific Ocean. Searching from east to west, we find signs of uplift and divergence in the ridges of the Indian Ocean, the rift valleys of Eurasia and Africa, the Mid-Atlantic Ridge, and the Great Basin of North America. Nothing impressive in the way of sinking and convergence is found from the Java trench westward to the Chile-Peru trench. Could this mean that plate motion is governed by a single global convection cell with limbs rising under the "land hemisphere" and sinking under the Pacific? Such a pattern seems highly unlikely in view of the layered nature of the earth's interior and the presence in the Pacific of the rapidly spreading East Pacific Rise. The alternative remains that cellular convection in the classical sense is inapplicable to global tectonics.

Penetrative convection, the rise of heated mantle materials into zones of weakness in the lithosphere, has long been regarded as an important process by scientists who stress the predominance of vertical over horizontal motion in crustal tectonics. Today, some advocates of plate tectonics who have abandoned the idea of rolling convection cells suggest that the configuration of ridges and rifts depends less on forces active in the mantle than on zones of weakness in the lithosphere; that mantle materials, always close to their melting point, will liquify and rise wherever they are subject to a decrease in pressure; and that the ridges are essentially crack-fillings.

Another proposal raised in 1963 by J. Tuzo Wilson and again in 1972 by Jason Morgan of Princeton University is that the mantle includes several "permanent hot-spots" which have maintained stationary plumes of rising materials beneath the sliding lithosphere. Morgan calculates that the plumes are roughly 150 kilometers in diameter, have their source at or near the core-mantle boundary, and rise at a rate of about 2 meters per year. Once injected into the asthenosphere, this material spreads horizontally and initiates plate motion. Although some of the risen material is erupted onto the surface to form aseismic ridges and island chains, the remainder eventually sinks at a rate that causes an overturn of the entire mantle in two billion years.

Still another view advocated in recent years is that mantle convection is a self-limited process—that once heated materials have risen toward the surface and cooled they invert to a denser phase, sink, and accrete as an inert residue. Thus, the actively convecting layer in the mantle is growing progressively thinner and finally will choke itself out of existence. If this occurs at some future date all plate motion will cease and the continents and oceans of that time will be frozen into position to become permanent features of the earth's crust.

The downward plunge of lithospheric plates may

be attributed to the force of gravity with or without any assistance from convective motion in the mantle. New insight into the density stratification in the suboceanic crust and mantle was developed in 1969 by Frank Press of Massachusetts Institute of Technology. The low strength of the asthenosphere reflects a lower density than that of the overlying crust, a situation which is gravitationally unstable and facilitates the sliding of heavy lithospheric plates into the asthenosphere. Speculating on this relationship, R. A. Daly, in 1926, guessed that the base of the lithosphere would have a density of 3.0 grams per cubic centimeter and the top of the asthenosphere would have a density of 2.85 grams per cubic centimeter. Press' figures are higher. The values he determined by seismic evidence are as follows: in grams per cubic centimeter, the density of oceanic lithosphere at depths of 10 to 70 kilometers is 3.2 to 3.4; at depths of 70 to 125 kilometers the density is 2.5 to 3.6; and at depths of 125 to 300 kilometers the density is 3.4 to 3.5. To explain the density inversion at the boundary between the lithosphere and the asthenosphere, Press suggests that the upper mantle consists mainly of eclogite which on partial melting generates basalt plus a residue of peridotite. The hot, light basaltic liquid rises buoyantly through zones of weakness in the oceanic lithosphere and extrudes on the surface to form the ridges. As the plates move sideways, the new basalt added to their margins cools and part of it transforms back to eclogite. At shallow depths this change is only about 50 percent efficient, but at depths of 70 to 100 kilometers the transformation is complete and produces a heavy eclogitic layer in the base of the lithosphere.

Does the new global view of tectonics explain the existence of continents, the existence of the land and water hemispheres, and the periodicity of mountain-building? It suggests that continents exist because sial is generated over sinking slabs of oceanic lithosphere. If this is true, then the earliest crust of the earth consisted of oceanic sima which destroyed itself in the process of subduction but left behind granitic rocks such as those recently dated as the oldest samples of the crust. Plate theory would not predict one land and one water hemisphere or the antipodal distribution of continents and oceans. Both features of the earth's surface must therefore be viewed as a coincidence of the present time. The theory accounts for folded mountain ranges by collisions between sialic plate margins. Over geologic time this should probably be a random process; no rule suggests that continental collisions should occur simultaneously in all parts of the earth. Does one plate collision somehow bring on another? Is the periodicity of mountain-building real? Were those few geologists who denied the existence of geological revolutions and said that orogenesis was occurring somewhere throughout most of earth history prophets after all? Or is plate motion governed by some factor which propagates planet-wide tectonic pulses?

Perhaps plate motions are governed in part by global deformations related to the rotation and precession of the planet. Changes in the vector of plate motion may be initiated if a redistribution of load on the earth's surface causes new portions of the lithosphere to be pressed into the equatorial bulge. The idea that the configuration of the earth's surface features may be altered drastically by large-scale deformations of the globe is one of the oldest concepts in geology and cannot be dismissed on any grounds known today.

The search continues for more knowledge of the nature of the earth's interior and the mechanisms causing plate motion. Meanwhile, the acceptance of the general model, together with an admission of uncertainty with respect to the mechanism, is a refreshing change in geophysics. Evidence is no longer ignored or summarily dismissed for lack of a computed causal force. The motion of lithospheric plates has been compared (by D. P. McKenzie) with that of ice floes on a lake. The plates can move horizontally, rotate, and collide, but who can reconstruct the complex pattern of currents and eddies beneath them? "The plate motions are no longer believed to be closely related to the mantle motions below so that there is no longer any difficulty in understanding how the African and Antarctic plates can be almost entirely surrounded by ridges," wrote McKenzie in 1972.

A year earlier, in an interview with *Science News*, Walter Elsasser commented: "I avoid the term 'convective cells' as much as people in the Middle Ages avoided the term 'devil.' " (Elsasser did not add that the people of the Middle Ages avoided speaking of the devil for fear that he might appear.)

Voices of Dissent

"I think that a time has come when some protest must be made against the biased propaganda in favor of continental drift," wrote Harold Jeffreys in a letter to *Nature* dated May 17, 1969. And in the fifth edition (1970) of *The Earth* Jeffreys reiterated his earlier opinion that the evidence for matching continental margins and faunal zones is very dubious. Above all, he rejected the notion that the surface features of the earth are controlled by convection currents in the mantle. "How Soft is the Earth?" Jeffreys asked in an article of that title in 1964. As an answer, he stated: "The position may be summarized: there is evidence for slight imperfections of elasticity at small strains, and for large imperfections at large strains, but neither is of a type that would permit convection."

In the same article Jeffreys cited the uncertainties involved in the paleomagnetic method and in sorting out the supposed evidence for continental drift from that for polar wandering, the latter phenomenon being one toward which he felt far more tolerant. He also undertook to reply to some of his critics, evidently including S. Warren Carey. Referring to the fit of Africa and South America, Jeffreys said:

> A moment's look at a globe shows that there is a misfit of at least 10°. . . . One writer said that opponents call the agreement a coincidence; I simply deny that there is an agreement. Another said that the fit would be better if the boundaries were taken at the 1000 fathom line instead of at the coast but this would . . . leave the disagreement much as it was. The continued insistence that the agreement is good makes one wonder whether advocates of continental drift have doubts about the rest of their case. In any case the rest of the supposed fits are far worse.

The recent articles in favor of continental drift, Jeffreys said, "are remarkable for fallacious data, misinterpretations of the data, and omission to mention any objections."

"We have to agree with Sir Harold Jeffreys' opinion," wrote E. N. Lyustikh, of the Institute of Physics of the Earth in Moscow, in 1967. Lyustikh attacked several different convection theories, exposing their contradictions. In particular he opposed the idea that the midocean ridges with all their offset segments mark the rising limbs of convection cells. His most unique contribution, however, is a diagram (Figure 88) of fifteen different shorelines drawn to the same scale and showing "amazingly exact" fits, none of which could have any relationship to continental drift. Looking at the diagram of Lyustikh leads one to question whether a computer might not find a pattern of strikingly concordant matches between the margins of Terra Australis and the other continents on a map such as that of Abraham Ortelius shown in Figure 10.

In an article titled "The Position against Continental Drift" that was published in December 1970, Paul Wesson of Cambridge University wrote as follows:

> All work aimed at fitting continents together is bedevilled by ambiguities and inconsistencies. . . . It is to be deplored that part of the evidence is seized upon with avidity, while other parts, usually those that do not find an easy interpretation in terms of drift, are often ignored or left by the wayside as incongruities.

But misfitting continents was only one of many problems that Wesson discussed. He also exposed ambiguities and inconsistencies in the paleomagnetic method, paleoclimatic studies, and, above all, in the various convection theories that have been put forward by advocates of drift. Wesson concluded that no compelling evidence has been presented and no adequate mechanism proposed for continental drift or sea-floor spreading. As a possible alternative he suggested a history of polar wandering combined with earth expansion.

Three views of paleoclimates

To reconstruct the history of the earth's climatic zones by mapping the changing distributions of temperature-sensitive deposits or organisms would seem, at first glance, to be a fairly straightforward project. Climates, however, depend only partly on latitude; also important are worldwide fluctuations in temperatures and the distribution of landmasses with respect to changes in global and local patterns of winds and ocean currents. Perhaps it is small wonder

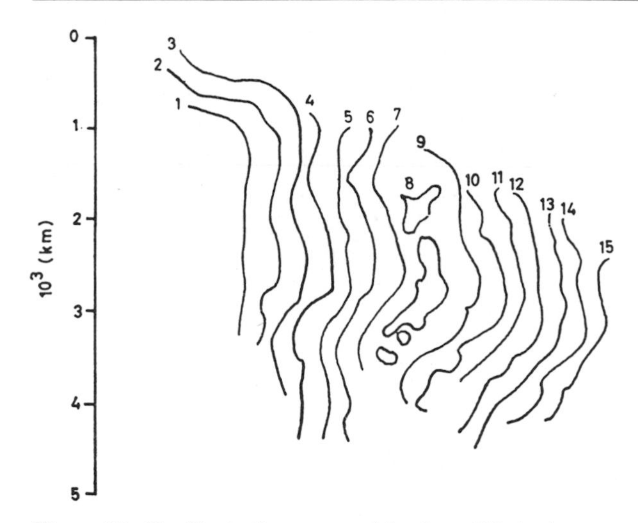

Figure 88. The illustration prepared by Lyustikh to demonstrate the striking similarities of numerous coastlines not recognized by advocates of continental drift. Lyustikh's caption reads: "Coastal lines of different continents on equal scales: 1, the western coast of South America from 15 to 40° S; land is to the right. 2, the western and northwestern coast of Australia from 35° S to 133° E; the south is at the top, land is to the left. 3, the western coast of North America from 175° W to 35° N; land is to the right. 4, the southeastern coast of South America from 5 to 35° S; land is to the left. 5, the north-eastern coast of South America from 35 to 60° W; the east is at the top, land is to the right. 6, the western coast of Africa from 4° N to 30° S; land is to the right. 7, the eastern coast of Africa from 0 to 24° S; land is to the left. 8, the Islands of Japan. 9, the western coast of South America from 5° N to 18° S; the south is at the top, land is to the left. 10, the southern coast of Australia from 117 to 144° E; the west is at the top; land is to the right. 11, the eastern coast of Australia from 17 to 38° S; land is to the left. 12, the eastern coast of Greenland from 84 to 60° N; land is to the left. 13, the eastern coast of Africa from 15° S to 12° N; the south is at the top, land is to the right. 14, the western coast of Africa from 5 to 25° N; the south is at the top, land is to the left. 15, the eastern coast of the Hindustan Peninsula from 8 to 22° N; the south is at the top, land is to the right." (Used with permission of E. N. Lyustikh and of the Royal Astronomical Society.)

that specialists disagree on the paleoclimatic evidence for polar wandering and continental drift.

In 1963, George Bain, of Amherst College, published a set of maps on which he had plotted the climatic zones for every geological period since the early Precambrian. He depicted a pattern of extensive polar wandering with no significant continental drift. In the Precambrian he shows the earth with the continents nearly upside down from their present orientation; throughout the Paleozoic they are tilted about 90° so that the equator passes lengthwise through North and South America. Of special interest is his map of the Permian (Figure 89), which accounts for the Gondwana glaciations by a tilt that places the tillites of southern Africa near the south pole while those of India and Brazil lie close to the (Permian) Tropic of Capricorn. With respect to paleomagnetic data Bain pointed out that the present magnetic axis is curved so that the magnetic poles are separated by an angle of 161° across the Pacific and 199° across the Atlantic. A straight line connecting the poles would be tangent to the core element some 960 kilometers from the center. A fundamental error is therefore introduced into paleopole positions computed on the basis of a geocentric dipole. An eccentric dipole should be used, but there is no assurance that the eccentricity has not varied significantly over geologic time. Bain therefore distrusted paleomagnetic polar wandering curves.

More recently Bain has written that he is fully aware of the documented evidence for translations—approaching "drift" magnitude—of large crustal blocks relative to one another. In 1972 he cited as examples the movement exceeding 3.5° on the Alpine fault of New Zealand, the Dead Sea Rift of Jordan, and the San Andreas Fault of California. And he believes that, since the Permian, India has moved about 10° and Antarctica, the Falkland Islands, and southern South America have moved about 20° from an "anchor" in South Georgia. However, he argues that these motions are of ordered tectonic types and that

> Much of what is referred to as Continental Drift produces extreme disorder, and I believe that the measurements used to support it can be accounted for better in other ways . . . some fundamental basement features of pre-Permian age which fail to match when nucleated according to Continental Drift are accounted for better by ocean basin foundering. I regard most of the evidence used to support drift of South America away from Africa as due to a climatic belt displaced by polar migration. Certainly if Australia, India, and Africa were nucleated with South America during Triassic time, the Triassic strata of Australia should have some fossils of the

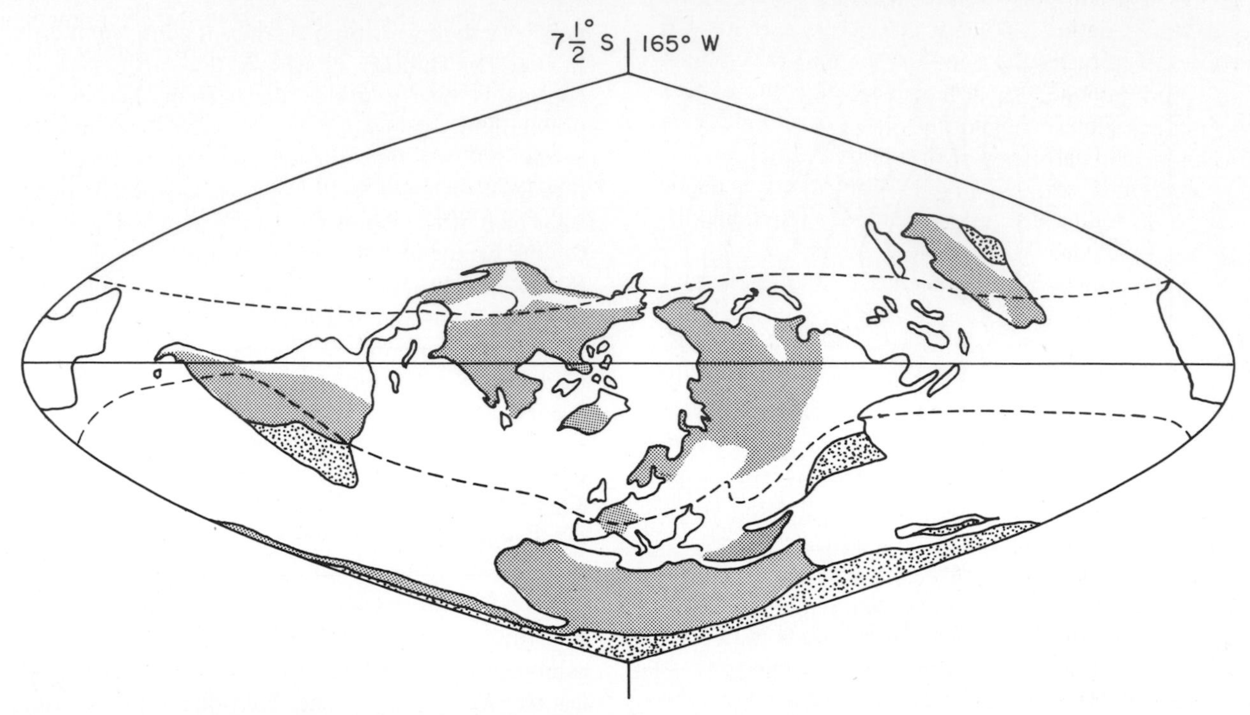

Figure 89. A simplified drawing of George W. Bain's 1963 map of the distribution of landmasses and climatic zones in the Permian. As a result of polar wandering without significant continental drift, the north pole is located in the Pacific Ocean at 7½° south, 165° west, and the equator passes lengthwise through the Americas and eastern Asia. Gray areas represent Permian land; dark stippling marks Gondwana glaciations, all of which occurred in regions isolated from warm ocean currents. (Used with permission of G. W. Bain and the Society of Economic Paleontologists and Mineralogists.)

land living mammal-like reptiles of the Madagascar, Africa, South America fauna.

With no apparent prejudice for or against plate theory, Francis G. Stehli, of Case Western Reserve University, tested paleontological evidence against paleomagnetic evidence for polar wandering. He studied the global distribution of marine invertebrates and found that at present the diversity of groups reaches a maximum in warm waters near the equator and declines steadily in the cooler waters to the north and south. He found a duplication of this pattern among Permian brachiopods of the northern hemisphere. The Permian brachiopods show a maximum diversity at the *present* equator and no obvious relationship at all to the paleomagnetic Permian equator. This evidence against polar wandering was so persuasive that it prompted an editorial note in *Nature* (August 1970) titled "Could Palaeomagnetism Be Wrong? " Certainly Stehli's fossil diversity gradient, which is subject to rigorous checks against sampling and other errors, and the paleomagnetic pole position for the Permian cannot both be right. Stehli concludes that the Permian magnetic field was not an axial dipole. Stehli's exhaustive research is the type of effort for which Chester Longwell issued an appeal in 1943 when he cited the need for many different specialists, each working in his own field, to build a body of dependable, significant data.

In the January 1970 issue of the *Journal of Geology* A. A. Meyerhoff of the American Association of Petroleum Geologists published the first of numerous articles challenging various aspects of plate theory. The first article focused on paleoclimates and included a series of maps showing the earth's climatic zones—as indicated by the distribution of evaporites, carbonates, coals, and tillites—oriented essentially parallel to the present equator but with a persistent offset toward the north for every period back to the middle Precambrian. Meyerhoff rejected both continental drift and polar wandering. He argued that the global circulation of wind and ocean currents has remained virtually the same for the past 800 million to 1,000 million years, although episodic fluctuations in average global temperatures have caused alternate widening and narrowing of the climatic zones.

Meyerhoff set out to demolish several climatic myths, including one that coal swamps develop in the tropics and one that glaciers do not. From the Devonian, when it first appeared in the stratigraphic record, coal has formed in fairly high latitudes where heavy moisture enhances organic growth but cool climates inhibit its decay. Equatorward from the coal belts, according to Meyerhoff, are found marine evaporites, indicative of drier climates, and below these lie the windblown sandstones of the horse-latitude deserts. Meyerhoff believed that the Gondwana glaciations are more easily explained on the basis of present continental distributions than they would be on a supercontinent at the south pole. He stated that all of the known Permocarboniferous tillites lie either in highland areas or near present shorelines. He described India, in the Permian, as a rugged peninsula with 1,000 to 3,000 meters of relief. Previous to the Tertiary uplift of the Himalayas, summer monsoons from the Arabian Sea and winter monsoons from interior Asia both dropped moisture on India. Given a worldwide drop in average temperature, Permian glaciation at this site as well as in South Africa, South America, Australia, and Antarctica is understandable.

Why, then, were there no contemporaneous glaciations in Canada, Greenland, Siberia? According to Meyerhoff, Permian glaciations did occur in several highland regions of Eurasia and a north polar ice pack also existed. However, he believed that the submarine rise between Greenland and the Faeroe Islands had not formed as early as the Permian and therefore the warm waters of the Gulf stream-North Atlantic drift then flowed freely into the Arctic, ameliorating the climate of Greenland and other northern lands.

Meyerhoff also described structural and stratigraphic evidence that Paleozoic redbeds and evaporites can be traced in deep wells from central Rajasthan to West Pakistan and are correlated with the Salt Series that extends to Iran and Afghanistan. India, he argues, was always part of Asia.

In Meyerhoff's view the paleomagnetic evidence for polar wandering is at least as poor as if not poorer than the climatic evidence. He questioned the validity of the dynamo theory of origin of the geomagnetic field, the assumption of a permanently dipolar axis, and the statistical "double-averaging" procedure used to determine pole positions for continents or large geological provinces. First, the measurements on all samples from a single formation at one locality are averaged to determine a pole for that locality; second, all poles of a given age from a continent are averaged to determine a paleopole for that continent. Meyerhoff stated that on large-scale maps these poles seem to cluster satisfactorily; however, when the details are closely examined "it is not unusual to discover that the pole for continent 'X' was determined from say, fifteen single-locality positions with a spread of 4,000 km, that the pole for continent 'Y' was determined from twenty-two positions also having a 4,000-km spread, and that the two spreads have considerable overlap." The spread of "reliable" poles is, in fact, often wider than the Atlantic Ocean, where drift is supposed to have taken place.

Meyerhoff concluded that the significance of paleomagnetic data is unknown because of the wide scatter of poles and the uncertainties of the method and so cannot be used to "prove" continental drift or polar wandering. In later articles Meyerhoff documented his objections to other arguments for plate theory, but first let us hear from V. V. Beloussov of Moscow University and the Institute of the Physics of the Earth.

A complete fiasco for ocean-floor spreading

"It is evident that not a single aspect of the ocean-floor spreading hypothesis can stand up to criticism," wrote Beloussov in *Tectonophysics* in 1970. Beloussov had argued the structural and petrological case against continental drift for many years. For a time, proponents of the new global tectonics suggested that scientists of the Soviet Union, unlike those of maritime nations, had perhaps restricted their attentions too much to the landmass of Eurasia to appreciate the significance of the ocean basins. Then, in 1970, Beloussov published a penetrating critique entitled "Against the Hypothesis of Ocean-Floor Spreading."

Reviewing the now-familiar principles of plate

tectonics, Beloussov found them incomprehensible when applied to the geology of the real earth. In his view the oceanic ridges are rugged topographic features without the least resemblance to the margins of rigid plates built of a succession of vertical dikes. A section of the Mid-Atlantic Ridge, for example, has an axial valley from 25 to 50 kilometers wide that is bordered on either side by high mountains which appear to be faulted into a series of longitudinal ridges and valleys with 1 kilometer of relief. These mountains extend outward for about 50 kilometers and are flanked by plateaus that are about 100 kilometers wide and have about 700 meters of relief. Beyond the plateaus the surface slopes toward the abyssal plain, to which it descends in three steps separated by steep ledges 300 meters high. How, Beloussov asked, does a plate margin that begins as a rugged mountain range, 50 kilometers broad, change into a plateau that is 100 kilometers broad, and this into a broken slope which finally becomes a flat abyssal plain, simply by moving horizontally?

An intense magnetic anomaly marks the axial valley of the ridge and may be caused by a near-surface basaltic body from 6 to 10 kilometers thick. Anomalies of various breadths and shapes occur over the flanks but are so weak that they must arise from bodies less than 2 kilometers thick residing in layer 2. As the plates move sideways, how does a thick, strongly magnetized body (dike) become a thin, broad, weakly magnetized body?

If the magnetic strip anomalies truly record reversals of the geomagnetic field imprinted upon them during spreading, their widths should reflect the relative lengths of time between reversals. The present (Bruhnes) epoch has lasted 0.7 million years; and the preceding Matuyama and Gauss epochs lasted 1.7 and 1.1 million years, respectively. No earlier reversals, Beloussov stated, have ever been established by radiometric dating of rock samples. Taking the present epoch as 1.0, the ratio for the other two epochs is 1.0:2.4:1.6. According to Beloussov, the ratios of anomaly widths along several ridge segments are as follows:

Juan de Fuca Ridge	1.0:2.6:1.2
Carlsberg Ridge	1.0:2.5:1.2
South Atlantic Ridge	1.0:2.9:1.1
Reykjanes Ridge	1.0:0.5:0.4
North Pacific Ocean	1.0:0.7:1.3

Beloussov also listed six different published estimates for spreading rates of segments of the South Pacific Ridge. He commented:

> It is readily seen that, strictly speaking, not a single measurement is in accord with the hypothesis of uniform ocean-floor spreading. The closest are the anomalies on Juan de Fuca Ridge, Carlsberg Ridge, and in the South Atlantic. Actually, these are the three whales that support the hypothesis.

The other ratios, Beloussov pointed out, are not even remotely what they should be—not even over the Reykjanes Ridge, south of Iceland, where the anomalies make a pattern so beautiful that it is often printed in six rainbow colors. Indeed, over vast distances along the ridges, such as in the Atlantic south of latitude 35° south and in the southwestern Indian Ocean, the anomaly pattern shows no symmetry at all, a fact which most proponents of sea-floor spreading mention only in passing. (Refer to Figure 73.)

Beloussov was skeptical of the idea that, in a dynamic system driven by thermal energy, ridges can stop and start moving sporadically, that the ridges can themselves migrate horizontally, or that volcanoes can grow on a moving floor and maintain contact with their feeder canals. Following the example of many advocates of plate tectonics, he examined Iceland as the best available example of an exposed ridge segment. "The result," he said, "is unexpected." [Unexpected to supporters of plate theory.]

The central anomaly of the Reykjanes Ridge is aligned with a strip of young Pleistocene basalts less than a million years old occurring on the island. Two anomalies lie on either side of the ridge crest at distances corresponding to about eight million years of spreading; and they are also aligned with belts of Pleistocene and Recent volcanism! "This result blasts all conclusions concerning 'fossilized magnetic epochs' in the structure of mid-ocean ridges," concluded Beloussov.

Rejecting as both inadequate and misguided the evidence for the dilation of Iceland, Beloussov added:

> We may conclude that a comparison of the structure of the submarine ridge and its land surface continuation results in a complete fiasco for the hypothesis of ocean-floor spreading. . . .
>
> Let us not forget that we still have only a very approximate idea of the composition of the second layer of the oceanic crust and know nothing at all

about the composition of the third layer. When dealing with magnetic anomalies, we still do not know with what rocks they are associated, how they occur and what their direction or remanent magnetization is. To this very day not a single oriented rock sample has been taken from the ocean floor!

Beloussov finds the deep ocean trenches as unsuited for their role in plate tectonics as the ridges. Like Hess, he would expect to find great heaps of oceanic sediments caught up in the bulldozer and scraped off the surface of the descending slab against the opposing wall. Nothing of this sort is ever seen. Most trenches are empty of sediments or are floored by a flat-lying series which nowhere is more than 4 kilometers thick.

The Aleutian trench merits special attention, for here the sea floor is represented as sliding downward at an oblique angle, with younger magnetic strips preceding older ones (Figures 74, 81). At the same time, diagrams show a northwest-trending ridge axis pitching into the trench under the Aleutian arc. The consequences, one would think, should be dramatic in the extreme. Yet, Beloussov said, "the trench continues to exist as if nothing were happening." Beloussov concluded that this trench must have nothing whatever to do with resorption of the lithosphere. [One might argue, on the other hand, that the quiet swallowing of a ridge, hot dikes and all, *is* resorption—par excellence.] Beloussov continued: "It is hardly worthwhile criticizing this whole situation from the position of elementary conceptions of hydrodynamics. Simple common sense will suffice." And he concluded that

This hypothesis is based on a hasty generalization of certain data whose significance has been monstrously overestimated. It is replete with distortions of actual phenomena of nature and with raw statements. It brought into the earth sciences an alien rough schematization permeated by total ignorance of the actual properties of the medium.

Beloussov favors an interpretation of earth history that emphasizes the evolution of crustal rocks and tectonic structures as a result of thermal and chemical changes at depth operating mainly in a vertical sense. Light materials differentiate from the mantle and rise in the manner of salt domes; heavy residues sink. The sial may or may not have been continuous over the earth but in the past it has been far more extensive than it is at present. He believes that continental sial is "oceanized" by the intrusion of ultramafic mantle materials and by the formation of eclogite at deep levels until slabs of it become heavy, sink, and are resorbed.

Beloussov postulates that the destruction of continental crust has taken place over the entire area of the oceans, with the Atlantic and Indian oceans being relatively new features, and that the same process has opened the Mediterranean Sea, Black Sea, China Sea, and other waters. The Beloussov hypothesis of oceanization accounts for the youth of the ocean floors and their thin cover of sediments. The ridges are seen as young features built up of basalt flows arranged like tiles on a roof, with older ones projecting from beneath the younger ones. The magnetic anomalies record the succession of flows, which often are not symmetrical, on the two flanks. The ocean floor is not migrating, however. Like the more ancient continents, it responds to energy changes from the subjacent interior. As to the parallelism of continental coastlines, perhaps they are controlled by ancient fractures in the crust that have a persistent symmetry.

Here is a modern, highly developed theory of the earth with aspects reminiscent of the ideas of Eduard Suess and Joseph Barrell. Our truncated abstract cannot begin to do it justice. Needless to say, it is not in favor today with any quorum of earth scientists outside the Soviet Union.

The case for permanence: 1972

The February 1972 issue of the *American Association of Petroleum Geologists Bulletin* was devoted to plate tectonics and included two articles co-authored by A. A. Meyerhoff and H. A. Meyerhoff. The articles are a powerful restatement of the case for permanence not only of continents and ocean basins but also of the position of the earth's rotation axis, which, the authors argue, has remained constant for at least the past 1.6 billion years. In support of their view the Meyerhoffs have documented evidence from meteorology, biology, paleontology, physical geology, geophysics, and oceanography. Like many an earlier scientist they charge that continental margins do not match and that fossil floras and faunas show none of the uniformity they should within any given climatic zone if the continents had ever been joined. With V. V. Beloussov they charge that oceanic ridges do not spread, trenches do not subduct, transform faults do not transform, and magnetic stripes are neither linear nor symmetrical, nor, indeed, are they present at all along some 12,500 kilometers of the ridge system. The Meyerhoffs point out that if the Atlantic Ocean first broke along a line of weakness in Pangaea and the offsets in the Mid-Atlantic Ridge are primary features, then the "transform" faults should continue from the oceanic into the continental lithosphere; but they do not. If, on the other hand, the Americas are overriding the Pacific floor the two types of crust are wholly decoupled and the great fracture zones of the Pacific floor should not continue into the continents; but, according to the Meyerhoffs, they do.

In the Meyerhoffs' view the ocean basins are ancient Precambrian features. Numerous specimens of Precambrian and early Paleozoic rocks have been dredged from the Atlantic and Indian oceans in the vicinity of the ridge crests. These specimens are commonly dismissed as erratics picked up by ice caps on the continents and floated to sea by drifting icebergs—although this explanation strains credulity for the tropical waters overlying the Carlsberg ridge. The Meyerhoffs cite the results, published in 1969, of an expedition on which 64 (75 percent) of 84 rock specimens dredged from three sites on the Mid-Atlantic Ridge near latitude 45° north were discarded because they were too old to fit the theoretical age of oceanic bedrock. Such a proportion of glacial erratics in the ocean was described by the scientists on the expedition as a "remarkable phenomenon." Earlier, in 1949, specimens containing Cambrian trilobites were dredged from the Atlantic north of the Azores. This occurrence, which is very puzzling in the light of sea-floor spreading, was recently explained away as due to either the mislabeling of specimens or the recovery of ballast dropped by old steamers out of Cardiff, Wales (ancient Cambria). The Meyerhoffs quoted from a letter on this subject written to them by an ardent advocate of the new global tectonics: " 'I am amazed that you regard such a dredge haul as having any significance. After all, the dredge predates 1960, and nothing done before that date is of importance in modern geology.' " Such advocates, say the Meyerhoffs, "do not claim that they are impartial, but . . . do insist that they are scientific."

The JOIDES program has recovered drill cores from the Atlantic which show a thickening of sediments from the ridge flanks toward both continental margins together with a progressive increase in age from Recent to early Jurassic of the sedimentary layers overlying basement rock. But what is meant by basement rock? The deep drilling is brought to an end after the cores strike basalt. Are the basalts the authentic crystalline bedrock of the ocean floors or are they flows or sills of lava that are interlayered with the sediments? The Meyerhoffs argue the latter point of view. They state that some of the sediments have been baked by contact with the basalt which was therefore intruded into the preexisting sediments. Such a relationship was demonstrated conclusively in one instance where the basalt underlying 150-million-year-old Cretaceous sediment was shown by K/Ar dating to be only 15 million years old. Deeper drilling, the Meyerhoffs argue, would penetrate a long succession of basalts and sediments leading back to the Precambrian. Whether or not this hypothesis will be tested and with what results only the future will tell. Meanwhile, to question the work of the ship *Glomar Challenger*, from which the drilling program is carried on, is to lay siege to one of the most elegant new artifacts in earth science. That has not been done before.

The Meyerhoffs defend a permanent configuration of continents and ocean basins. They believe that the continental shields began forming in situ in the early Precambrian and have acted as thick, relatively rigid units which have grown by accretion. They suggest that the chemical differentiation of sial scavenges radioactive elements from the sima, leaving the mantle beneath the continents colder and stronger than that beneath the ocean basins. In their view the earth is probably contracting and the weakest portions of the crust (oceanic areas) are buckling into ridges under tangential compression. The zones of weakness are controlled by the distribution of rigid shields to which the oceanic ridges are approximately concentric. Continued buckling causes rejuvenation of the ridges and the bleeding of fresh basalts to the ridge crests, which therefore are kept young and free of sediments. As most marine sediments are derived from the continents, the beds are naturally older and thicker toward the continental margins.

According to the Meyerhoffs, the magnetic anom-

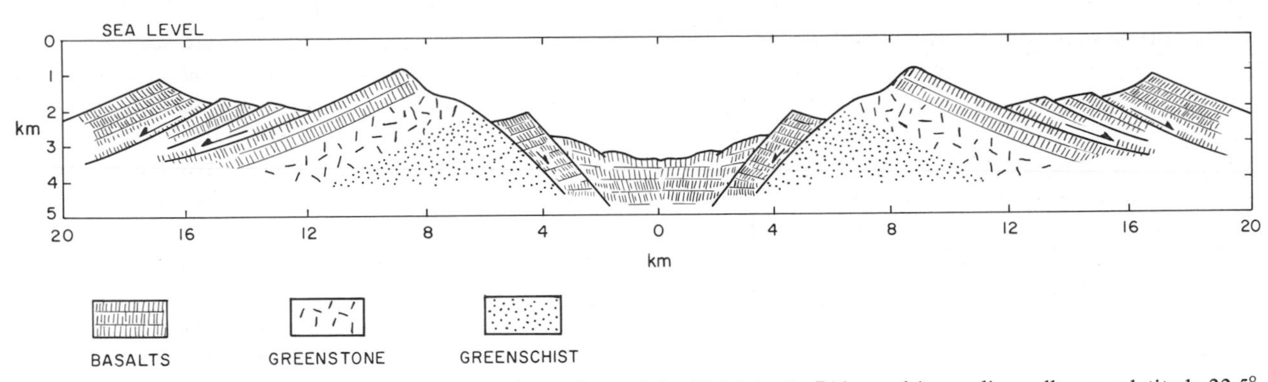

Figure 90. An interpretation of the structure and petrology of the Mid-Atlantic Ridge and its median valley near latitude 22.5° north, by Tjeerd van Andel and C. O. Bowin. (Used with permission of Tjeerd van Andel and the American Geophysical Union.)

aly belts are very complex igneous and metamorphic features, some of which record field reversals while others reflect major differences in the magnetic susceptibility of the rocks. These features are not limited to ridge flanks. The Meyerhoffs state that in fifteen places the belts dive beneath continental margins of great age and that some of the belts therefore originated in the early Precambrian and the others have been emplaced at intervals up to the present.

Figure 90 depicts a cross section through the axial valley and crestal range of the Mid-Atlantic Ridge near latitude 22.5° north. In this interpretation by Tjeerd H. van Andel and C. O. Bowin, the range consists of a series of tilted fault blocks composed of three types of bedrock. According to an analysis by N. D. Watkins and A. Richardson, published in 1968, differences in magnetization of rocks in such a structure could account for linear magnetic anomalies on ridge flanks without sea-floor spreading. This is a type of interpretation favored by the Meyerhoffs and by V. V. Beloussov.

Other voices of dissent have been raised since the founding of the new global tectonics. Many of them are directed specifically at the problem of the mechanism and the grotesque inconsistencies between convection in the classical sense and the inferred characteristics of lithospheric plates. Others present hard data that appear inconsistent with the theory. In no cases can the dissenters be accused of simply trying to impede progress; they are asking very seriously whether the new view of the earth coincides more accurately with reality and will serve better than the old for solving geological problems.

Scientific Revolutions

The story of continental drift as a geologic concept, with its slow, tentative beginnings and violent controversy, followed by the spectacular bandwagon effect which has swept up the majority of earth scientists, bears out in dramatic fashion a thesis developed by Thomas S. Kuhn in his book *The Structure of Scientific Revolutions*, first published in 1962. Kuhn rejected the widespread notion that, unlike other, less-exacting disciplines, science progresses in a linear manner by the steady increment of shared knowledge. He perceived that, in common with art, politics, and other human endeavors, science develops through a succession of tradition-bound periods separated by decisive breaks during which a fundamental revision of concepts takes place. Recalling the examples of Copernicus and Galileo, he showed that the breaks are precipitated not so much by great new discoveries or compendia of improved data as by the buildup to crisis proportions of anomalies, inconsistencies, and unsolved problems within the traditional framework. Only when such a point is reached are more than a few scientists willing to abandon a familiar system of thought for a wholly new one. Once established, the new conceptual system gradually grows into a tradition encompassing its own anomalies and unsolved problems.

Plate tectonics has brought a new way of viewing the earth, a new array of assumptions, intuitions, and problems, and new criteria for evaluating observational data and making predictions. It has not been possible to embrace it piecemeal. Acceptance has required the rejection of an older corpus of belief centered upon the concept of an essentially stable configuration of the earth's surface features. Kuhn points out that to change one matrix of beliefs for another is tantamount to undergoing a conversion experience. Such a change can be accomplished only with the greatest difficulty by any serious thinker and, as scientific integrity does not admit of facile changes of reference, it is most difficult of all for highly productive scientists who have spent a lifetime creating order out of the "chaos" they found when they began their careers. For this reason, older scientists generally do not join the ranks of the converted. However, successful scientific revolutions depend only partly on converts. Presently, young students who have never faced any problems are trained from the beginning to perceive questions and seek out answers within the new framework. Scientists, according to Kuhn, are particularly prone to rewrite history. New textbooks tend to omit all serious discussion of outmoded views. Indeed, although they would be totally helpless without the knowledge of the earth gathered over the last century, some geologists already regard as comic all ideas held before 1960.

At this point in time earth science is pervaded by a powerful sense of euphoria. The new global tectonics is almost universally accepted as a fact. However, the theory in its original pure form has already undergone considerable modification and it clearly must be modified further to account for deformations within plates, such as the massive amounts of uplift and downwarping recorded on the continents, and the occurrence of earthquakes and active volcanism deep within continental interiors. The true test of the theory will come when geologists shift their attention from those features that are most easily explained to those that are more difficult to reconcile with the principles of plate tectonics.

Meanwhile, the objections raised by the dissenters should, at the very least, inspire a critical attitude on the part of readers threading their way through the volumes of new articles published weekly in the scientific and news journals of the world. The computer-fit shows a perfect matching of coastlines; the paleomagnetic polar wandering curves correlate with those of paleoclimates; the magnetic strips measure the spreading rates. But whose computer-fit? Whose paleomagnetic curves? Whose climatic zones? Which magnetic strips? Selected by what criteria? Checked against what independent evidence?

Possibly we should find it reassuring that the hoary debate on the origin of continents and oceans has lost nothing of its elemental vigor or polemical style with the changing of majority opinion.

Patterns of Continental Drift

The tectonic history of the earth is not, of course, a matter of opinion. The planet has followed only one course of evolution—one which has produced the tectonic framework of cratons and mountain belts represented on the map of Figure 91 and a layered interior of approximately the character indicated in the two sections in Figure 92, both within the time scale specified in Figure 93. The evidence of the various stages and mechanisms involved in the earth's evolution is, however, so fragmentary, so conflicting, so inconclusive that its interpretation remains controversial.

The theory of plate tectonics, despite the objections of its critics, opens the most significant and fruitful new approach to understanding earth history that we have seen in many decades. How do the continental reconstructions, inferred by plate motions, compare with those postulated by Alfred Wegener and the other early advocates of continental drift? To answer this question we must begin at the oceanic ridges and imagine plate motion thrown into reverse. The great slabs of lithosphere must pull out of the trenches, migrate up the ridge flanks, and melt down into the mantle beneath the ridge crests. The new oceans will close, and the continental margins will converge to form one great protocontinent, Pangaea, or two protocontinents, Laurasia and Gondwanaland. Possibly there were more such continents. We have arrived back in Permian time, 270 million years ago.

Have we also arrived at a configuration of continents that grew in place around stable nuclei through the earlier part of earth history? Such an idea is unthinkable, say the uniformitarians. If plates have been moving since the Mesozoic, then plates have always moved. Oceans were opening and closing and continents were splitting and regrouping with great mountain ranges rising on their margins all through the Paleozoic and Precambrian.

Yet the alternative remains that the planet may have slowly evolved from one tectonic pattern to another. Precambrian and Paleozoic events may have been dominated by the differentiation of new sial from the mantle and the growth of one or more large continents. By the early Triassic, the chemical and thermal regime of the earth's interior may have changed sufficiently for the lithosphere to begin rupturing, for the first time, into moving plates.

What do our reversed plate motions indicate about world geography at the beginning of the Triassic? Laurasia and Gondwanaland can be reconstructed with relatively little difficulty. Major uncertainties attend the match between the northern and southern continents. Before attempting to solve the problem of Pangaea, let us examine the latest evidence with respect to Laurasia and Gondwanaland.

Laurasia

If Laurasia was a single landmass, or a compact group of landmasses, it might have looked something like the reconstruction shown in Figure 94, which is a simplified version of an illustration published in 1971 by P. M. Hurley. The map, compiled from data on bedrock ages, shows the Precambrian areas of the northern hemisphere surrounded and intruded by younger mountain belts and island arcs. The central position of the shields suggests that Permian Laurasia as a whole was the final product of continental accretion uninterrupted by plate motion.

Several lines of contrary evidence, however, indicate that Permian Laurasia was a mosaic of formerly separate fragments, that the North Atlantic existed in the lower Paleozoic, and that seaways separated parts of Siberia and North America. We may examine the problem of Laurasia by beginning in the region of the Atlantic Ocean and working our way eastward around the northern hemisphere.

We have heard much about the similarities between late Paleozoic fossil assemblages on both sides of the North Atlantic. But as we noted in our section on Grabau's Paleozoic Pangaeas, the earlier Cambrian and Ordovician strata on either side carry fundamentally different assemblages—with two exceptions: typically American fossils populate the early Paleozoic formations along the west coasts of Norway, northern Scotland, and Ireland; and European fossils occur in the early Paleozoic rocks of New England, Nova Scotia, and Newfoundland (Figure 95). In 1966 J. Tuzo Wilson accounted for these

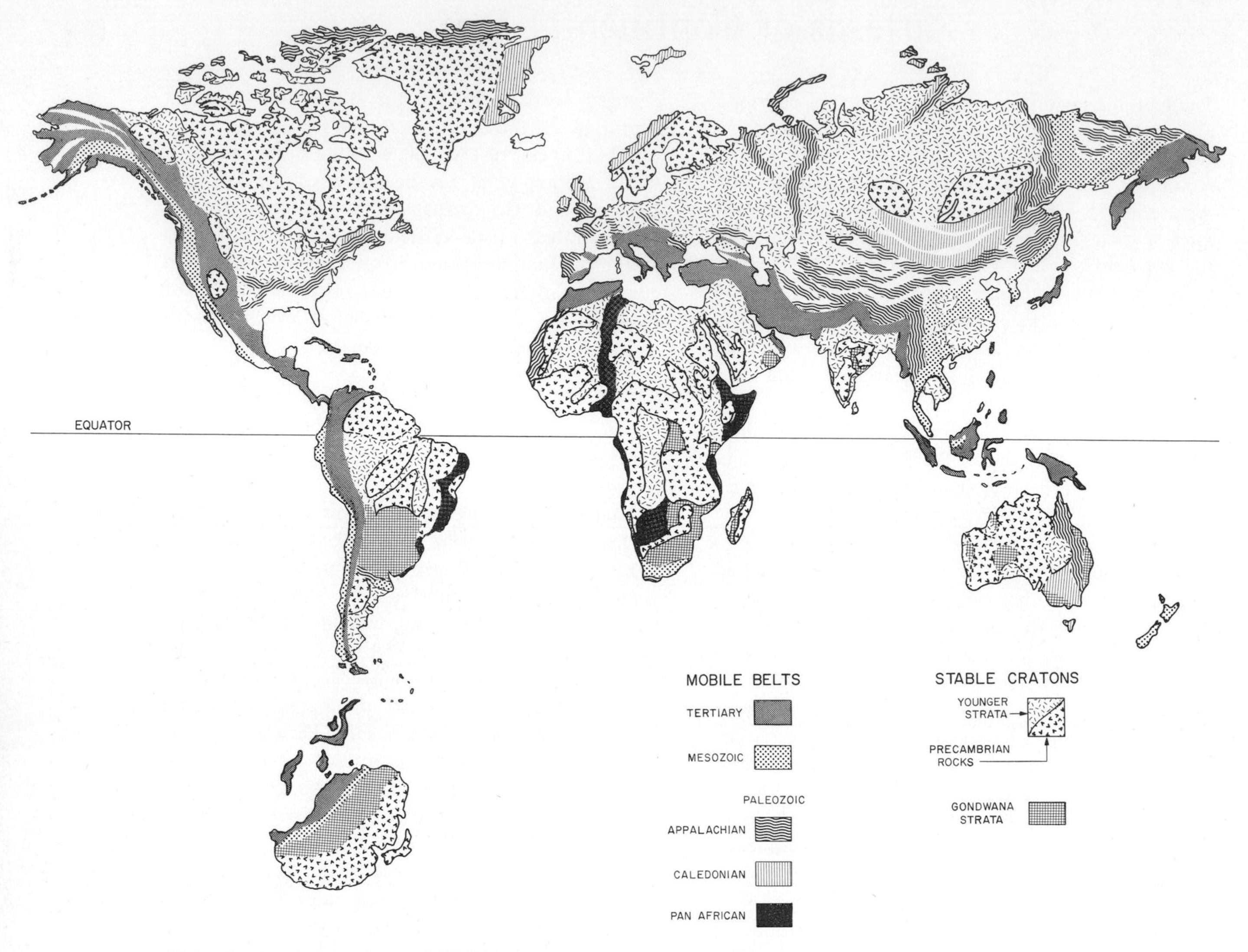

Figure 91. A tectonic map of the continents. (Compiled from numerous sources.)

anomalous slices of misplaced Cambrian and Ordovician sediments by suggesting that the North Atlantic was open at the end of the Precambrian; that it was closed by the Devonian; and that it began opening again in the Cretaceous along a line slightly different from the earlier suture.

We will remember that Alfred Wegener was severely taken to task for claiming that the structural provinces of Newfoundland match those of the British Isles and for fitting together the Devonian redbeds of North America, the British Isles, and northern Europe. However, in a symposium on North Atlantic geology and continental drift held in Gander, Newfoundland, in 1967, Marshall Kay described three metamorphic belts in Newfoundland with counterparts in the British Isles. He dismissed as "trivial" the probability that their similarities are accidental. During the past 25 years many similarities have been revealed in the freshwater fish genera of the Devonian redbeds. Indeed, 57 percent of the genera of North America also occur in the redbeds of Europe, and much the same order of correspondence persists among fish, amphibians, and reptiles from the Devonian through the Permian, when a decline in shared genera indicates a rupture. Alfred S. Romer, professor emeritus of vertebrate paleontology at Harvard University, wrote in 1968 that such evidence had finally transformed him into a drifter after a lifetime as an orthodox disciple of permanence.

One of the most controversial puzzles of the North Atlantic is a landmass that was not part of Laurasia at all. It is Iceland, a young, Tertiary island and the only large area in the world where an oceanic ridge is built above sea level. Because of its elevation, it was long assumed to have a sialic basement, and in 1965 Arthur Holmes stated: "Iceland may be cited as an outstanding example of a large continental crustal block that has been left stranded in the formerly 'dead' space between deep-seated diverging currents."

The occurrence of numerous inclusions of granitic rock in the Icelandic basalts and also in the fresh eruptions of Surtsey (the volcanic island immediately southeast of Iceland) is taken by some geologists as confirmation of sialic bedrock. Others, however, interpret the inclusions as products of differentiation from a basaltic magma. One specialist in rift volcanics who favors this view is Ian Gibson, of the University of London, who remains convinced that Iceland is a strictly oceanic island and an integral part of the Mid-Atlantic Ridge.

The geologic history of the island as outlined in 1967 by John Haller of Harvard University began with the formation in the Eocene or Oligocene of a basaltic plateau which, in the Miocene, was fractured under tension into an array of tilted blocks. Subsequently, horizontal bands of onyx were deposited in joints and vesicles in the lavas and most of these bands remain essentially horizontal. Iceland was eroded almost to a level plain in the Pliocene before a renewal of tectonic activity caused the downfaulting of a central valley. Volcanism, which began again at that time and still continues, has covered most of the Pliocene erosion surface. During the Pleistocene many extrusions occurred under a thick cap of ice.

Haller described the central valley of Iceland as a zone of active tension-fracturing that apparently is a landward extension of the Mid-Atlantic Ridge rift. Among the more unusual features of this valley are open fissures, some of which are up to 50 meters wide. These gaping cracks occur in parallel swarms that occupy a zone from 1 to 4 kilometers wide and

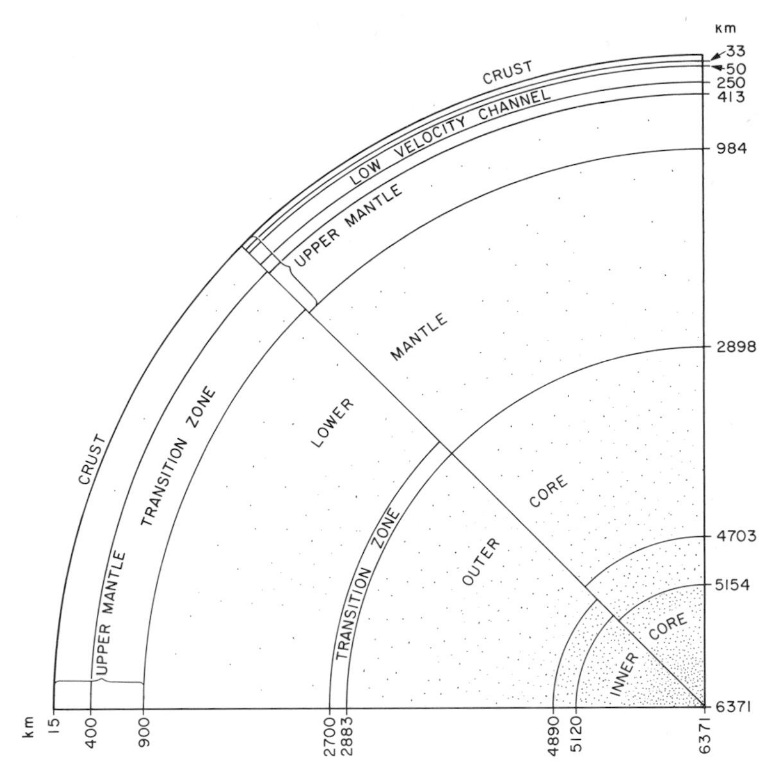

Figure 92. Two sections of the earth's interior. No one interpretation of the evidence for the nature of the earth's interior can be demonstrated to be the unique solution to the problem. Of these two slightly different versions, the upper section is from F. S. Birch, 1965 (with permission of the author and The Geological Society of America); the lower section is from F. D. Stacey, 1969 (*Physics of the Earth*, John Wiley & Sons, Inc., by permission).

10 to 30 kilometers long. As the average age of the surface rocks cut by these fractures is about 5,000 years, it is possible to estimate the rate of extension represented by the fissures. Haller cited an investigation made in 1943 by F. Bernauer who concluded that for the past 5,000 years each kilometer of width of the zone occupied by fractures has opened at a rate of 3.56 meters per 1,000 years, or 0.356 centimeters per year. This rate applies only to the zone of open tension fissures where direct measurements can be made. Extrapolations, which Haller regards as of dubious value, have subsequently been made for spreading rates of the central valley and, indeed, of the North Atlantic as a whole.

Geodetic measurements reported in 1971 by

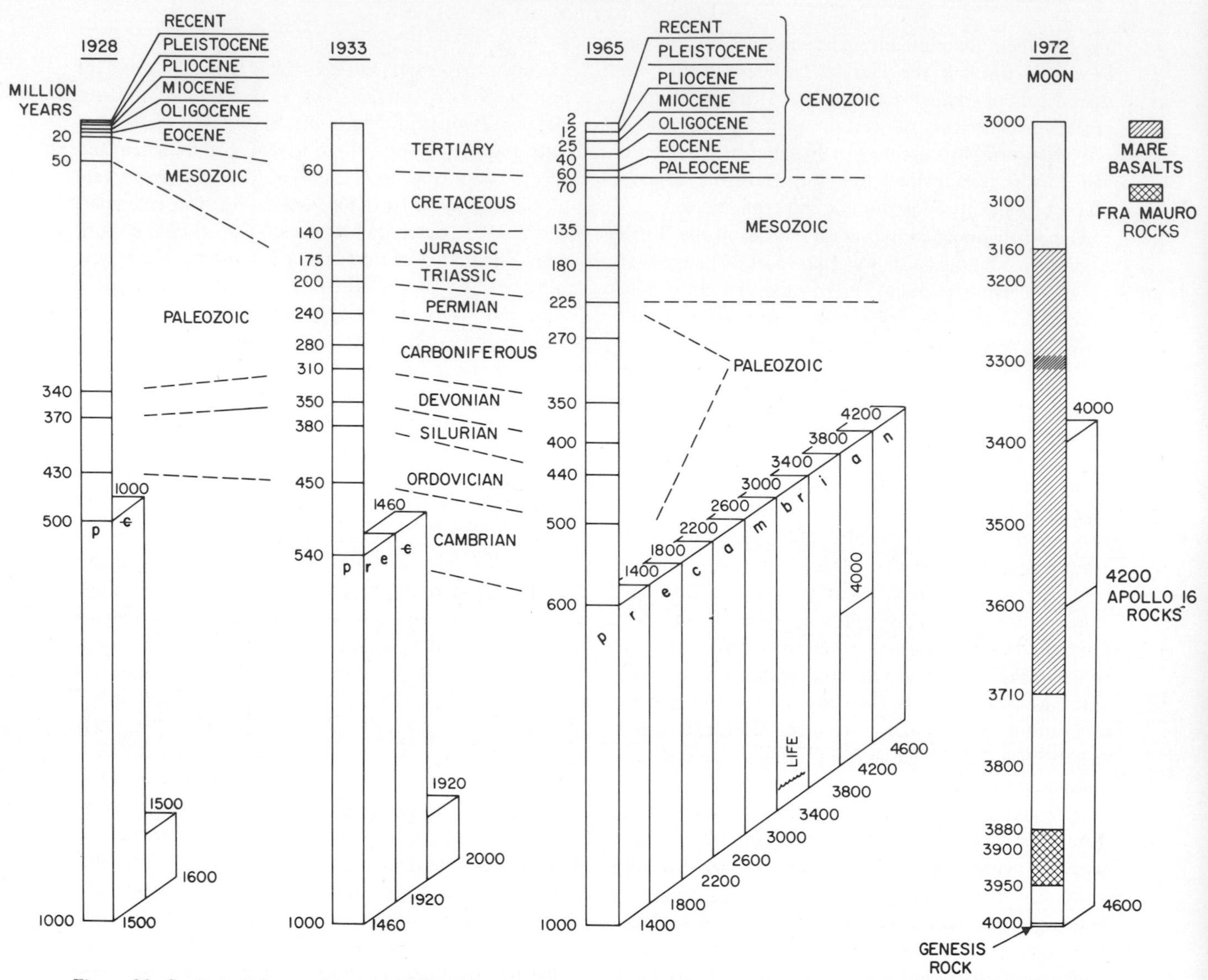

Figure 93. Geological time scales circa 1928, 1933, and 1965, and a lunar time scale circa 1972. The 1928 scale was compiled from Cenozoic and Mesozoic dates listed by Wegener in 1929 (fourth edition) and earlier dates listed by Holmes in *The Age of the Earth*, 1928. The 1933 scale was compiled from the writings of Charles Schuchert; the 1965 scale, mainly from those of Holmes. The lunar scale, which includes only the first 1.6 billion years of the moon's history, was constructed from reports presented at the Fourth Lunar Science Conference, Houston, Texas, March 1973.

Robert W. Decker of Dartmouth College, Páll Einarsson of Lamont-Doherty Geological Observatory, and Paul A. Mohr of Smithsonian Astrophysical Observatory show that an extension of 6 centimeters occurred between 1967 and 1970 across a survey line north of Hekla volcano. This period, however, was marked by fissure eruptions from the volcano, so the very rapid rate of spreading is probably exceptional for this sector and all other sectors of the Icelandic rift.

Although Iceland may provide some clues to the nature of midocean ridges, the site it occupies is far from being a typical ridge segment. Iceland stands at the intersection of the Mid-Atlantic Ridge and a rise called the Faeroes-Greenland Sill or the Brito-Arctic Rise, which stretches east to west across the North

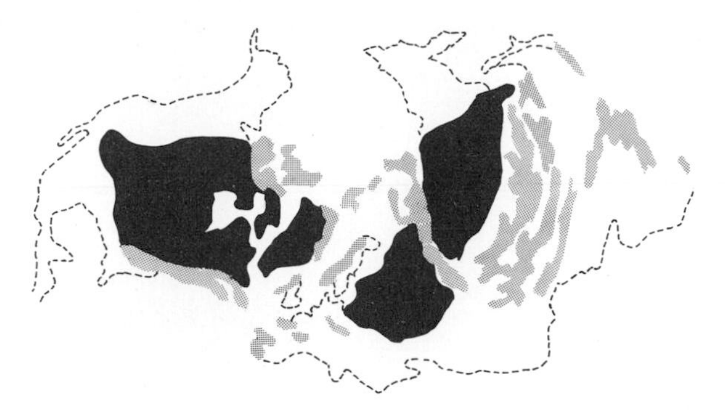

Figure 94. A simplified version of P. M. Hurley's proposed reconstruction of Laurasia. Stable areas more than 800 million years old (black) are bordered by orogenic belts 250 to 800 million years old (stippled). (From *Earth and Planetary Science Letters*, volume 8, 1970; used with permission of author and North-Holland Publishing Company.)

Atlantic. Transoceanic rises were formerly interpreted by antidrifters as sunken land bridges and by drifters as streaks of sial that were fused at depth and pulled out, like strands of taffy, between diverging continents—a process that would seemingly threaten the preservation of matching shorelines. A newer suggestion is that transverse rises are built on diverging oceanic plates by the continuous eruption of lavas from a permanent "hot spot" on a ridge. The precise nature of the Faeroes-Greenland Sill remains uncertain.

At present Iceland is regarded by A. A. Meyerhoff as a structural unit dominated by compression and uplift, by some geologists as a complex island with a sialic basement subject to tension and overwhelmed by basaltic volcanism, and by numerous others as a spreading ridge.

Evidence of a former north-south seaway separating parts of the Eurasian continent was described in 1970 by Warren Hamilton, who believes that the Ural Mountains were formed by the convergence of two continental plate margins during the Devonian. Farther east we are faced with the possibility that a large portion of China and of western North America may have been joined in the late Paleozoic. Among the faunal and floral distributions described in the 1920s, by J. W. Gregory, Chester Longwell, and others, were numerous instances wherein the organisms of northwestern North America match those of eastern Asia more closely than they do those of eastern North America. This is reported of present-day earthworms, for example, and is well established with respect to late Paleozoic land plants. When land plants first appeared, in the Silurian, they were worldwide in distribution and uniform in character. By the Permocarboniferous they had diversified into four distinctive groups: the *Glossopteris* flora, now preserved in the southern continents and India; a European flora in Europe, eastern North America, and northern Africa; an Angaran flora in Siberia; and a Cathasian flora in China and western North America. The matching biota of America and Asia have been largely ignored except by opponents of continental drift. Such a matching could only be lamented by advocates of the classical ideas of drift in which the Americas moved toward and not away from the Pacific.

In 1966, R. Melville of the Royal Botanic Gardens at Kew proposed that the floral groups reflect the existence of four late Paleozoic continents. Eastern Asia and western North America formed one such continent—which Melville called Pacifica—that was ruptured during the Cretaceous by the initiation of a northwest-southeast-trending ridge. The two fragments of Pacifica were subsequently rafted to their present sites by the spreading sea floor. Finally, after a lifetime of some 100 million years, the ridge subsided to form the deeply submerged Darwin Rise of the western Pacific Ocean. According to Melville, the seamline where part of Pacifica is welded to North America extends from the delta of the McKenzie River to that of the Mississippi, thus all of the mountainous American West and Northwest once belonged to Pacifica. Some investigators reject the idea that North America includes any block that was shared with Paleozoic Cathay. Others, including J. Tuzo Wilson, favor such an interpretation for a smaller area centered approximately on the State of Washington.

In the northern Pacific Ocean the JOIDES drilling program has produced results suggesting the progressive increase in the age of the crystalline basement on either side of the East Pacific Rise shown in Figure 96. The results are broadly similar to those inferred from magnetic anomalies shown in Figure 81. The predominant trend from Recent to Cretaceous rocks westward from North America is the reverse of what one would expect if the North

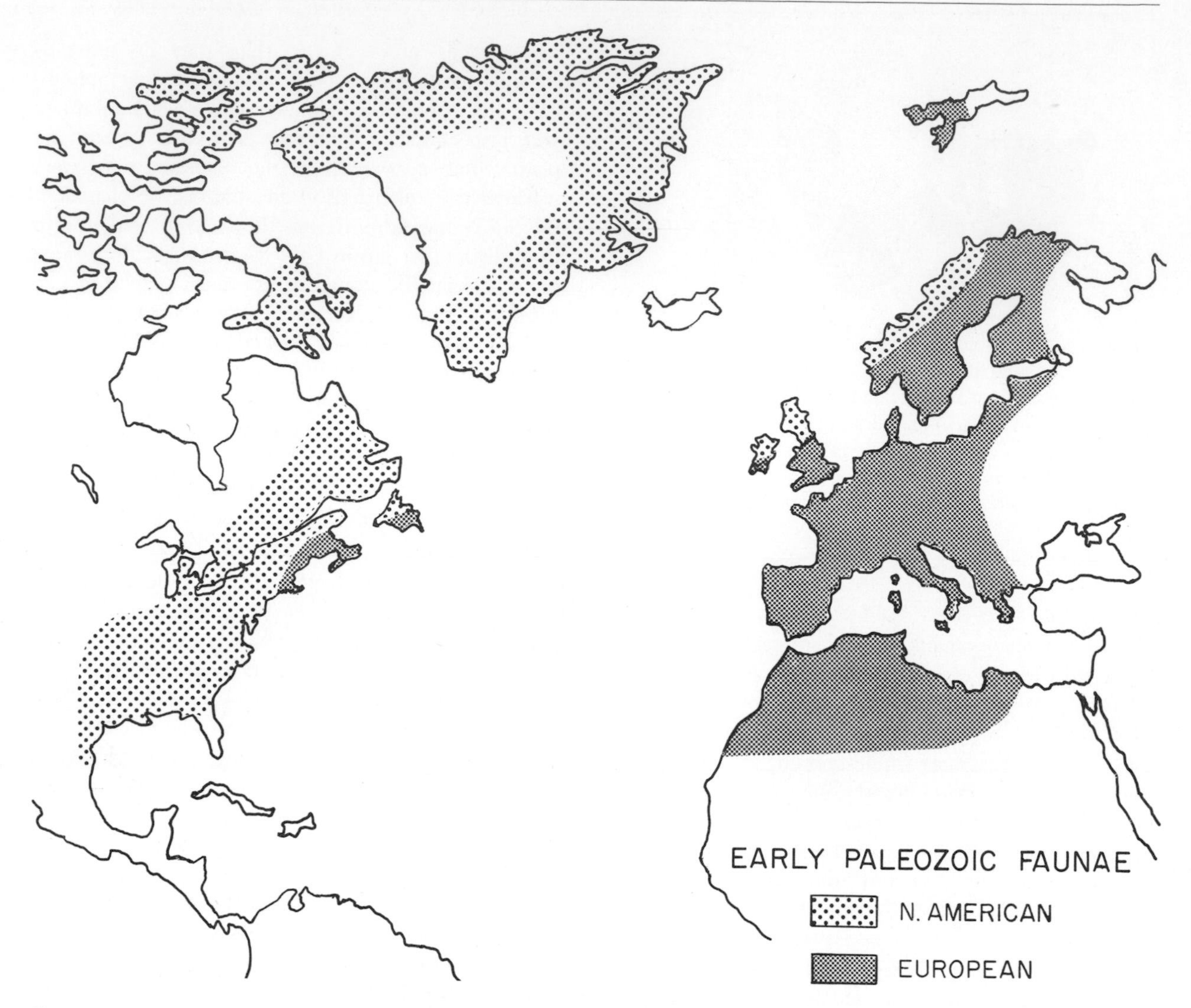

Figure 95. The distribution of early Paleozoic fossil faunae on both sides of the North Atlantic Ocean. (After J. Tuzo Wilson, *Nature*, volume 211, 1966; used with permission of the author and *Nature*.)

Pacific Ocean opened in the late Mesozoic and grew wider as North America and China were rafted to either side as suggested by Melville.

Western North America is surely one of the most complex tectonic regions of the world. At present there is no trench off the coast of California, and no active Benioff zone plunging beneath the continental margin from Baja California to Alaska. A broad zone of the continent is, however, seismically active and characterized by recent volcanism. In 1972 James Gilluly pointed out that even without an underthrusting oceanic plate this continental margin is about as unstable as any such margins having Benioff zones and probably is as active as it was during the Tertiary when a Benioff zone may have existed. Gilluly also called attention to the very long span of time during which the western cordilleran region has been a site of ore deposition. Minerals rich in copper and molybdenum have been emplaced along this zone from Precambrian times (as at Jerome, Arizona) to the Oligocene (as at Bingham, Utah). Enrichment in silver also has a long history. The mantle beneath the western portion of the continent has clearly retained a distinct signature throughout much of geologic

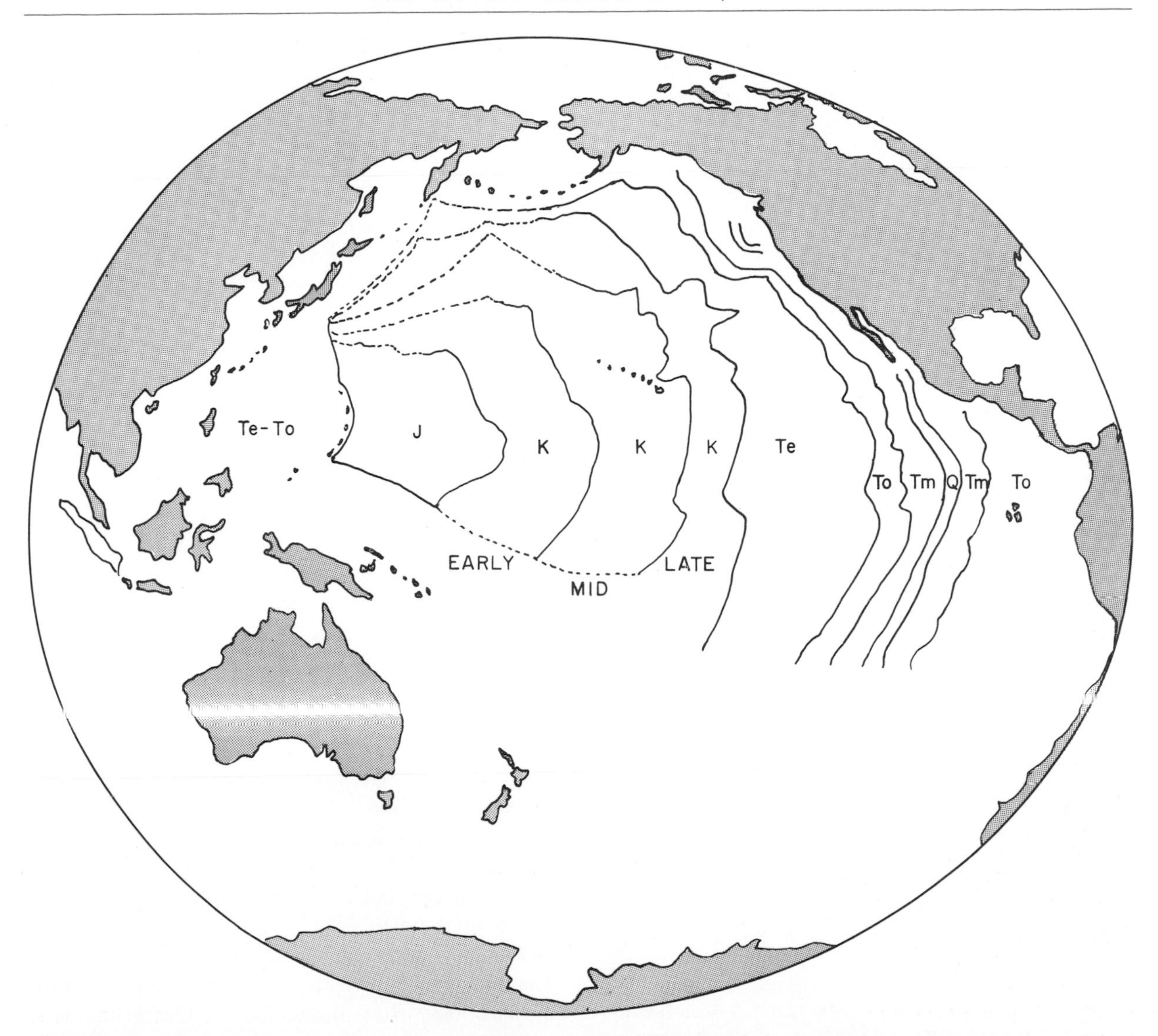

Figure 96. Ages of ocean floor basalts in the North Pacific inferred from the drilling results of the *Glomar Challenger*: **J**, Jurassic; **K**, Cretaceous; **T**, Tertiary; **e**, Eocene; **o**, Oligocene; **m**, Miocene; **Q**, Quaternary. (From A. G. Fischer, et al., *Science*, volume 168, number 3936, page 1211. Copyright 1970 by the American Association for the Advancement of Science. Used with permission of principal author and the publisher.)

time—one that differs completely from that of the mantle under the Great Plains. Thus, cautions Gilluly, we must assume that the ores are derived from within the lithospheric plates at depths of less than about 70 kilometers and that each segment of the continental crust has remained permanently coupled with the sample of uppermost mantle directly beneath it.

Now that we have explored the possibilities of former landmasses in the North Atlantic, North Pacific, and former seaways separating parts of Siberia and North America, we leave Laurasia, temporarily, for a perusal of Gondwanaland.

Gondwanaland

The problem of Gondwanaland is much more complicated than that of Laurasia. Here, five large continents and several islands must be reassembled

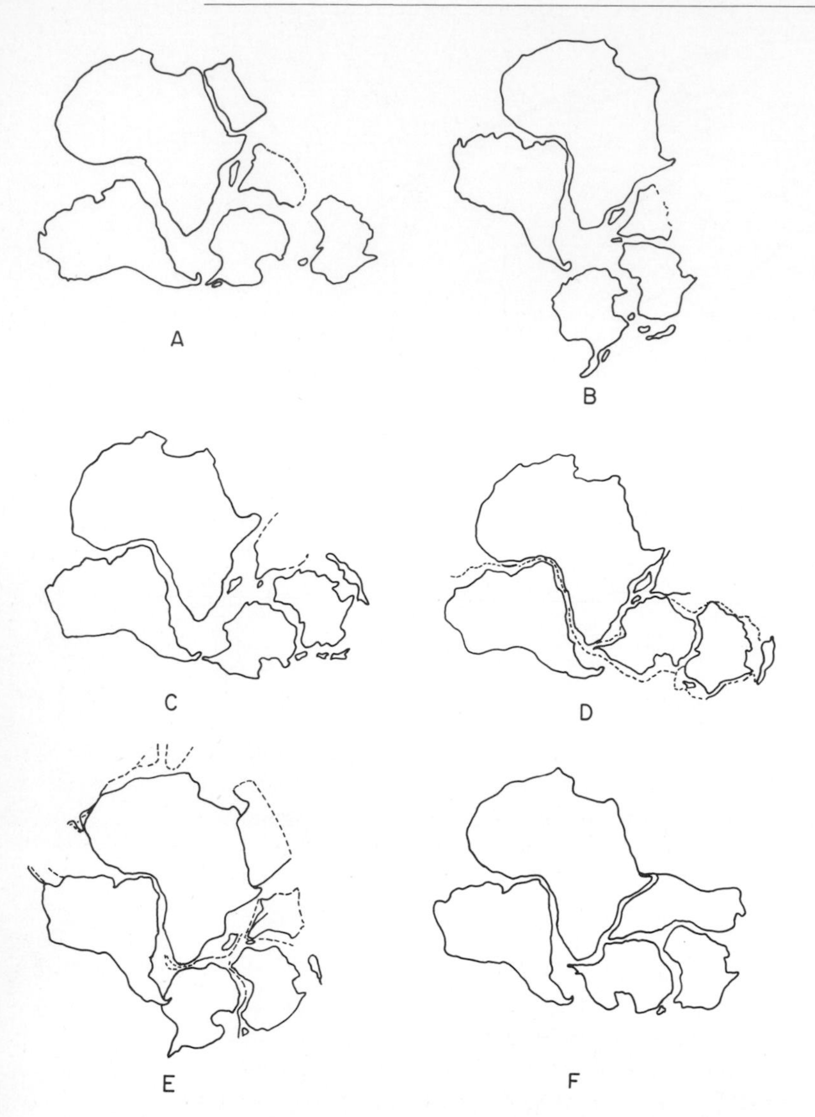

Figure 97. Six proposed reconstructions of Gondwanaland: **A**: Lester King, 1958. **B**: S. W. Warren Carey, 1958. **C**: Arthur Holmes, 1965. **D**: A. G. Smith, A. Hallam, and a computer, 1970. **E**: D. H. Tarling, 1971. **F**: P. M. Hurley, 1971. (Used with permission of authors and publishers.)

along oceanic ridges. The magnitude of the problem is illustrated by Figure 97, which includes six of the reconstructions published within the past fifteen years. Each of the six was carefully based on evidence for matching continental margins, structures, stratigraphic formations, climatic zones, and floral or faunal distributions. Where data were available, dated rock provinces, polar-wandering curves, and magnetic stripes on the ocean floors also were used. Nevertheless, there was no genuine agreement even on so simple a point as the distance and angle at which South America should be joined to Africa—the factor that originally inspired Alfred Wegener with the idea of continental drift. Possibly Harold Jeffreys had been examining such maps as these when he stated once again, in the fifth edition of *The Earth* (1970), that the match is really a misfit of about 12°.

In general, however, the evidence for former contiguity between Africa and South America has been strengthened in recent years. The matching of contacts between dated rock provinces was followed by the discovery in 1969 by G. O. Allard and V. J. Hurst of a Precambrian structure called the Propria geosyncline that extends from Sergipe, Brazil, to the Gabon in West Africa. In addition, small basins of Mesozoic sediments, truncated by both coasts, contain identical species of freshwater fish and ostracods. Salt beds of apparently shallow-water marine origin attest to the Cretaceous age of the first invasions of seawater along this portion of the Atlantic margin.

Although extensive field work has been done in recent decades in Permian strata, the fossilized skeletons of the little reptile *Mesosaurus* are still found only in the lower Gondwana-type beds of South America and South Africa.

With one or two notable exceptions, discussed in our section on "voices of dissent," geologists around the world are most favorably impressed with the results of the JOIDES program. Drill cores taken in 1970 along a traverse between Dakar and Rio de Janeiro show a progressive increase in age with distance away from the midocean ridge of the basal sediment overlying basalt. The youngest sediments lap highest on the ridge flanks, leaving the crestal zone bare. The oldest bottom sediments, dated by microfossils, lie near the continental margins and are late Jurassic. The sedimentary succession shows variations in thickness and distribution that have been attributed to changes in elevation of the ridge. This, in turn, suggests occasional cooling and subsidence of the ridge followed by elevation and renewed spreading.

Outside the Atlantic, the proper fit of fragments of Gondwanaland is very much open to debate. The Gondwana Series, with its basal tillites, *Glossopteris* flora, continental sediments, and basalt flows, still constitutes some of the best evidence of former contact between these landmasses.

The paleomagnetic data, developed in the laboratory of P. M. S. Blackett, supports the idea that India lay in the southern hemisphere as recently as 80 million years ago when the Deccan basalt flows crystallized. Most of the recent reconstructions (such as those in Figure 97) show India as a small triangular wedge of uncertain northern boundaries. This wedge is described as occupying an oceanic plate which moved rapidly northward as its leading edge plunged into a subduction zone beneath central Asia. That pattern would render impossible any of the Precambrian and Paleozoic structural and stratigraphic continuities between India and Asia that are described by A. A. Meyerhoff. However, Alfred Wegener's solution to the problem (Figures 29 and 31) may suggest a possible means of reconciliation. Another version of the relations between India and Asia was recently suggested by P. M. Hurley, who found from his compilation of data on the ages of bedrock samples that a large block consisting of China and Korea would fit better in Gondwanaland that it does in Laurasia. Having considered this radical move, Hurley found that it could also be supported by structural trends and fossil distributions. This suggestion of Hurley's is still strictly tentative but it calls attention to the fact that the former position of China may be due for serious reconsideration. We have already seen that R. Melville and J. Tuzo Wilson favor a Paleozoic contact between parts of China and North America. Hurley is the first advocate of continental drift to place China with the southern continents, but Neumayr's map of the Jurassic world (Figure 19), which originally inspired the idea of Gondwanaland, might provide justification from paleontology.

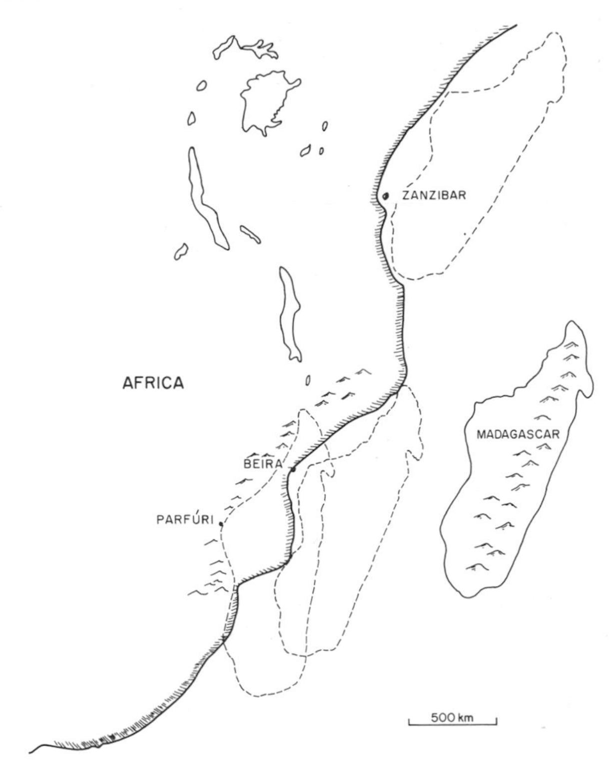

Figure 98. Three proposed locations for Madagascar in the late Cretaceous period. Drilling results from the Mozambique channel, reported late in 1972, suggest that Madagascar was not adjacent to Africa and may have occupied its present position in the Cretaceous (*Geotimes*, November 1972, pages 21–24).

The reconstructions in Figure 97 show a striking lack of agreement on a "straightforward" case such as the former position of Madagascar. Figure 98 illustrates how some geologists fit the island to Tanzania at Zanzibar and others to Mozambique at Beira; still others dispose of the wedge of post-Paleozoic sediments in southern Mozambique and weld Madagascar to the interior highlands at Parfurí. The latter match was suggested in 1971 in an informal discussion of the problem by Ian Gibson of the University of London.

Of all the southern continents, Antarctica is the most troublesome in the reconstructions. It fits poorly because in whatever way it is rotated the prominent hook of the Palmer Peninsula tends to overlap some other landmass. However, the case for joining Antarctica with the other Gondwana continents was strengthened in 1967 by the discovery in the central Transantarctic Mountains of a jawbone fragment of a labyrinthodont, a Triassic freshwater amphibian. Shortly thereafter the remains of *Lystrosaurus*, a cold-blooded reptile 1.5 meters long, also were found. *Lystrosaurus* is the key index fossil by which lower Triassic strata are identified in Africa. As a land-based reptile he presumably could not cross wide marine barriers; as a cold-blooded creature he probably enjoyed climates warmer than those prevailing today in the vicinity of the pole. *Lystrosaurus* thus seems to tell of both continental drift and polar wandering. But not necessarily so, warned Harold

Figure 99. *Lystrosaurus* contemplates a crossing.

Jeffreys in the fifth edition of *The Earth*. Perhaps *Lystrosaurus* spread through the South Sandwich arc, which formerly may have been a land bridge.

The discovery of *Lystrosaurus* in Antarctica was described by D. H. Elliot, of Ohio State University, and four other scientists in a September 1970 issue of *Science*. That report elicited a letter to *Science* in November of the same year from George Gaylord Simpson at the University of Arizona. Simpson stated that the discovery was an important event for paleontology and geology, and he agreed that it lent strong support to drift theory. However, he questioned the idea, favored by Elliot and his colleagues, that the new evidence proved the former existence of Gondwanaland as a separate landmass. He pointed out that *Lystrosaurus* also lived in Sinkiang, China. Thus, it would seem that either that region of Asia must be added to Gondwanaland (an idea suggested on geochronological evidence by P. M. Hurley a year later) or Gondwanaland must be connected with the northern continents.

In June 1972 Simpson reviewed some of the old and new evidence bearing on continental drift:

. . . as regards my paper of 1943 on *Mammals and the Nature of Continents* . . . I meant my title literally and was primarily talking about Cenozoic mammals, not fossils in general. My study was a reaction to the fact that proponents of continental drift, notably Wegener, Du Toit, and Joleaud, had claimed that the distribution of Cenozoic mammals supported their views, especially as regards the South America-Africa fit. That was not true when they wrote and in fact it still is not true today. . . . Arguments for continental drift based on Cenozoic mammals were wrong, and to that extent it still appears that I was right in 1943. The Atlantic Ocean existed and was an effective barrier to land mammals well before the beginning of the Cenozoic, according to all present students of plate tectonics. If there was any spread of Cenozoic mammals from Africa to South America—still a possibility—it was certainly by sweepstakes dispersal across an ocean. There are certain dubious points still. It seems likely now that there was a Paleocene-early Eocene connection direct to Europe, not through Asia until late Eocene and Oligocene, but that has little or no clear implication for drift. It is also possible that Australia was linked with Antarctica into the early Cenozoic, but there is no fossil evidence whatever bearing on that.

It is unfortunate that I concluded the 1943 paper with the statement that, "The distribution of mammals definitely supports the hypothesis that continents were essentially stable throughout the whole time involved in mammalian history." . . . M. C. McKenna points out that as mammals appear in the late Triassic and continents were definitely not stable in the Jurassic, that statement cannot now be accepted. I agree, but it is a curious fact that what we *know* about Mesozoic mammals still does not support continental drift. The probability is that we just don't know enough about Mesozoic mammals (and perhaps not enough about drift, either). My whole study was explicitly based on Cenozoic mammals and "the whole history" was just a careless slip.

In 1943 I also pointed out that the South American Triassic reptilian fauna as then known and published did not resemble that of Africa as much as would be consistent with the union of the two continents. That was decidedly true at that time. Since then knowledge of the South American Triassic fauna has increased enormously. Faunal resemblance to Africa is said by some students of the new materials to have been increased thereby (e.g., Romer in *Gondwana Stratigraphy*, 1970). . . . I have to take their word for it, but the evidence has not been well published and is not clear. As far as I have been able to learn, the crucial *Lystrosaurus* fauna now known in Antarctica, Africa, and Central Asia is not known in South America. The real evidence for Triassic union of South America and Africa still seems to be nonpaleontological. . . . The whole subject of plate tectonics is very exciting and has revivified geology. I think it is great. But as inevitably happens, some students get over-excited. . . . Now *everything* has to relate to plate tectonics. I can't go into detail here, but it is rather amusing that currently there are *four completely different* plate tectonic "explanations" of the early distribution of marsupials, none of them based on a reasonable balance of evidence.

A type of evidence that is independent of the stratigraphic record and that supported the idea of two Permian supercontinents rather than one such continent was published in 1969 by P. M. Hurley and J. R. Rand. These authors plotted the distribution of Precambrian rocks on the reconstruction of Bullard, Everett, and Smith and discovered that the oldest rocks cluster in two compact groups. Such a configuration would be unexpected indeed if the various continental fragments had drifted far and wide before coming together at the end of the Paleozoic. From this map pattern, together with measurements of the strontium isotopic composition of continental bedrock, Hurley and Rand concluded that Laurasia and Gondwanaland grew as separate units, neither of which began to rupture into fragments that were rafted apart by plate motion until the Mesozoic. They calculated that the growth of continents has been caused by the generation of new sial from the mantle that has occurred at rates increasing from 45,000 cubic meters per million years in the Precambrian to 120,000 cubic meters per million years in the Tertiary.

Pangaea

The postulate of two early continents of about equal area, one in the north and one in the south, appeals to mankind's ancient love of symmetry. Partly for this reason and partly because the northern and southern continents meet in a zone of profound complexity and instability, some advocates of plate tectonics favor the concept of Laurasia and Gondwanaland as proposed by Alexander du Toit, Lester King, and Patrick Hurley. However, a majority of the continental reconstructions published in recent years depict a single late Paleozoic world landmass. All modern advocates of Pangaea have departed from the scheme of fitting Europe to Africa and North America to South America as proposed by Alfred Wegener in 1922 (Figure 31). Instead, they have placed northern Africa against North America (Figure 69). Wegener himself used this configuration in 1915 in the first edition of his book (see Figure 29A) but he abandoned it shortly afterward. This aspect of the fit is highly controversial but it retains an immense appeal because it "looks so good," an argument of dubious scientific value as was amply illustrated by E. N. Lyustikh in Figure 88.

In the reconstruction by Bullard, Everett, and Smith (Figure 69) the margin of northwestern Africa abuts the present edge of the North American continental shelf. According to some advocates of plate tectonics, the central and southern portion of the Appalachian mountain chain was formed by a late Permian collision of Africa with North America. By this hypothesis, parts of these mountains consist of African territory left behind when the two continents broke apart again in the late Mesozoic. If this is so, then the contact with Africa should lie at least 200 kilometers farther inland in North America than is shown on the map of Bullard, Everett, and Smith. This fact was pointed out in 1972 by James Gilluly, who also pointed out that Spain, in the Cretaceous, lay about 400 kilometers farther north than it appears on that map. Further errors are clearly apparent in the region of Central America and the Caribbean where the map simply eliminates much territory where Paleozoic formations are well known. For these and numerous other reasons, Gilluly, while agreeing that the east-to-west matches of the lands across the Atlantic are surely meaningful, has questioned whether the contact between the northern and southern continents shown on the map of Bullard, Everett, and Smith may not be equal in geological significance to the resemblance of the map of Italy to a boot.

On the other hand, some evidence of matching stratigraphy has been reported between the northern and southern continents. In 1970, for example, Paul A. Mohr described similar sequences of fossils in manganese carbonate beds of Wales, Newfoundland, and Mauretania. The full significance of these occurrences awaits further geological field work in Mauretania.

Perhaps it is both unknown and unknowable whether the Permian landscape consisted of one great continent or of two or more continents. In any case, the rifting apart of the northern and southern sectors is one of the first Triassic events postulated by supporters of Pangaea. One version of the development of oceanic ridges, the breakup of continental fragments, the opening of trenches, and the coalescence of sialic areas to achieve our present world map was outlined in 1970 by Robert S. Dietz and John C. Holden. Dietz and Holden even extended plate motions into the future to depict how the world should look 50 million years from now. While some geologists dispute the details of their reconstruction (how unprecedented a situation if they did not!), the broad principles of continental drift in terms of plate motions are vividly represented on their maps (Figures 100, 101).

Dietz and Holden began with Pangaea as it appeared in the Jurassic and translated and rotated the fragments northward as well as eastward and westward. Unlike Alfred Wegener, Dietz and Holden tied plate motions to fixed coordinates, which they did not choose arbitrarily. In searching out a clue to the original position of Pangaea on the globe, they seized upon a "permanent hot spot" which they believe has erupted lava onto the sea floor ever since the end of the Jurassic. This hot spot is the Walvis thermal center of the South Atlantic, which lies near the island group of Tristan da Cunha where the Mid-Atlantic Ridge is intersected by both the Walvis and Rio Grande rises. Dietz and Holden assume that the thermal center has remained at the same latitude and longitude since the Mesozoic while the adjacent plates have spread away from it to either side and all of the other plates of the world have shifted in concord. Tristan da Cunha is an eruptive volcanic island group from which all 262 human inhabitants were evacuated to safety in 1961. The islands are sufficiently remote to have been chosen as a nesting ground by the wandering albatross, the rockhopper penguin, the giant fulmar, and the Tristan great shearwater, a bird which leaves in prodigious flocks each spring to slowly circle clockwise once around the Atlantic—to

Figure 100 (facing page). Pangaea—at the end of the Paleozoic and at the end of the Triassic. A: The Dietz-Holden reconstruction of Pangaea as it appeared at the end of the Paleozoic, 200 million years ago. The continents are fitted at the 1,000-fathom isobath. Hatched crescents indicate the present positions of the Antilles arc and the Scotia arc, between South America and Antarctica. B: By the end of the Triassic, after 20 million years of drift, Pangaea has split into two continents, Laurasia and Gondwana. Plates are migrating from spreading oceanic ridges toward deep trenches (hatched lines). (From "The Break-up of Pangaea" by Robert S. Dietz and John C. Holden.)

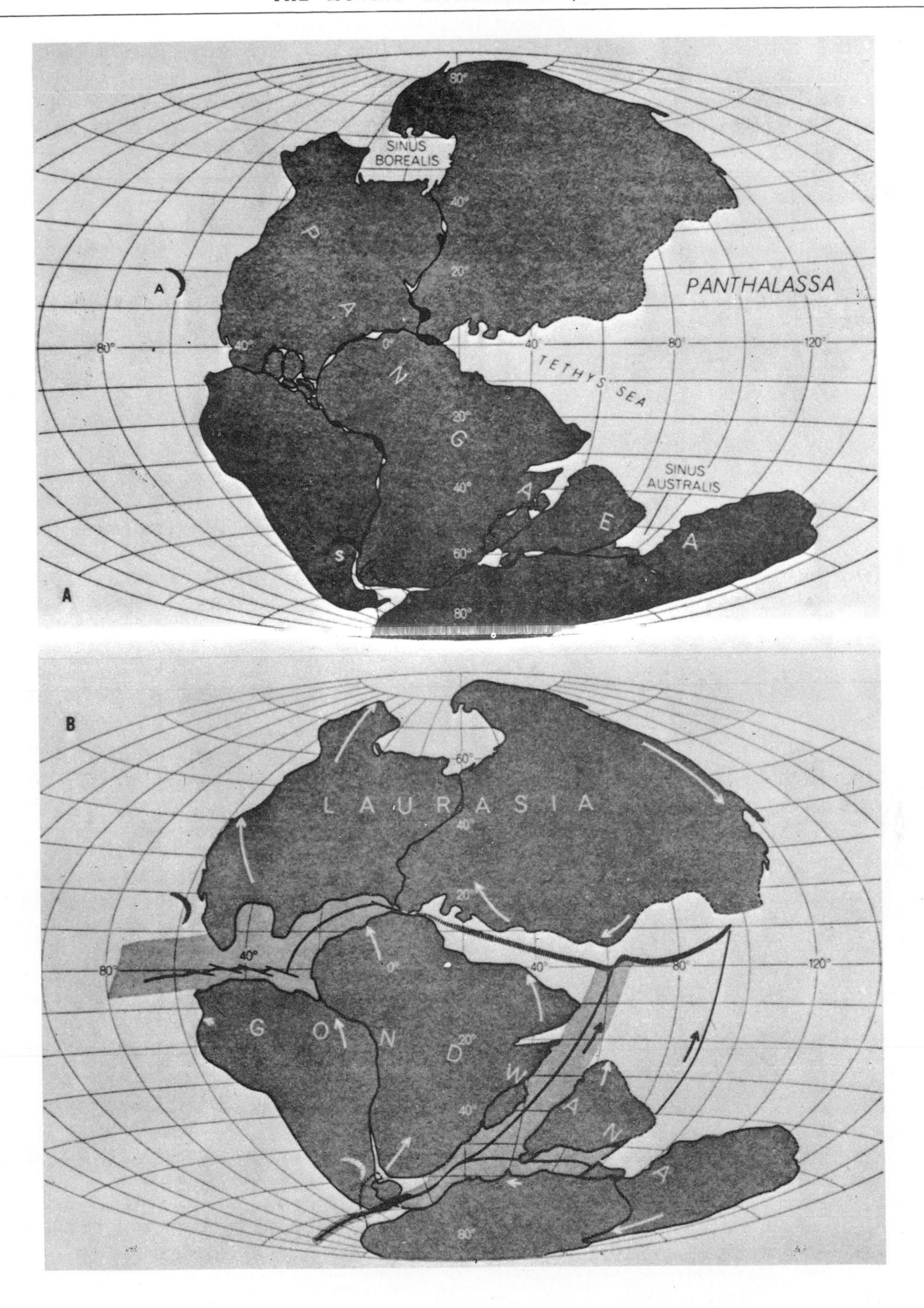
A
SINUS
BOREALIS
PANTHALASSA
TETHYS SEA
SINUS
AUSTRALIS
P A N G A E A
B
L A U R A S I A
G O N D W A N A

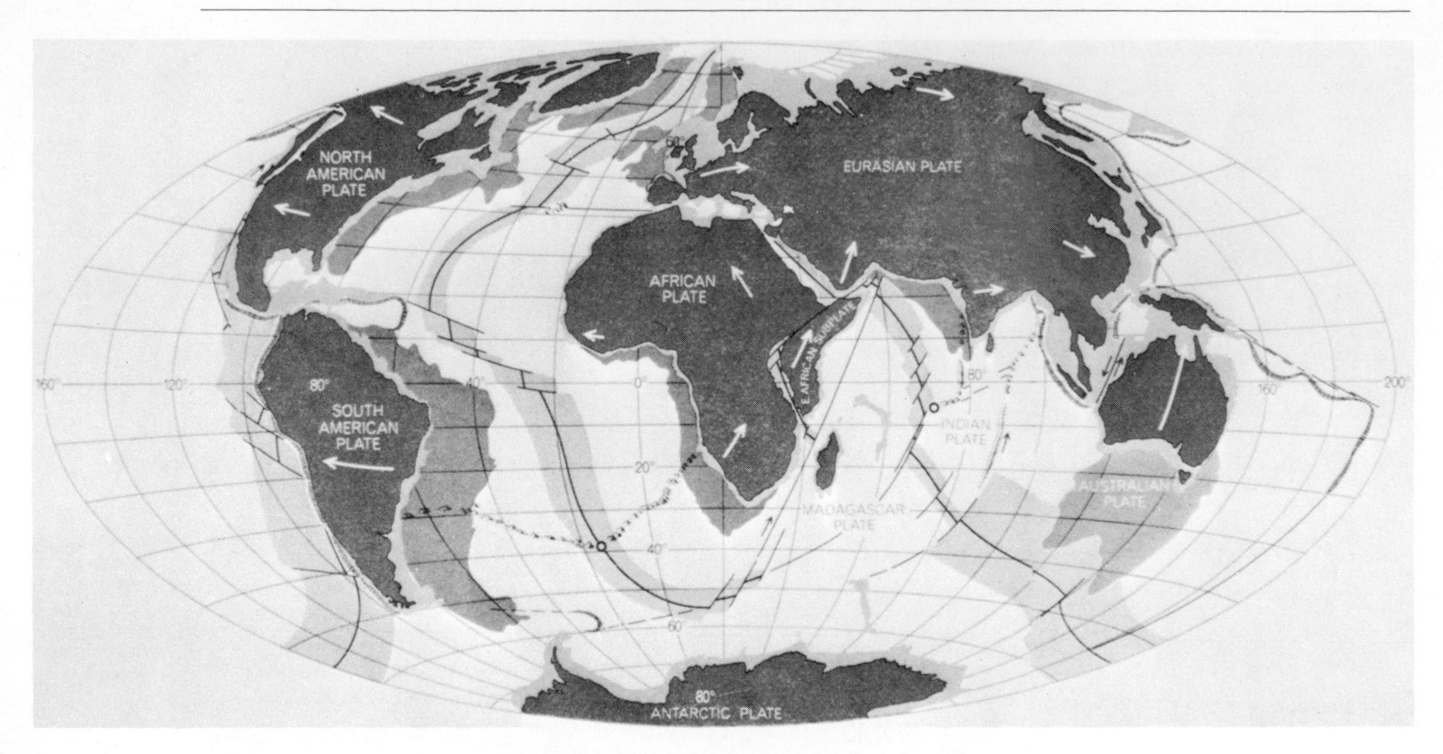

Figure 101. The distribution of continental fragments 50 million years in the future as predicted by Dietz and Holden. The smaller oceans are still growing at the expense of the Pacific. The permanent reference point of the globe is the Walvis thermal center at the site of the Tristan da Cunha island group in the South Atlantic. Lavas extruded continuously from this center are depicted as having built the aseismic ridges on the sea floor as the coasts of Africa and South America have drifted apart. (From "The Break-up of Pangaea" by Robert S. Dietz and John C. Holden. Copyright 1970 by Scientific American, Inc. All rights reserved.)

Brazil, Florida, Nova Scotia, England, Spain, Senegal, and southern Africa. Did the birds begin these yearly rounds when the Atlantic was a narrow rift? Tristan da Cunha, in this conception, is the one fixed reference point of the globe, the "Greenwich Observatory" of Pangaea.

Postscript

Why do the continents exist at all? Why are most of them triangular in shape and crowded into one hemisphere? We began with these questions and we have glanced through 23 centuries of thoughts and observations on continents and oceans. If we have not achieved the final answers, we have at least seen real, if unsteady, progress toward the solutions. The progress has been marked by dramatic changes in opinion. The latest of these changes occurred in 1966, the year which saw the publication of the *Source Book of Geology: 1900-1950*. That book, a compilation of the significant ideas advanced in the first half of this century, includes no mention of Alfred Wegener or of the hypothesis of continental drift. Today, conferences and symposia on continental drift have become so numerous as to constitute, in the phrase of one observer, "a major sociological phenomenon."

Leaving out Alfred Wegener was a very serious oversight indeed [commented K. F. Mather, compiler of the *Source Book*, in the autumn of 1971]. Today I would certainly include him. One can think of justifications that seemed valid only a few years ago. But in retrospect we see that Wegener had one of the most fertile geological imaginations of this century.

Other scientists are clearly in accord with such a reevaluation, for in 1970 the International Astronomical Union named one of the craters of the moon for Alfred Wegener. It will come as no surprise to a reviewer of the vicissitudes of the hypothesis of continental drift that the Wegener crater lies on the moon's far side, permanently turned away from the earth. There was no room left on the near side where each of the prominent craters was already named for a scientist, philosopher, or explorer. Lunar features have been named for many other figures who contributed to the history outlined in these pages: Eratosthenes, Posidonius, Strabo, and Ptolemy; Leonardo da Vinci, Nicolaus Copernicus, René Descartes, Isaac Newton, and Edmond Halley; Christopher Columbus, Ferdinand Magellan, James Cook, and James Clark Ross; Martin Behaim and Gerhard Mercator; Pierre Bouguer and R. J. Boscovitch; Alexander von Humboldt, Abraham Werner, John Playfair, Charles Darwin, Charles Lyell, and Lord Kelvin. Galileo, the condemned heretic, was denied a crater when the lunar naming began in the 17th century, but eventually this genius, who was the first man to turn a telescope toward the moon and recognize the presence there of a variegated landscape, was assigned an inconspicuous nearside feature. James Hutton, the first great uniformitarian, was awarded a crater on the back of the moon along with Alfred Wegener in 1970. And so too were Nicolaus Steno, the Comte de Buffon, and J. B. Lamarck; R. Eötvös, Ernst Weichert, and Andreiji Mohorovičić; Henri Becquerel and Marie Sklodowska Curie; T. C. Chamberlin and Felix A. Vening Meinesz; the historian of science George Sarton and the progenitor of sea-floor spreading, Harry H. Hess. Alfred Wegener is memorialized in good company.

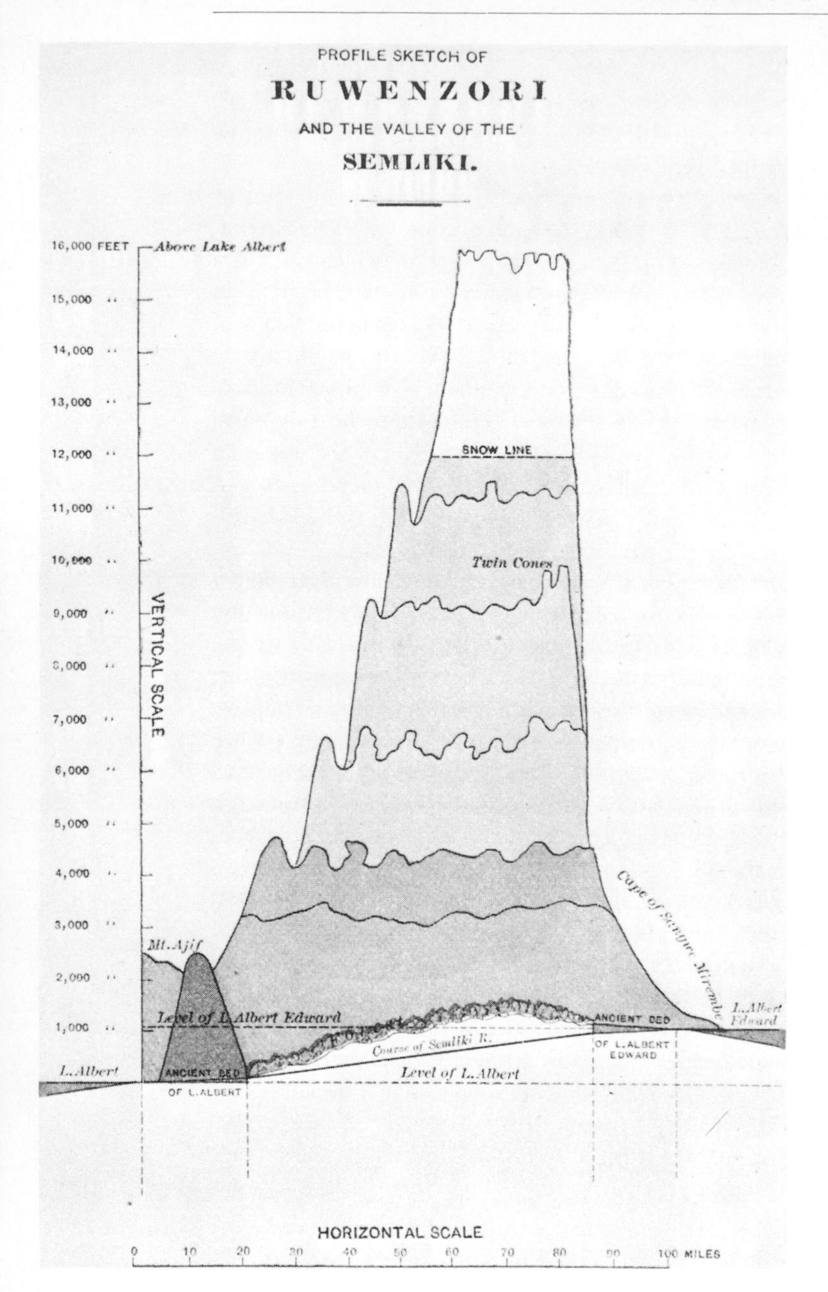

Figure 102. The fabled Ruwenzori. A topographic profile drawn to scale (with some vertical exaggeration) by H. M. Stanley (1890).

The investigations carried out by these scientists and many others made it possible in 1969 for men to visit the moon and return in safety. The dynamical problems were fully solved within some two million days of that early time when man first perceived the principle of the wheel. Still remaining unsolved are the geological problems relating to the origin and evolution of the earth and moon. The rock samples returned from the moon are yielding some of our most valuable evidence of the early history of the earth-moon system. Meanwhile, we are still seeking to unravel the visible record of the continents. To examine a specific problem we may turn to Africa.

The oldest homeland of mankind now known to us lies in the rift valleys of East Africa. For more than two million years man has lived in the rift along with the antelope, zebra, and giraffe; the lion, leopard, and cheetah; the elephant and rhinoceros; the ostrich and ibis—all of which evolved with him during the Tertiary. Geographical explorers became aware of the rift in the 19th century and began to map and describe it. They were followed by an illustrious succession of geologists. Among the early ones now familiar to us were J. W. Gregory, who invented the term "rift valley," Bailey Willis, and Alexander du Toit. Scores of more recent investigators have sampled the rocks and surveyed the structures by means of specialized devices—magnetometers, geodimeters, gravimeters, seismometers—and stations have been erected to sense crustal motions via the laser tracking of geodetic satellites.

To this day, nevertheless, the genesis of the African rift system is not well understood. Is it a tensional valley where the walls have pulled apart and a block of continental crust has dropped like a fallen keystone? Is it a deep furrow held down by compression? Is it a zone of crustal stretching and thinning? Is it a fissure opening toward the mantle and filling at depth with oceanic basalts? All of these interpretations have been proposed, each supported by some individually persuasive line of evidence.

Wherever the sense of past motion is apparent, the walls of the rift are seen to have slipped upward while the floor has remained at a low elevation. This suggests that the valley began as a down-warped trough in a low-lying plain and stayed depressed while highlands rose along steep fault scarps on either side. A few geologists are convinced that the faults date back to the Precambrian. Most believe that the structures began to develop in the Tertiary but that in some sections they follow a grain of the Precambrian basement rock.

From the air the rifts look like an arcuate system of tension fractures. Gravity measurements, first made by Sir Edward Bullard in the 1930s, show a negative anomaly over the rift and rift shoulders, signifying that the valley is underlain by too-little mass and should be rising. That the floor shows no sign of rising has been used as an argument that it is held down by compression. Recent gravity determinations show a narrow positive strip along the axis of the valley, and this has been interpreted as being due to injection of dense magma into the lower crust in response to tensional stretching and thinning.

Magnetic anomalies paralleling the rift axes have not been observed. But in October 1972, in a special issue of *Tectonophysics* devoted to the East African rift system, J. Wohlenberg and N. V. Bhatt announced the discovery of magnetic anomalies trending northwest-southeast across portions of the Kenya rift where the valleys strike north-south. This oblique pattern bears no relationship to the surface structures or igneous rocks and renders perplexing indeed the meaning of magnetic anomaly belts.

Volcanic rocks occur sporadically and unpredictably along the rift, both on the floor and on the bordering plateaus. Unpredictably because in some of the most deeply faulted sections—as, for example, where the floor of Lake Tanganyika plunges to depths of 655 meters below sea level—the rifts are almost completely free of volcanism. The extrusive rocks typically are enriched in alkalis; and in their extreme manifestations, as in the active volcano Ol Doinyo Lengai in Tanzania, are largely composed of sodium and calcium carbonates. In this the rift volcanics differ markedly from the common types of oceanic basalt and island arc andesite. Petrologists remain in disagreement on whether continental rift extrusives are primary differentiates or partial fusion products from the mantle, and on the importance of contamination by crustal sial.

Where the western fork of the rift—bordering Uganda and the Republic of Zaire—crosses the equator, a chain of deep lakes is interrupted by the massive Ruwenzori (Figure 102), a mountain range that towers to altitudes of over 5,000 meters. The Ruwenzori—the legendary Mountains of the Moon—was regarded as fabulous until the morning of May 25, 1888, when Africans pointed out to H. M. Stanley, the indefatigable explorer, a vista of snow-capped peaks shining in the sun.

Geologically speaking, the Ruwenzori remains fabulous. Unlike the great cones of Kilimanjaro, Kenya, Ras Dejen, and Elgon, all of which rise from the plateaus, the Ruwenzori is not volcanic. It is an upraised block of the ancient Precambrian shield of Africa. The existence of the Ruwenzori would not be predicted by any present theory of rift tectonics. Explanations in terms of transcurrent shearing, local up-warping, and isostatic uplift may all share a grain of truth, but all leave the Ruwenzori unexplained.

These are a few of the problems associated with one continental rift where mankind has lived from the beginning and scientists have worked for a century. How soon should we conclude that we fully understand the oceanic rifts which none of us has ever seen? In learning about the earth—a most fundamental preoccupation of man—geologists are limited to direct examination of its outermost surface and that of its nearest neighbor, the moon. All knowledge of the deep interior of our own planet and of the nature of other planets in the solar system is gathered by remote sensing devices. Characteristically, the signals from these devices are interpreted differently by different scientists and every new advance tends to raise new problems, leaving us acutely aware of the limitations of our knowledge. Nevertheless, like the people along the sand, described in a poem by Robert Frost, earth scientists persist in probing the unknown:

> They cannot look out far
> They cannot look in deep.
> But when was that ever a bar
> To any watch they keep?

Sources of Quotations

Introduction

1 "If we are to believe": Chamberlin (1928), page 87.
1 "Can we call geology a science": Chamberlin, page 83.
2 "since further discussion": Willis (1944), page 509.
2 "Scientists who are not geologists": Willis (1944), page 510.
3 "If we imagine an observer": Suess, Sollas translation, volume 1 (1904), page 8.
5 "A moving continent": Lake (1922), page 338.

Geographical Speculations

10 "hath made of one blood": Acts 17:26.
11 "verily, their sound": Romans 10:18.
11 "your reasonings are capricious": Cosmas, McCrindle translation (1897), page 136.
11 "Cease, O ye wiseacres!": Cosmas, page 14.
12 "The triumph of Christianity": Sarton (1927), page 11.
12 "It does not lie in the destinies": Humboldt, *Cosmos*, Otté translation, volume 2 (1867), page 229.
13 "consider natural processes": Roger Bacon, Burke translation, volume 1 (1928), page 313.
14 "Although he was not worthy": Roger Bacon, page 381.
17 Letters of Peter Martyr. Translations based on those of Thacher, volume 1 (1903): Letter of May 14, 1493, page 54; September 13, 1493, page 58; November 1, 1493, page 60; October 31, 1494, page 65.
18 "Or like stout Cortez": John Keats, *On First Looking into Chapman's Homer.*
19 "Moreover, beyond [these] three parts": Wroth (1944), page 81.
21 "I have seen to my great satisfaction": Buffon (1778), page 355.
21 "Even in the very figure of the world": Francis Bacon, Kitchin translation (1855), page 182.
23 "We revolve about the sun": Copernicus, Rosen translation (1939), page 59.
24 "I, Galileo": from Santillana (1955), page 312.
24 "The opinion of the earth's motion": Inchofer, translation of passage in Williams (1930), page 222.
25 "great break-up of a continent": Gilbert, Mottelay translation (1893), page 241.
25 "if the season of the Yeare permit": Halley, *Correspondence*, MacPike edition (1937), appendix 11, page 243.

Geological Speculations

27 "all the fountains": Genesis 7:11.
27 "Fifteen cubits upward": Genesis 7:20.
28 "The poor world": *As You Like It*, act 4, scene 1.
29 "may perhaps to some seem not impossible": Hooke (1705), page 321.
31 "is chiefly intended to refute": Halley (1715), page 300.

32 "It was no sooner suspected": Buffon [1749], Smellie translation, volume I (1785), page 484.
33 "for, in Ireland": Buffon [1749], Smellie translation, page 507.
34 "That which we call the Atlantic": Humboldt (1801), page 33.
34 "It would be very interesting": Humboldt (1801), page 34.
35 "endless and as nothing": Hutton (1795), volume I, page 15.
35 "no powers [are] to be employed": Hutton, volume 2, page 547.
36 "all told, Hutton's presentation": Davies (1969), page 178.
36 "nor does it appear that their language": Playfair (1802), page 126.
37 "Sir Charles Lyell probably accomplished": Mather and Mason (1939), page 263.
38 "The increasing recognition of such structures": French, in French and Short, editors 1968), page 8.
39 "The great depressions occupied by the oceans": Dana (1846), page 353.
39 "The great ocean basins": Willis (1910), page 243.
40 "What we are witnessing": Suess, volume 1 (1885), page 604.
40 "would at the same time change": Owen (1857), page 65.
42 "Some things are very interesting": Emerson (1900), page 96.

Geophysical Speculations

46 "arising no doubt from": Pratt (1855), page 95.
46 "ought to have been anticipated": Airy (1855), page 101.
46 "the downward projection": Airy, page 103.
46 "It is supposed that the crust is floating": Airy, page 104.
46 "thus leading to a law": Pratt, "On the Deflection of the Plumb-Line in India" (1859), page 747.
48 "Mr. Hall's hypothesis": Dana (1866), page 210.
51 "As I have no time to discuss": Dutton (1889), page 51.
52 "the wound of the earth": Ball (1882), page 103.
52 "This would make the Atlantic": Fisher (1882), page 244.
52 "But the difficulties surrounding": Fisher (1882), page 244.
52 "Who it was that first suggested": Pickering (1907), page 30.
53 "A curious feature of the Atlantic": Pickering, page 28.
53 "When the moon separated from us": Pickering, page 30.
57 "The crust of the earth gives way": Suess, Sollas translation (1904-1909), volume 2, page 537.
57 "The land has become habitable": Suess, volume 2, page 553.
57 "The whole southern border of Eurasia": Suess, volume 1, page 596.
58 "In the face of these open questions": Suess, volume 4, page 673.
58 "The theory is here entertained": Willis (1907), page 124.
58 "Their vast depressions have been oceanic basins": Willis (1907), page 125.
59 "The modern seismograph": Oldham (1906), page 165.
59 "if not also differing": Oldham, page 166.
62 "Led by Hutton and Playfair": Kelvin (1898), page 341.
62 "I am as incapable of estimating": Kelvin, page 341.
62 "and probably much nearer 20 than 40": Kelvin, page 345.

Continental Drift

63 "It is probably much nearer the truth": Taylor (1910), page 218.
64 "The earth never was a molten globe": Taylor (1926), page 66.

Alfred Wegener: The Hypothesis

66 "But now this similarity": Köppen, in Runcorn, editor (1962), page 310.
66 "When one cannot follow": Wegener, in Runcorn, editor (1962), page 313.
68 "In the whole of geophysics": Wegener, *The Origin of Continents and Oceans*, Skerl translation, 3rd edition (1924), pages 30-31.

72 "The result is therefore proof": Wegener, Biram translation, 4th edition (1966), page 30.
76 "Where the ocean basins are involved": Wegener, Biram translation, page 98.
79 "Polar wandering is a geological idea": Wegener, Biram translation, page 148.
79 "Let us assume that the inertial pole": Wegener, Biram translation, page 157.
82 "The Newton of the drift theory": Wegener, Biram translation, page 167.
82 "Continental drift, faults and compressions": Wegener, Biram translation, page 179.

The Reaction

82 "The trend of a mountain range": Coleman (1916), page 190.
83 "In examining ideas so novel": Lake (1922), page 338.
83 "Since this is not the case": Wegener, *The Origin of Continents and Oceans*, Skerl translation, 3rd edition (1924), page 30.
83 "Since, so far as I know, this is not the case": Lake (1923), page 181.
83 "There is the force of gravity": Lake (1922), page 339.
84 "It is easy to fit the pieces": Lake (1923), page 183.
84 "[Wegener] does not think that the flattening": Lake (1922), page 341.
84 "No doubt he is right": Lake (1922), page 343.
85 "geological features of the two sides": Lake (1923), page 187.
85 "It may seem surprising": Lake (1923), page 188.
85 "But also I can remember": Lake (1923), page 189.
85 "Now, not for the first time": Lake (1923), page 191.
86 "it would be a very long business": Lake (1923), page 192.
86 "The impression left on my mind": Lake (1923), page 193.
86 "Some theory of this kind": Lake (1923), page 193.
87 "small force can not only produce": Jeffreys (1924), page 261.
88 "Far from responding to shocks": Birch (1965), page 142.
89 "The mere fact that a group of American geologists": Longwell, "Some Physical Tests" (1928), page 145.
89 "Perhaps the very completeness": Longwell, page 145.
89 "Wegener's theory, which is easily grasped": Chamberlin (1928), page 83.
90 "Wegener's hypothesis in general": Chamberlin, page 87.
90 "That hypothesis, instead of being detached": Chamberlin, page 87.
91 "Is there a geologist anywhere": Schuchert (1928), page 114.
91 "deals a crushing blow": Schuchert, page 125.
91 "Facts are facts": Schuchert, page 139.
92 "We are on safe ground": Schuchert, page 140.
92 "the geophysicists will in time": Schuchert, page 142.
92 "a beautiful dream": Schuchert, page 140.
92 "How is it, then": Willis (1928), page 78.
93 "The verdict on Professor Wegener's theory": Gregory (1925), page 255.
94 "I think continental movement is not improbable": Joly (1928), page 88.
94 "The hypothesis of continental drift": Wegener (1928), page 100.
95 "Yet they exist": Waterschoot van der Gracht (1928), page 205.

Alternative Hypotheses

97 "Less widely published": Daly (1926), page 260.
97 "The continents appear to have slid downhill": Daly, page 263.
98 "face to face with a principal mystery of nature": Daly, page 305.
98 "thick, hot, basic, rigid yet weak shell": Barrell (1927) [1917], page 287.
101 "We need not conclude": Lake (1933), page 121.
101 "Yet Suess, Haug, and others": Willis (1929), page 286.
101 "Thus, in the twentieth century": Willis (1929), page 289.
103 "Holmes was an anachronism": John S. Dickey, Jr., personal communication, April 1971.
103 "The impression that remains with me": Holmes (1928), page 433.

104 "[in] masses in that state": Hopkins (1839), page 381.
106 "Splendid sledging weather": Sorge, in E. Wegener (editor), Dean translation (1939), page 182.
107 "One thing is certain": Bucher (1933), page 459.
107 "This little map": Bucher, page 458.
108 "du Toit's book made an extraordinary impression": Wegener, *The Origin of Continents and Oceans*, Biram translation, 4th edition (1966), page 72.
108 "with many tremblings": Du Toit (1937), page 305.
110 "In pointing out the difficulties": Du Toit (1937), page 6.
110 "Most persons view": Du Toit (1937), page 54.
110 "The submarine shelves": Wegener, *The Origin of Continents and Oceans*, Skerl translation, 3rd edition (1924), page 8.
110 "The apparent changes in the positions of the poles": Du Toit (1937), page 18.
110 "Throughout this discussion": Du Toit (1937), page 272.
111 "the great primitive shoreless sea": Grabau (1940), page 7.
111 "Finally the theory that the Pacific basin": Grabau, page 9.
114 "There are few subjects": Simpson (1943), page 1.
114 "It must be almost unique": Simpson (1943), page 2.
115 "But no fossil land faunas": Simpson (1943), page 21.
115 "Resemblances of this degree": Simpson (1943), page 21.
115 "Such looseness of thought": Simpson (1943), page 14.
116 "seems to me the climax": Simpson (1943), page 26.
116 "The plain fact is": Simpson (1943), page 26.
116 "sometimes erroneous": Simpson (1943), page 11.
117 "The undoubted likenesses": Du Toit (1937), page 294.

Renaissance in Earth Science

121 "There are no old men practicing the art": Wasserburg (1966), page 432.
124 "It seems we now should admit": Patterson (1956), page 230.
124 "Some problems in geology": Wetherill, personal communication, December 1972.
128 "It is true that the transformation": Read (1957), page xiii.
129 "no other theory can explain": Wilson (1959), page 23.
131 "If isostasy is to be maintained": Gilluly (1955), page 16.
135 "The greatest depths": Menard (1964), page 311.
140 "Obviously very little of the annual supply": Holmes (1965), page 1026.
140 "Now let us follow the career": Shapley (1967), page 46.
148 "This is not the time for a reappraisal": Runcorn (1962), page vii.

Intimations of Change

149 "All will agree without argument": Carey (1955), page 199.
149 "If continental drift should be rejected": Carey (1955), page 196.
149 "should never be used against it again": Carey (1955), page 199.
150 "Wegener worked out the broad outlines": Carey (1958), page 53.
150 "All that can be said": Irving (1958), page 53.
154 "I shall consider this paper": Hess (1962), page 599.
155 "The great advantage of serpentine": Hess, in lecture at American Geophysical Union meeting, April 1967.
160 "It is suggested that the whole history": Vine (1966), page 229.

The Moving Lithosphere

175 "It seems now to be well established": Holmes (1965), page 1013.
177 "This change in thinking": Dickinson (1971), page 166.
179 "So far as I know": Gilluly (1971), page 2383).
181 "The plate motions": McKenzie (1972), page 324.
181 "I avoid the term": Elsasser, in Frazier (1970), page 75.

182 "I think that a time has come": Jeffreys (1969), page 706.
182 "The position may be summarized": Jeffreys (1964), page 19.
182 "A moment's look at a globe": Jeffreys (1964), page 16.
182 "are remarkable for fallacious data": Jeffreys (1964), page 16.
182 "We have to agree": Lyustikh (1967), page 351.
182 "All work aimed at fitting continents": Wesson (1970), page 315.
183 "Much of what is referred to": Bain, in letter to author, June 7, 1972.
185 "It is not unusual": Meyerhoff (1970), page 8.
185 "It is evident": Beloussov (1970), page 505.
186 "It is readily seen": Beloussov (1970), page 494.
186 "The result is unexpected": Beloussov (1970), page 502.
186 "This result blasts all conclusions": Beloussov (1970), page 502.
186 "We may conclude": Beloussov (1970), page 504.
186 "Let us not forget": Beloussov (1970), page 506.
187 "the trench continues to exist": Beloussov (1970), page 498.
187 "It is hardly worthwhile criticizing": Beloussov (1970), page 499.
187 "This hypothesis is based": Beloussov (1970), page 505.
188 "I am amazed": Meyerhoff and Meyerhoff (1972), page 316.
188 "do not claim that they are impartial": Meyerhoff and Meyerhoff (1972), page 316.
193 "Iceland may be cited": Holmes (1965), page 1001.
200 "as regards my paper of 1943": Simpson, in letter to author, June 8, 1972.

Postscript

205 "a major sociological phenomenon": Phinney (1968), page 23.
205 "Leaving out Alfred Wegener": Mather, personal communication, April 1971.

Bibliography

AIRY, G. B. "On the Computation of the Effect of Attraction of Mountain-Masses, as Disturbing the Apparent Astronomical Latitude of Stations in Geodetic Surveys." *Philosophical Transactions of the Royal Society of London*, vol. 145, pp. 101-104, 1855.

ALBRITTON, Claude C., Jr., editor. *The Fabric of Geology*. Stanford: Freeman, Cooper and Co., 1963.

ALLARD, G. O., and V. J. HURST. "Brazil-Gabon Geologic Link Supports Continental Drift." *Science*, vol. 163, pp. 528-532, 1969.

ANDERSON, D. L., C. SAMMIS, and T. JORDAN. "Composition and Evolution of the Mantle and Core." *Science*, vol. 171, pp. 1103-1112, 1971.

ARGAND, E. "La Tectonique de l'Asie." *Compte-Rendu du 13e Congrès Géologique International, Belgique*, pp. 171-372, 1922.

ATWATER, Tanya. "Test of New Global Tectonics: Discussion." *Bulletin of the American Association of Petroleum Geologists*, vol. 56, pp. 385-388, 1972.

__________ . "Implications of Plate Tectonics for the Cenozoic Tectonic Evolution of Western North America." *Bulletin of the Geological Society of America*, vol. 81, pp. 3513-3536, 1970.

BABBAGE, C. *The Ninth Bridgewater Treatise*. 2nd edition. London, 1838.

BACON, Francis. *Novum Organum* [1620]. Translated by G. W. Kitchin. London: Oxford University Press, 1855.

BACON, Roger. *Opus Majus* [1268]. Translated by Robert Belle Burke. Vol. 1, Mathematics. Philadelphia: University of Pennsylvania Press, 1928.

BAIN, G. W. "Climatic Zones throughout the Ages." In A. C. Munyan, editor, *Polar Wandering and Continental Drift* (pp. 94-130). Tulsa: Society of Economic Paleontologists and Mineralogists, 1963.

BAKER, H. B. *The Atlantic Rift and Its Meaning*. Detroit, 1932.

BALL, Robert. "A Glimpse through the Corridors of Time." *Nature*, vol. 25, pp. 79-82, 103-107, 1881.

BARAZANGI, M., and J. DORMAN. "World Seismicity Maps Compiled from ESSA, Coast and Geodetic Survey, Epicenter Data, 1961-1967." *Bulletin of the Seismological Society of America*, vol. 59, pp. 369-380, 1969.

BARRELL, J. "On Continental Fragmentation and the Geologic Bearing of the Moon's Surficial Features." *American Journal of Science*, vol. 13, pp. 283-314, 1927. [Written in 1917.]

__________ . "The Strength of the Earth's Crust." *Journal of Geology*, vol. 22 (1914), pp. 28-48, 145-165, 209-236, 289-314, 441-468, 537-555, 655-683, 729-741; vol. 23 (1915), pp. 27-44, 425-443, 499-515.

__________ . "Review of 'The Place of Origin of the Moon—The Volcanic Problem' by W. H. Pickering." *Journal of Geology*, vol. 15, pp. 503-507, 1907.

BARRETT, P. J., R. J. BAILLIE, and E. H. COLBERT. "Triassic Amphibian from Antarctica." *Science*, vol. 161, pp. 460-462, 1968.

BELOUSSOV, V. V. "Against the Hypothesis of Ocean-Floor Spreading." *Tectonophysics*, vol. 9, pp. 489-511, 1970.

_________. "An Open Letter to J. Tuzo Wilson." *Geotimes*, vol. 13, no. 10, pp. 17-19, 1968.

_________. "Against Continental Drift." *Science Journal*, vol. 3, no. 1, pp. 56-61, 1967.

_________. "Modern Concepts of the Structure and Development of the Earth's Crust and the Upper Mantle of Continents." *Quarterly Journal of the Geological Society of London*, vol. 122, pp. 293-314, 1966.

_________. *Basic Problems in Geotectonics.* Translated by P. T. Broneer. New York: McGraw-Hill, 1962.

BENIOFF, H. "Seismic Evidence for Crustal Structure and Tectonic Activity." In A. Poldervaart, editor, "Crust of the Earth" (pp. 61-74). *Geological Society of America Special Paper*, no. 62, 1955.

_________. "Orogenesis and Deep Crustal Structure–Additional Evidence from Seismology." *Bulletin of the Geological Society of America*, vol. 65, pp. 385-400, 1954.

BERRY, E. W. "Paleobotany: A Sketch of the Origin and Evolution of Floras." *Annual Report . . . of the Smithsonian Institution . . . 1918*, pp. 289-407, 1920.

BIRCH, F. "Speculations on the Earth's Thermal History." *Bulletin of the Geological Society of America*, vol. 76, pp. 133-154, 1965.

BLACKETT, P. M. S., Sir Edward BULLARD, and S. K. RUNCORN. "A Symposium on Continental Drift." *Philosophical Transactions of the Royal Society of London*, series A, volume 258, no. 1088, 1965.

BLACKETT, P. M. S., J. A. CLEGG, and P. H. S. STUBBS. "An Analysis of Rock Magnetic Data." *Proceedings of the Royal Society of London*, series A, vol. 256, pp. 291-322, 1960.

BOWIE, W. "The Yielding of the Earth's Crust." *Annual Report of the . . . Smithsonian Institution . . . 1921*, pp. 235-247, 1922.

BROWN, F. B. H. "Cornaceae and Allies in the Marquesas and Neighboring Islands." *Bulletin of the Bernice P. Bishop Museum*, no. 52, pp. 1-22, 1928.

BROWN, Lloyd A. *The Story of Maps.* Boston: Little Brown and Co., 1950.

BUCHER, W. H. *The Deformation of the Earth's Crust.* Princeton: Princeton University Press, 1933.

BUFFON, Comte de. *Supplément a l'Histoire naturelle*, vol. 5. Paris: 1778.

_________. *Natural History*. Vol. I. *Theory of the Earth* [1749]. Translated by W. Smellie. London, 1785.

BULLARD, E. C. "Heat Flow through the Floor of the Eastern North Pacific Ocean." (Letter.) *Nature*, vol. 170, p. 200, 1952.

BULLARD, Sir Edward, J. E. EVERETT, and A. G. SMITH. "The Fit of the Continents around the Atlantic." In P. M. S. Blackett, Sir Edward Bullard, and S. K. Runcorn, "A Symposium on Continental Drift" (pp. 41-51). *Philosophical Transactions of the Royal Society of London*, series A, vol. 258, no. 1088, 1965.

CAREY, S. W., convener. *Continental Drift–A Symposium*. Hobart: University of Tasmania, 1958.

_________. "The Tectonic Approach to Continental Drift." In S. W. Carey, convener, *Continental Drift–A Symposium* (pp. 177-355). Hobart: University of Tasmania, 1958.

_________. "Wegener's South America-Africa Assembly, Fit or Misfit?" *Geological Magazine*, vol. 92, pp. 196-200, 1955.

CHAMBERLIN, R. T. "Some of the Objections to Wegener's Theory." In W. A. J. M. van Waterschoot van der Gracht, et al., *Theory of Continental Drift* (pp. 83-87). Tulsa: American Association of Petroleum Geologists, 1928.

CLARK, S. P., K. K. TUREKIAN, and L. GROSSMAN. "Model for the Early History of the Earth." In E. C. Robertson, editor. *The Nature of the Solid Earth* (pp. 3-18). New York: McGraw-Hill, 1972.

CLEGG, J. A., M. ALMOND, and P. H. S. STUBBS. "The Remanent Magnetism of Sedimentary Rock in Britain." *Philosophical Magazine*, vol. 45, pp. 583-598, 1954.

COLEMAN, A. P. "Permo-Carboniferous Glaciation and the Wegener Hypothesis." (Letter.) *Nature*, vol. 115, p. 602, 1925.

_________. "Dry Land in Geology." *Bulletin of the Geological Society of America*, vol. 27, pp. 175-192, 1916.

COPERNICUS, Nicolaus. *The Commentariolus* [1514]. In *Three Copernican Treatises*. Translated by Edward Rosen. New York: Columbia University Press, 1939.

COSMAS [An Egyptian Monk]. *The Christian Topography*. Translated from the Greek by J. W. McCrindle. London: The Hakluyt Society, 1897.

COX, A., R. R. DOELL, and G. B. DALRYMPLE. "Reversals of the Earth's Magnetic Field." *Science*, vol. 144, pp. 1537-1543, 1964.

CREER, K. M. "A Review of Paleomagnetism." *Earth Science Reviews*, vol. 6, pp. 369-466, 1970.

_________. "A Synthesis of World-Wide Paleomagnetic Data." In S. K. Runcorn, editor, *Mantles of the Earth and Terrestrial Planets* (pp. 351-382). New York: Interscience Publishers, 1967.

DALY, R. A. *Our Mobile Earth*. New York: Charles Scribner and Sons, 1926.

_________. *Strength and Structure of the Earth*. New York: Prentice-Hall, 1940.

DANA, J. D. "On Some Results of the Earth's Contraction from Cooling, Including a Discussion of the Origin of Mountains, and the Nature of the Earth's Interior. Part I." *American Journal of Science*, vol. 5, pp. 423-443, 1873.

_________. "Observations on the Origin of Some of the Earth's Features." *American Journal of Science*, vol. 42, pp. 205-211, 1866.

_________. "On the Volcanoes of the Moon." *American Journal of Science*, vol. 2, pp. 335-355, 1846.

DARWIN, G. H. "The Precession of a Viscous Spheroid and the Remote History of the Earth." *Philosophical Transactions of the Royal Society of London*, vol. 170, part 2, pp. 447-538, 1879.

DAUBRÉE, M. A. *Expériences synthétiques relatives aux météorites*. Paris, 1868.

DAVIES, G. L. *The Earth in Decay*. New York: American Elsevier Publishing Co., 1969.

DAVIS, W. M. *The Coral Reef Problem*. New York: American Geographical Society, 1928.

DECKER, R. W., P. EINARSSON, and P. A. MOHR. "Rifting in Iceland: New Geodetic Data." *Science*, vol. 173, pp. 530-533, 1971.

DELEVORYAS, T. "Glossopterid Leaves from the Middle Jurassic of Oaxaca, Mexico." *Science*, vol. 165, pp. 895-896, 1969.

DEWEY, J. F., and J. M. BIRD. "Origin and Emplacement of the Ophiolite Suite: Appalachian Ophiolites in Newfoundland." *Journal of Geophysical Research*, vol. 76, pp. 3179-3206, 1971.

_________. "Mountain Belts and the New Global Tectonics." *Journal of Geophysical Research*, vol. 75, pp. 2625-2647, 1970.

DICKINSON, W. R. "Plate Tectonic Models of Geosynclines." *Earth and Planetary Science Letters*, vol. 10, pp. 165-174, 1971.

_________. "Relations of Andesites, Granites, and Derivative Sandstones to Arc-Trench Tectonics." *Reviews of Geophysics and Space Physics*, vol. 8, pp. 813-860, 1970.

_________. "Circum-Pacific Andesite Types." *Journal of Geophysical Research*, vol. 73, pp. 2261-2269, 1968.

DIETZ, R. S. "Geosynclines, Mountains and Continent-Building." *Scientific American*, vol. 226, no. 3, pp. 30-38, 1972.

_________. "Continent and Ocean Basin Evolution by Spreading of the Sea Floor." *Nature*, vol. 190, pp. 854-857, 1961.

DIETZ, R. S., and J. C. HOLDEN. "The Breakup of Pangaea." *Scientific American*, vol. 223, no. 4, pp. 30-41, 1970.

_________. "Reconstruction of Pangaea: Breakup and Dispersion of Continents, Permian to Present." *Journal of Geophysical Research*, vol. 75, pp. 4939-4956, 1970.

DONN, W. L., B. D. DONN, and W. G. VALENTINE. "On the Early History of the Earth." *Bulletin of the Geological Society of America*, vol. 76, pp. 287-306, 1965.

DOTT, R. H., Jr. "Squantum 'tillite,' Massachusetts-Evidence of Glaciation or Subaqueous Mass Movements?" *Bulletin of the Geological Society of America*, vol. 72, pp. 1289-1306, 1961.

DOTT, R. H., Jr., and R. L. BATTEN. *Evolution of the Earth*. New York: McGraw-Hill, 1971.

DOUGLAS, G. V., and A. V. DOUGLAS. "Note on the Interpretation of the Wegener Frequency Cycle." *Geological Magazine*, vol. 60, pp. 108-111, 1923.

DU TOIT, A. L. "Tertiary Mammals and Continental Drift." *American Journal of Science*, vol. 242, pp. 145-163, 1944.

_________. *Our Wandering Continents*. Edinburgh: Oliver and Boyd Ltd., 1937.

_________. "A Geological Comparison of South America with South Africa." (With a palaeontological contribution by F. R. Cowper Reed.) *Carnegie Institution of Washington Publication*, no. 381, 1927.

DUTTON, C. "On Some of the Greater Problems of Physical Geology." *Bulletin of the Philosophical Society of Washington*, vol. 11, pp. 51-64, 1889.

ELLIOT, D. H., E. H. COLBERT, W. J. BREED, J. A. JENSEN, and J. S. POWELL. "Triassic Tetrapods from Antartica: Evidence for Continental Drift." *Science*, vol. 169, pp. 1197-1201, 1970.

EMERSON, B. K. "The Tetrahedral Earth and Zone of the Intercontinental Seas." *Bulletin of the Geological Society of America*, vol. 11, pp. 61-106, 1900.

ENGEL, A. E. J. "Geologic. Evolution of North America." *Science*, vol. 140, pp. 143-152, 1963.

ENGEL, A. E. J., H. L. JAMES, and B. F. LEONARD, editors. *Petrologic Studies: A Volume to Honor A. F. Buddington*. New York: Geological Society of America, 1962.

EVISON, F. F. "Rock Magnetism in Western Europe as an Indicator of Continental Growth." *Geophysical Journal of the Royal Astronomical Society*, vol. 4, pp. 320-335, 1961.

EWING, J., and M. EWING. "Sediment Distribution of the Mid-Ocean Ridges with Respect to Spreading of the Sea Floor." *Science*, vol. 156, pp. 1590-1592, 1967.

FISCHER, A. G., B. C. HEEZEN, R. E. BOYCE, D. BUKRY, R. G. DOUGLAS, R. E. GARRISON, S. A. KLING, V. KRASHENINNIKOV, A. P. LISITZIN, and A. C. PIMM. "Geological History of the Western North Pacific." *Science*, vol. 168, pp. 1210-1214, 1970.

FISHER, O. "On the Physical Cause of the Ocean Basins." *Nature*, vol. 25, pp. 243-244, 1882.

_________. *Physics of the Earth's Crust*. London: Macmillan and Co., 1881.

FISHER, R. L., and R. REVELLE. "The Trenches of the Pacific." *Scientific American*, vol. 193, no. 5, pp. 36-41, 1955.

FRAZIER, K. "The Unfathomed Forces Driving Earth's Plates." *Science News*, vol. 98, pp. 74-76, 1970.

FRENCH, Bevan M., and N. M. SHORT, editors. *Shock Metamorphism of Natural Materials*. Baltimore: Mono Book Corporation, 1968.

FROST, Robert, "Neither Far Out Nor in Deep." *The Poetry of Robert Frost*, edited by Edward Connery Lathem. New York: Holt, Rinehart and Winston, Inc., 1969.

GARLAND, G. D., editor. *Continental Drift*. Royal Society of Canada Special Publication No. 9, Toronto: University of Toronto Press, 1966.

GASKELL, T. F., editor. *The Earth's Mantle*. New York: Academic Press, 1967.

GEORGI, J. "Memories of Alfred Wegener." In S. K. Runcorn, editor, *Continental Drift* (pp. 309-324). New York: Academic Press, 1962.

_________. *Mid-Ice. The Story of the Wegener Expedition to Greenland*. Translated by F. H. Lyon. New York: E. P. Dutton and Co., 1935.

GILBERT, G. K. "The Origin of Hypotheses, Illustrated by the Discussion of a Topographic Problem." *Science*, vol. 3, pp. 1-13, 1896.

GILBERT, William. *On the Loadstone and Magnetic Bodies and on the Great Magnet the Earth* [1600]. Translated by P. Fleury Mottelay. New York: John Wiley and Sons, 1893.

GILLULY, J. "Plate Tectonics and Magmatic Evolution." *Bulletin of the Geological Society of America*, vol. 82, pp. 2383-2396, 1971.

_________. "Geologic Contrasts between Continents and Ocean Basins." In A. Poldervaart, editor, "Crust of the Earth" (pp. 7-18). *Geological Society of America Special Paper*, no. 62, 1955.

_________. "Oceanic Sediment Volumes and Continental Drift." *Science*, vol. 166, pp. 992-993, 1969.

GOLD, T. "Instability of the Earth's Axis of Rotation." *Nature*, vol. 175, pp. 526-529, 1955.

GOLDREICH, P., and A. TOOMRE. "Some Remarks on Polar Wandering." *Journal of Geophysical Research*, vol. 74, pp. 2555-2567, 1969.

Gondwana Stratigraphy. Symposium, International Union of Geological Sciences, Buenos Aires, 1-15 October 1967, Paris: UNESCO, 1969.

GRABAU, A. *The Rhythm of the Ages, Earth History in the Light of the Pulsation and Polar Control Theories*. Peking: Henri Vetch, 1940.

GREEN, W. L. *Vestiges of the Molten Globe as Exhibited in the Figure of the Earth, Volcanic Action, and Physiography*. Part 1, London: Edward Stanford, 1875. Part 2, Honolulu: Hawaiian Gazette Publishing Co., 1887.

GREGORY, J. W. "The Geologic History of the Atlantic Ocean." *Quarterly Journal of the Geological Society of London*, vol. 85, pp. xviii-cxxi, 1929.

__________. "Continental Drift." [A review of the English translation of the 3rd edition of Wegener's book.] *Nature*, vol. 115, pp. 255-257, 1925.

GUTENBERG, B. "Structure of the Earth's Crust and the Spreading of the Continents." *Bulletin of the Geological Society of America*, vol. 47, pp. 1587-1610, 1936.

__________. "Die Veränderungen der Erdkruste durch Fliessbewegungen der Kontinentalscholle." *Gerlands Beiträge zur Geophysik*, vol. 16, pp. 239-247, vol. 18, pp. 281-291, 1927.

HALL, James, *Natural History of New York*. Paleontology, vol. 3, part 1 (text). Albany, 1859.

HALLEY, Edmond. "A Short Account of the Cause of the Saltiness of the Oceans, and of Several Lakes that Emit no Rivers; with a Proposal, by Help Thereof, to Discover the Age of the World." *Philosophical Transactions of the Royal Society of London*, vol. 29, pp. 296-300, 1715.

__________. *Correspondence and Papers of Edmund Halley*. Arranged and edited by Eugene F. MacPike. History of Science Society Publications, new series II. Oxford University Press, 1932. Reissued by Taylor and Francis, Ltd., London, 1937.

HAMILTON, W. "The Uralides and the Motion of the Russian and Siberian Platforms." *Bulletin of the Geological Society of America*, vol. 81, pp. 2553-2576, 1970.

HARLAND, W. B. "Review of 'The Origin of Continents and Oceans.' " *Geological Magazine*, vol. 106, pp. 100-104, 1969.

HARRISON, C. G. A. "Formation of Magnetic Anomaly Patterns by Dyke Injection." *Journal of Geophysical Research*, vol. 73, pp. 2137-2142, 1968.

HAUG, E. *Traité de Géologie*. Vol. 1, Paris: Librairie Armand Colin, 1907.

HAYES, D. E., and W. C. PITMAN III. "Magnetic Lineations in the North Pacific." In J. D. Hays, editor, "Geological Investigations of the North Pacific" (pp. 291-314). *Geological Society of America Memoir*, no. 126, 1970.

HAYS, J. D., editor. "Geological Investigations of the North Pacific." *Geological Society of America Memoir*, no. 126, 1970.

HAYS, J. D., and N. D. OPDYKE. "Antarctic Radiolaria, Magnetic Reversals, and Climatic Change." *Science*, vol. 158, pp. 1001-1011, 1967.

HEIRTZLER, J. R. "Sea-Floor Spreading." *Scientific American*, vol. 219, no. 6, pp. 60-70, 1968.

HEIRTZLER, J. R., G. O. DICKSON, E. M. HERRON, W. C. PITTMAN III, and X. LE PICHON. "Marine Magnetic Anomalies, Geomagnetic Field Reversals, and Motions of the Ocean Floor and Continents." *Journal of Geophysical Research*, vol. 73, pp. 2119-2136, 1968.

HERODOTUS. *The Persian Wars*. Translated by George Rawlinson. New York: The Modern Library, 1942.

HESS, H. H. "History of Ocean Basins." In A. E. J. Engel, H. L. James, and B. F. Leonard, editors. *Petrologic Studies: A Volume to Honor A. F. Buddington* (pp. 599-620). New York: Geological Society of America, 1962.

__________. "Drowned Ancient Islands of the Pacific Basin." *American Journal of Science*, vol. 244, pp. 772-791, 1946.

HILL, M. N. "A Median Valley of the Mid-Atlantic Ridge." *Deep-Sea Research*, vol. 6, pp. 193-205, 1960.

HOLMES, A. *Principles of Physical Geology*. New York: Ronald Press, 1965.

__________. "Radioactivity and Earth Movements." *Transactions of the Geological Society of Glasgow*, vol. 18, part 3, pp. 559-606, 1931.

__________. "Continental Drift." [A review of the symposium volume of 1928 by Waterschoot van der Gracht and others.] *Nature*, vol. 122, pp. 431-433, 1928.

HOOKE, Robert. *The Posthumous Works of Robert Hooke*. London, 1705.

HOPKINS, W. "Researches in Physical Geology; Preliminary Observations on the Refrigeration of the Globe." *Philosophical Transactions of the Royal Society of London*, vol. 129, pp. 381-385, 1839.

HOYLE, Fred. *Astronomy*. New York: Crescent Books, Inc., 1962.

HUMBOLDT, A. von. "Esquisse d'un tableau géologique de L'amérique méridionale." *Journal de Physique, de Chemie, d'Histoire Naturelle*, vol. 53, pp. 30-60, 1801.

_________. *Cosmos*. (Stuttgart: vols. 1-4, 1845-1858.) Translated by E. C. Otté. New York: Harper and Brothers, 1867-1868.

HURLEY, P. M. "Distribution of Age Provinces in Laurasia." *Earth and Planetary Science Letters*, vol. 8, pp. 189-196, 1970.

_________. "The Confirmation of Continental Drift." *Scientific American*, vol. 218, no. 4, pp. 52-64, 1968.

_________, editor. *Advances in Earth Science*. Cambridge, Massachusetts: Massachusetts Institute of Technology Press, 1966.

HURLEY, P. M., F. F. M. De ALMEIDA, G. C. MELCHER, U. G. CORDANI, J. R. RAND, K. KAWASHITA, P. VANDOROS, W. H. PINSON, and H. W. FAIRBAIRN. "Test of Continental Drift by Comparison of Radiometric Ages." *Science*, vol. 157, pp. 495-500, 1967.

HURLEY, P. M., and J. R. RAND. "Pre-Drift Continental Nuclei." *Science*, vol. 164, pp. 1229-1242, 1969.

HUTTON, James. *Theory of the Earth with Proofs and Illustrations*. Vols. 1 and 2. Edinburgh, 1795. (Reprinted at Codicote, England: Wheldon and Wesley Ltd., 1959.)

IHERING, H. von. *Die Geschichte des Atlantischen Ozeans*. Vienna: Verlag Gustav Fischer, 1927.

_________. *Archhelenis und Archinotis, Gesammelte Beiträge zur Geschichte der Neotropischen Region*. Leipzig: 1907.

INCHOFER, Melchior, S. J. *Tractatus Syllepticus, in quo quid de Terrae Solisque Motu vel Statione Secundum Sacram Scripturam Sentiendum Ostenditur*. Rome, 1633.

IRVING, E. *Paleomagnetism*. New York: J. Wiley and Sons, 1964.

_________. "Rock Magnetism: A New Approach to the Problems of Polar Wandering and Continental Drift." In S. W. Carey, convener, *Continental Drift—A Symposium* (pp. 24-61), Hobart: University of Tasmania, 1958.

IRVING, E., and W. A. ROBERTSON. "Test for Polar Wandering and Some Possible Implications." *Journal of Geophysical Research*, vol. 74, pp. 1026-1036, 1969.

ISACKS, B. "Distribution of Stresses in the Descending Lithosphere from a Global Survey of Focal-Mechanism Solutions of Mantle Earthquakes." *Reviews of Geophysics and Space Physics*, vol. 9, pp. 103-174, 1971.

ISACKS, B., J. OLIVER, and L. R. SYKES. "Seismology and the New Global Tectonics." *Journal of Geophysical Research*, vol. 73, pp. 5855-5899, 1968.

JACOBY, W. R. "Instability in the Upper Mantle and Global Plate Movements." *Journal of Geophysical Research*, vol. 75, pp. 5671-5680, 1970.

JAMIESON, T. F. "On the Cause of the Depression and Re-Elevation of the Land during the Glacial Period." *Geological Magazine*, decade II, vol. 9, pp. 400-407, 457-466, 1882.

_________. "On the History of the Last Geological Changes in Scotland." *Quarterly Journal of the Geological Society of London*, vol. 21, pp. 161-203, 1865.

JANE, Cecil, translator and editor. *Select Documents Illustrating the Four Voyages of Columbus*. 2 vols. London: The Hakluyt Society, 1930 (vol. 1), 1933 (vol. 2).

JEFFREYS, H. "Imperfections of Elasticity and Continental Drift." *Nature*, vol. 225, pp. 1007-1008, 1970.

_________. "Continental Drift." (Letter.) *Nature*, vol. 222, p. 706, 1969.

_________. "How Soft Is the Earth?" *Quarterly Journal of the Royal Astronomical Society*, vol. 5, pp. 10-22, 1964.

_________. *The Earth: Its Origin, History, and Physical Constitution*. 1st edition, 1924; 2nd edition, 1929; 3rd edition, 1952; 4th edition, 1959; 5th edition, 1970. Cambridge: Cambridge University Press.

JOLY, J. "Continental Movement." In W. A. J. M. van Waterschoot van der Gracht, et al., *Theory of Continental Drift . . .* (pp. 88-89). Tulsa: American Association of Petroleum Geologists, 1928.

_________. *The Surface History of the Earth*. Oxford: Clarendon Press, 1925.

JONES, T. R. "The Past and Present Aspects of Geology." *The Geological Magazine*, no. 1, pp. 1-4, 1864.

KARIG, D. E. "Ridges and Basins of the Tonga-Kermadec Island Arc System." *Journal of Geophysical Research*, vol. 75, pp. 239-254, 1970.

KAULA, W. M. "A Tectonic Classification of the Main Features of the Earth's Gravitational Field." *Journal of Geophysical Research*, vol. 74, pp. 4807-4826, 1969.

_________. *An Introduction to Planetary Physics*. New York: John Wiley and Sons, 1968.

KAY, M., editor, "North Atlantic–Geology and Continental Drift." *American Association of Petroleum Geologists Memoir*, no. 12, 1969.

_________. "The Origin of Continents." *Scientific American*, vol. 193, no. 3, pp. 62-66, 1955.

_________. "North American Geosynclines." *Geological Society of America Memoir*, no. 48, 1951.

KEHLE, R. O. "Analysis of Gravity Sliding and Orogenic Translation." *Bulletin of the Geological Society of America*, vol. 81, pp. 1641-1664, 1970.

KELVIN, LORD, "The Age of the Earth as an Abode Fitted for Life." *Annual Report of the . . . Smithsonian Institution . . . 1897*, pp. 337-351, 1898.

KING, L. *The Morphology of the Earth*. New York: Hafner Publishing Co., 1962.

_________. "A New Reconstruction of Laurasia." In S. W. Carey, convener, *Continental Drift, a Symposium* (pp. 13-23). Hobart: University of Tasmania, 1958.

_________. "The Origin and Significance of the Great Sub-Oceanic Ridges." In S. W. Carey, convener, *Continental Drift, a Symposium* (pp. 62-102). Hobart: University of Tasmania, 1958.

KNOPOFF, L. "The Upper Mantle of the Earth." *Science*, vol. 163, pp. 1277-1287, 1969.

KÖPPEN, W. von, and A. WEGENER. *Die Klimate der geologischen Vorzeit*. Berlin: Verlag von Gebrüder Borntraeger, 1924.

KUHN, T. S. "The Structure of Scientific Revolutions." In *International Encyclopedia of Unified Science*, vol. 2, no. 2. Chicago: University of Chicago Press, 1970.

KUMMEL, B. *History of the Earth*. San Francisco: W. H. Freeman and Co., 1961.

LAKE, P. "Gutenberg's Fliesstheorie, a Theory of Continental Spreading." *Geological Magazine*, vol. 70, pp. 116-121, 1933.

_________. "Wegener's Hypothesis of Continental Drift." *Geographical Journal*, vol. 61, pp. 179-194, 1923.

_________. "Wegener's Displacement Theory." *Geological Magazine*, vol. 59, pp. 338-346, 1922.

LAMARCK, J. B. *Hydrogéologie* [1802]. Translated by A. V. Carozzi. Urbana: University of Illinois Press, 1964.

LANDSBERG, H. E., editor. *Advances in Geophysics*. Vol. 3. New York: Academic Press, 1956.

LAWSON, A. C. "Insular Arcs, Fore-Deeps, and Geosynclinal Seas of the Asiatic Coast." *Bulletin of the Geological Society of America*, vol. 43, pp. 353-382, 1932.

LE PICHON, X. "Sea-Floor Spreading and Continental Drift." *Journal of Geophysical Research*, vol. 73, pp. 3661-3697, 1968.

LLIBOUTRY, L. "Sea-Floor Spreading, Continental Drift and Lithosphere Sinking with an Asthenosphere at Melting Point." *Journal of Geophysical Research*, vol. 74, pp. 6525-6540, 1969.

LONGWELL, C. R. "Some Thoughts on the Evidence for Continental Drift." *American Journal of Science*, vol. 242, pp. 218-231, 1944.

_________. "Further Discussion of Continental Drift." *American Journal of Science*, vol. 242, pp. 514-515, 1944.

_________. "Herschel's View of Isostatic Adjustment." *American Journal of Science*, vol. 16, pp. 451-453, 1928.

_________. "Some Physical Tests of the Displacement Hypothesis." In W. A. J. M. van Waterschoot van der Gracht, et al., *Theory of Continental Drift . . .* (pp. 145-157). Tulsa: American Association of Petroleum Geologists, 1928.

LUYENDYK, B. P. "Dips of Downgoing Lithospheric Plates beneath Island Arcs." *Bulletin of the Geological Society of America*, vol. 81, pp. 3411-3416, 1970.

LYUSTIKH, E. N. "Criticism of Hypotheses of Convection and Continental Drift." *Geophysical Journal of the Royal Astronomical Society*, vol. 14, pp. 347-352, 1967.

MacDONALD, G. J. F. "The Deep Structure of Continents." *Reviews of Geophysics*, vol. 1, pp. 587-665, 1963.

MALKUS, W. V. R. "Precession of the Earth as the Cause of Geomagnetism." *Science*, vol. 160, pp. 259-264, 1968.

MASON, R. G. "A Magnetic Survey off the West Coast of the U.S. between Latitudes 32° and 36° N, Longitudes 121° and 128° W." *Geophysical Journal*, vol. 1, pp. 320-329, 1958.

MASON, R. G., and A. D. RAFF. "Magnetic Survey off the West Coast of North America, 32° N. Latitude–42° N. Latitude." *Bulletin of the Geological Society of America*, vol. 72, pp. 1259-1266, 1961.

MATHER, K. F., editor. *Source Book in Geology, 1900-1950*. Cambridge, Massachusetts: Harvard University Press, 1967.

MATHER, K. F., and S. L. MASON. *A Source Book in Geology*. New York: McGraw-Hill, 1939.

MAXWELL, A. E., R. P. Von HERZEN, K. J. HSÜ, J. E. ANDREWS, T. SAITO, S. F. PERCIVAL, Jr., E. D. MILOW, and R. E. BOYCE. "Deep Sea Drilling in the South Atlantic." *Science*, vol. 168, pp. 1047-1059, 1970.

McBIRNEY, A. R. "Oceanic Volcanism: A Review." *Reviews of Geophysics and Space Physics*, vol. 9, pp. 523-556, 1971.

McKENZIE, D. P. "Plate Tectonics." In E. C. Robertson, editor, *The Nature of the Solid Earth* (pp. 323-360). New York: McGraw-Hill, 1972.

_________. In K. Frazer, "The Unfathomed Forces Driving Earth's Plates." *Science News*, vol. 98, pp. 74-76, 1970.

_________. "Speculations on the Consequences and Causes of Plate Motions." *Geophysical Journal of the Royal Astronomical Society*, vol. 18, pp. 1-32, 1969.

McKENZIE, D. P., and R. L. PARKER. "The North Pacific, an Example of Tectonics on a Sphere." *Nature*, vol. 216, pp. 1276-1280, 1967.

MELVILLE, R. "Continental Drift, Mesozoic Continents and the Migrations of the Angiosperms." *Nature*, vol. 211, pp. 116-120, 1966.

MENARD, H. W. "Growth of Drifting Volcanoes." *Journal of Geophysical Research*, vol. 74, pp. 4827-4837, 1969.

_________. *Marine Geology of the Pacific*. New York: McGraw-Hill, 1964.

_________. "Deformation of the Northeastern Pacific Basin and the West Coast of North America." *Bulletin of the Geological Society of America*, vol. 66, pp. 1149-1198, 1955.

MEYERHOFF, A. A. "Continental Drift: I. Implications of Paleomagnetic Studies, Meteorology, Physical Oceanography, and Climatology." *Journal of Geology*, vol. 78, pp. 1-51, 1970.

MEYERHOFF, A. A., and H. A. MEYERHOFF. " 'The New Global Tectonics': Major Inconsistencies." *Bulletin of the American Association of Petroleum Geologists*, vol. 56, pp. 269-336, 1972.

_________. " 'The New Global Tectonics': Age of Linear Magnetic Anomalies of Ocean Basins." *Bulletin of the American Association of Petroleum Geologists*, vol. 56, pp. 337-359, 1972.

MITCHELL, A. H., and H. G. READING. "Evolution of Island Arcs." *Journal of Geology*, vol. 79, pp. 253-284, 1971.

MOHR, P. A. "Plate Tectonics of the Red Sea and East Africa." *Nature*, vol. 228, pp. 547-548, 1970.

MOORBATH, S., R. K. O'NIONS, R. J. PANKHURST, N. H. GALE, and V. R. McGREGOR. "Further Rubidium-Strontium Age Determinations on the Very Early Precambrian Rocks of the Godthaab District, West Greenland." *Nature*, vol. 240 (Physical Science, Nov. 27, 1972), pp. 78-82, 1972.

MORGAN, W. J. "Rises, Trenches, Great Faults, and Crustal Blocks." *Journal of Geophysical Research*, vol. 73, pp. 1959-1982, 1968.

_________. "Deep Mantle Convection Plumes and Plate Motions." *Bulletin of the American Association of Petroleum Geologists*, vol. 56, pp. 203-213, 1972.

MORISON, S. E. *The European Discovery of America. The Northern Voyages, A.D. 500-1600*. New York: Oxford University Press, 1971.

_________. *Admiral of the Ocean Sea, a Life of Christopher Columbus*. Boston: Little, Brown and Co., 1942.

MUNK, W. H., and G. J. F. MacDONALD. *The Rotation of the Earth, a Geophysical Discussion*. Cambridge: Cambridge University Press, 1960.

MUNYAN, A. C., editor. *Polar Wandering and Continental Drift*. Tulsa: Society of Economic Paleontologists and Mineralogists, 1963.

NEUMAYR, Melchior. *Erdgeschichte*. Volume 2. *Beschreibende Geologie*. 2nd edition. Leipzig und Wien: Bibliographisches Institut, 1895.

OLDHAM, R. D. "The Constitution of the Interior of the Earth, as Revealed by Earthquakes." *Philosophical Magazine*, vol. 12, pp. 165-166, 1906.

OPDYKE, N.D. "Paleomagnetism of Deep-Sea Cores." *Reviews of Geophysics and Space Physics*, vol. 10, pp. 213-249, 1972.

OROWAN, E. "Continental Drift and the Origin of Mountains." *Science*, vol. 146, pp. 1003-1010, 1964.

OWEN, Richard. *Key to the Geology of the Globe*. New York: A. S. Barnes & Co., 1857.

OXBURGH, E. R., and D. L. TURCOTTE. "Thermal Structure of Island Arcs." *Bulletin of the Geological Society of America*, vol. 81, pp. 1665-1688, 1970.

PANTOJA-ALOR, J., and R. A. ROBISON. "Paleozoic Sedimentary Rocks in Oaxaca, Mexico." *Science*, vol. 157, pp. 1033-1035, 1967.

PATTERSON, C. "Age of Meteorites and the Earth." *Geochimica et Cosmochimica Acta*, vol. 10, pp. 230-237, 1956.

PHINNEY, R. A., editor. *The History of the Earth's Crust*. Princeton: Princeton University Press, 1968.

PICKERING, W. H. "The Place of Origin of the Moon—the Volcanic Problem." *Journal of Geology*, vol. 15, pp. 23-38, 1907.

PIEL, G., editor. "Gondwanaland Revisited: New Evidence for Continental Drift." *Proceedings of the American Philosophical Society*, vol. 112, pp. 307-353, 1968.

PITMAN, W. C., III, and D. E. HAYES. "Sea-Floor Spreading in the Gulf of Alaska." *Journal of Geophysical Research*, vol. 73, pp. 6571-6580, 1968.

PLAYFAIR, John. *Illustrations of the Huttonian Theory of the Earth*. Edinburgh: Cadell and Davies, and William Creech, 1802. (Facsimile reprint with an introduction by George W. White, Urbana: University of Illinois Press, 1956. Reprinted by Dover Publications, New York, 1964.)

POLDERVAART, A., editor. "Crust of the Earth." *Geological Society of America Special Paper*, no. 62, 1955.

POWELL, J. W. *Exploration of the Colorado River of the West and Its Tributaries*. Washington: Government Printing Office, 1875.

PRATT, John Henry. "On the Indian Arc of Meridian." *Philosophical Transactions of the Royal Society of London*, vol. 151, pp. 579-594, 1861.

__________. "On the Deflection of the Plumb-Line in India, Caused by the Attraction of the Himalaya Mountains and of the Elevated Regions Beyond; and Its Modification by the Compensating Effect of a Deficiency of Matter below the Mountain Mass." *Philosophical Transactions of the Royal Society of London*, vol. 149, pp. 745-778, 1859.

__________. "On the Influence of the Ocean on the Plumb-Line in India." *Philosophical Transactions of the Royal Society of London*, vol. 149, pp. 779-796, 1859.

__________. "On the Attraction of the Himalaya Mountains and of the Elevated Regions beyond Them, upon the Plumb-Line in India." *Philosophical Transactions of the Royal Society of London*, vol. 145, pp. 53-100, 1855.

PRESS, F. "The Suboceanic Mantle." *Science*, vol. 165, pp. 174-176, 1969.

RAISZ, Erwin. *General Cartography*. New York: McGraw-Hill, 1938.

READ, H. H. *The Granite Controversy*. London: Thomas Murby and Co., 1957.

REID, H. F. "Drift of the Earth's Crust and Displacement of the Pole." *Geographical Review*, vol. 12, pp. 672-674, 1922.

REVELLE, R., and A. E. MAXWELL. "Heat Flow through the Floor of the Eastern North Pacific Ocean." (Letter.) *Nature*, vol. 170, pp. 199-200, 1952.

RINGWOOD, A. E. "On the Chemical Evolution and Densities of the Planets." *Geochemica et Cosmochemica Acta*, vol. 15, pp. 257-283, 1959.

ROBERTSON, E. C., editor. *The Nature of the Solid Earth*. New York: McGraw-Hill, 1972

ROMER, A. S. "Fossils and Gondwanaland." In G. Piel, editor, "Gondwanaland Revisited: New Evidence for Continental Drift" (pp. 335-343). *Proceedings of the American Philosophical Society*, vol. 112, pp. 307-353, 1968.

RONAN, Colin A. *Edmond Halley, Genius in Eclipse*. New York: Doubleday and Co., 1969.

RUBEY, W. W. "Geologic History of Sea Water." *Bulletin of the Geological Society of America*, vol. 62, pp. 1111-1148, 1951.

RUNCORN, S. K., editor. *Mantles of the Earth and Terrestrial Planets*. New York: Interscience Publishers, 1967.

_________. "Paleomagnetic Comparisons between Europe and North America." In P. M. S. Blackett, Sir Edward Bullard, and S. K. Runcorn, "A Symposium on Continental Drift" (pp. 1-11). *Philosophical Transactions of the Royal Society of London*, series A, vol. 258, no. 1088, 1965.

_________, editor. *Continental Drift*. New York: Academic Press, 1962.

_________. "Towards a Theory of Continental Drift." *Nature*, vol. 193, pp. 311-314, 1962.

SANTILLANA, Giorgio de. *The Crime of Galileo*. University of Chicago Press, 1955.

SARTON, George. *Introduction to the History of Science*. Vol. 1. Washington: Carnegie Institution of Washington, 1927.

SCHEIDEGGER, A. E. *Principles of Geodynamics*. New York: Academic Press, 1963.

SCHILLING, J. "Red Sea Floor Origin: Rare-Earth Evidence." *Science*, vol. 165, pp. 1357-1360, 1969.

SCHNEER, Cecil, J., editor. *Toward a History of Geology*. Cambridge, Massachusetts: Massachusetts Institute of Technology Press, 1969.

SCHOLL, D. W., R. VON HUENE, and J. B. RIDLON. "Spreading of the Ocean Floor: Undeformed Sediments in the Peru-Chile Trench." *Science*, vol. 159, pp. 869-871, 1968.

SCHUCHERT, C. "The Hypothesis of Continental Displacement." In W. A. J. M. Waterschoot van der Gracht, et al., *Theory of Continental Drift* . . . (pp. 104-144). Tulsa: American Association of Petroleum Geologists, 1928.

SCLATER, P. L. "On the General Distribution of the Members of the Class Aves." *Proceedings of the Linnean Society* (February 1858), pp. 130-145, 1858.

SHAPLEY, H. *Beyond the Observatory*. New York: Charles Scribner's Sons, 1967.

SIMPSON, G. G. "Drift Theory: Antarctica and Central Asia." (Letter.) *Science*, vol. 170, p. 678, 1970.

_________. "Mammals and the Nature of Continents." *American Journal of Science*, vol. 241, pp. 1-31, 1943.

_________. "Mammals and Land Bridges." *Journal of the Washington Academy of Sciences*, vol. 30, pp. 137-163, 1940.

SMITH, A. G., and A. HALLAM. "The Fit of the Southern Continents." *Nature*, vol. 225, pp. 139-144, 1970.

SNIDER, A. *La Création et ses mystères dévoilés*. Paris, 1858.

STACEY, F. D. *Physics of the Earth*. New York: John Wiley and Sons, 1969.

STANLEY, H. M. *In Darkest Africa*. New York: Charles Scribner's Sons, vol. 2, 1890.

STEHLI, F. G. "A Test of the Earth's Magnetic Field during Permian Time." *Journal of Geophysical Research*, vol. 75, pp. 3325-3342, 1970.

STENO, Nicolaus. *The Prodomus*. Translated by J. G. Winter. University of Michigan Studies, Humanistic Series, Volume 11. New York: Macmillan Co., 1916.

SUESS, E. "Asymmetry of the Northern Hemisphere." Translated by B. K. Emerson. *Bulletin of the Geological Society of America*, vol. 11, pp. 96-106, 1900.

_________. *Das Antlitz der Erde* [1885-1909]. Translated into English by Hertha B. C. Sollas under the direction of W. J. Sollas. 4 vols. Oxford: Clarendon Press, 1904-1909.

SYKES, L. R. "Mechanism of Earthquakes and Nature of Faulting on the Mid-Oceanic Ridges." *Journal of Geophysical Research*, vol. 72, pp. 2131-2153, 1967.

TAKEUCHI, H., S. UYEDA, and H. KANAMORI. *Debate about the Earth*. Revised edition. San Francisco: Freeman, Cooper and Co., 1970.

TARLING, D. H. "Gondwanaland, Palaeomagnetism and Continental Drift." *Nature*, vol. 229, pp. 17-21, 1971.

TARLING, D. H., and Maureen TARLING. *Continental Drift, a Study of the Earth's Moving Surface*. New York: Doubleday and Co., 1971.

TAYLOR, F. B. "Sliding Continents and Tidal and Rotational Forces." In W. A. J. M. van Waterschoot van der Gracht, et al., *Theory of Continental Drift* . . . (pp. 158-177). Tulsa: American Association of Petroleum Geologists, 1928.

_________. "Greater Asia and Isostasy." *American Journal of Science*, vol. 12, pp. 47-67, 1926.

_________. "Bearing of the Tertiary Mountain Belts on the Origin of the Earth's Plan." *Bulletin of the Geological Society of America*, vol. 21, pp. 179-226, 1910.

THACHER, J. B. *Christopher Columbus: His Life, His Work, His Remains, as Revealed by Original Printed and Manuscript Records together with an Essay on Peter Martyr of Anghera and Bartolome de las Casas, the First Historians of America*. 3 volumes, New York: G. P. Putnam's Sons, 1903.

TOOLEY, R. V., C. BRICKER, and G. R. CRONE. *Landmarks of Mapmaking*. Amsterdam and Brussels: Elsevier Publishing Co., 1968.

TURCOTTE, D. L., and E. R. OXBURGH. "Convection in a Mantle with Variable Physical Properties." *Journal of Geophysical Research*, vol. 74, pp. 1458-1474, 1969.

TUREKIAN, K. K., and S. P. CLARKE, Jr. "Inhomogeneous Accumulation of the Earth from the Primitive Solar Nebula." *Earth and Planetary Science Letters*, vol. 6, pp. 346-348, 1969.

UREY, H. C. "Primary and Secondary Objects." *Journal of Geophysical Research*, vol. 64, pp. 1721-1737, 1959.

_________. *The Planets*. New Haven: Yale University Press, 1952.

VACQUIER, V. "Measurement of Horizontal Displacement along Faults in the Ocean Floor." *Nature*, vol. 183, pp. 452-453, 1959.

VAN ANDEL, T. H., and C. O. BOWIN. "Mid-Atlantic Ridge between 22° and 23° North Latitude and the Tectonics of Mid-Ocean Rises." *Journal of Geophysical Research*, vol. 73, pp. 1279-1304, 1968.

VENING MEINESZ, F. A. *The Earth's Crust and Mantle*. Amsterdam and London: Elsevier Publishing Co., 1964.

VINE, F. J. "Proof of Ocean-Floor Spreading?" In "Abstracts for 1966" (p. 229). *Geological Society of America Special Paper*, no. 101, 1968.

_________. "Spreading of the Ocean Floor: New Evidence." *Science*, vol. 154, pp. 1405-1415, 1966.

VINE, F. J., and D. H. MATTHEWS. "Magnetic Anomalies over Oceanic Ridges." *Nature*, vol. 199, pp. 947-949, 1963.

WALLACE, A. R. *The Geographical Distribution of Animals*. Vol. 1. New York: Harper and Brothers, 1876.

WASSERBURG, G. J. "Geochronology, and Isotopic Data Bearing on the Development of the Continental Crust." In P. M. Hurley, editor, *Advances in Earth Science* (pp. 431-459). Cambridge: Massachusetts Institute of Technology Press, 1966.

WASSON, J. T. "The Chemical Classification of Iron Meteorites. III. Hexahedrites and Other Irons with Germanium Concentrations between 80 and 200 ppm." *Geochemica et Cosmochemica Acta*, vol. 33, pp. 859-876, 1969.

WATERSCHOOT VAN DER GRACHT, W. A. J. M. van. "Remarks Regarding the Papers Offered by the Other Contributors to the Symposium." In W. A. J. M. van Waterschoot van der Gracht, et al., *Theory of Continental Drift . . .* (pp. 197-226). Tulsa: American Association of Petroleum Geologists, 1928.

WATERSCHOOT VAN DER GRACHT, W. A. J. M. van, Bailey WILLIS, Rollin T. CHAMBERLIN, John JOLY, G. A. F. MOLENGRAAFF, J. W. GREGORY, Alfred WEGENER, Charles SCHUCHERT, Chester R. LONGWELL, Frank Bursey TAYLOR, William BOWIE, David WHITE, Joseph T. SINGEWALD, Jr., and Edward W. BERRY. In E. DeGolyer, compiler, *Theory of Continental Drift, a Symposium on the Origin and Movement of Land Masses both Inter-Continental and Intra-Continental, as Proposed by Alfred Wegener*. Tulsa: American Association of Petroleum Geologists, 1928.

WATKINS, N. D., and A. RICHARDSON. "Comments on the Relationship between Magnetic Anomalies, Crustal Spreading and Continental Drift." *Earth and Planetary Science Letters*, vol. 4, pp. 257-264, 1968.

WEGENER, Alfred. *Die Enstehung der Kontinente und Ozeane*. Braunschweig, Germany: Friedrich Vieweg & Sohns, 1st edition, 1915; 2nd edition, 1920; 3rd edition, 1922; 4th edition, 1924; 4th edition revised, 1929; 5th edition, revised by Kurt Wegener, 1936.

_________. *The Origin of Continents and Oceans*. Translated from the 3rd (1922) German edition by J. G. A. Skerl. London: Methuen and Company, 1924. Translated from the 4th (1929) German edition by John Biram. New York: Dover Publications, 1966.

_________. "Two Notes Concerning My Theory of Continental Drift." In W. A. J. M. van Waterschoot van der Gracht, et al., *Theory of Continental Drift . . .* (pp. 97-103). Tulsa: American Association of Petroleum Geologists, 1928.

_________. "Die Entstehung der Kontinente." *Petermanns Geographische Mitteilungen*, vol. 58, pp. 185-195, 253-256, 305-309, 1912.

_________. "Die Entstehung der Kontinente." *Geologische Rundschau*, vol. 3, pp. 276-292, 1912.

WEGENER, Else, editor, assisted by Fritz Loewe. *Greenland Journey*. Translated from the 7th German edition by Winifred M. Dean. London: Blackie and Son, Ltd., 1939.

WELLS, J. W. "Coral Growth and Geochronology." *Nature*, vol. 197, pp. 948-950, 1963.

WESSON, P. S. "The Position against Continental Drift." *Quarterly Journal of the Royal Astronomical Society*, vol. 11, pp. 312-340, 1970.

WHITE, D. A., D. H. ROEDER, T. H. NELSON, and J. C. CROWELL. "Subduction." *Bulletin of the Geological Society of America*, vol. 81, pp. 3431-3432, 1970.

WILLIAMS, Henry Smith. *The Great Astronomers*. New York: Simon and Schuster, 1930.

WILLIS, Bailey. "Continental Drift, Ein Märchen." *American Journal of Science*, vol. 242, pp. 509-513, 1944.

__________. "Isthmian Links." *Bulletin of the Geological Society of America*, vol. 43, pp. 917-952, 1932.

__________. "Continental Genesis." *Bulletin of the Geological Society of America*, vol. 40, pp. 281-336, 1929.

__________. "Continental Drift." In W. A. J. M. van Waterschoot van der Gracht, et al., *Theory of Continental Drift . . .* (pp. 76-82). Tulsa: American Association of Petroleum Geologists, 1928.

__________. "Principles of Palaeogeography." *Science*, vol. 31, pp. 241-260, 1910.

__________. *Research in China*. Vol. 2. *Systematic Geology*. Washington: Carnegie Institution of Washington, 1907.

WILSON, J. Tuzo. "A Revolution in Earth Science." *Geotimes*, vol. 13, no. 10, pp. 10-16, 1968.

__________. "Reply to V. V. Beloussov." *Geotimes*, vol. 13, no. 10, pp. 20-22, 1968.

__________. "Did the Atlantic Close and Then Re-Open?" *Nature*, vol. 211, pp. 676-681, 1966.

__________. "A New Class of Faults and Their Bearing on Continental Drift." *Nature*, vol. 207, pp. 343-347, 1965.

__________. "Geophysics and Continental Growth." *American Scientist*, no. 47, pp. 1-24, 1959.

WOHLENBERG, J., and N. V. BHATT. "Report on Airmagnetic Surveys of Two Areas in the Kenya Rift Valley." *Tectonophysics*, vol. 15, pp. 143-149, 1972.

WRIGHT, W. B. "The Wegener Hypothesis." *Nature*, vol. 111, pp. 30-31, 1923.

WROTH, Lawrence C. *The Early Cartography of the Pacific*. New York: The Bibliographical Society of America, 1944.

WYLLIE, P. J. *The Dynamic Earth: Textbook in Geosciences*. New York: Wiley, 1971.

Library of Congress Cataloging in Publication Data

Marvin, Ursula B
Continental drift
Bibliography: p.
1. Continental drift. I. Title.
QE511.5.M37 551.4'1 72-9575
ISBN 0-87474-129-7